Generative Adversarial Networks for Remote Sensing

Amol Dattatraya Vibhute
Symbiosis International University (Deemed), India

Rajesh Kumar Dhanaraj
Symbiosis International University (Deemed), India

Karbhari Vishwanth Kale
Dr. Babasaheb Ambedkar Technological University, Lonere, India

IGI Global
Scientific Publishing
Publishing Tomorrow's Research Today

Vice President of Editorial	Melissa Wagner
Managing Editor of Acquisitions	Mikaela Felty
Managing Editor of Book Development	Jocelynn Hessler
Production Manager	Mike Brehm
Cover Design	Phillip Shickler

Published in the United States of America by
IGI Global Scientific Publishing
701 East Chocolate Avenue
Hershey, PA, 17033, USA
Tel: 717-533-8845
Fax: 717-533-7115
Website: https://www.igi-global.com E-mail: cust@igi-global.com

Library of Congress Cataloging-in-Publication Data

LCCN: 2024050231 (CIP Data Pending)
ISBN13: 9798369369005
EISBN13: 9798369369029

British Cataloguing in Publication Data
A Cataloguing in Publication record for this book is available from the British Library.

Table of Contents

Detailed Table of Contents

Earth observation depends on the spatial-temporal data from satellites. Optical observations are often affected by random thin clouds. This cloud interference impacts the usefulness of satellite-based remote sensing in several application areas. This paper introduces a cloud-free reconstruction architecture based on a Generative Adversarial Network (GAN) that leverages spatial-attention mechanisms. The proposed Parallel Attention Guided Generative Adversarial Network for Efficient Thin Cloud Removal (PACR-GAN) integrates the benefits of the Convolutional Block Attention Module (CBAM) and the Coordinate Attention Module (CAM). When tested against the RICE-1 dataset, the proposed model demonstrated superior performance in terms of popular evaluation metrics when compared to existing methods. The model effectively reconstructed cloud-free images by focusing on critical features and spatial details, showing resilience to thin clouds.

In recent years, the intersection of machine learning (ML) and remote sensing has led to transformative advancements, with Generative Adversarial Networks (GANs) emerging as a powerful tool for generating synthetic images and enhancing data analysis. This chapter provides an overview of GANs' applications in remote sensing, starting with an introduction to their architecture, comprising a generator and a discriminator trained simultaneously. GANs are particularly valuable in augmenting remote sensing datasets, generating multispectral and hyperspectral images, enhancing image resolution, and enabling image-to-image translation. The chapter also explores GANs' role in data simulation for environmental monitoring and predictive modeling, especially in climate change research. Challenges such as training stability, computational demands, and ensuring geographic accuracy are discussed, along with future research directions, including integrating GANs with AI techniques like reinforcement and transfer learning. This chapter explores the potential and challenges of GANs in remote sensing

The advent of GANs by Generative AI has completely transformed ML profession. In addition to outlining the fundamental ideas & significant turning points that have influenced the development of GANs, this chapter offers a thorough overview of the technology. In addition to discussing different varieties of GANs, such as progressive, Wasserstein, and conditional GANs, & highlighting their special qualities & uses, the chapter delves into the basic architecture of GANs, which includes generator & discriminator networks. Further, it addresses methods to overcome training difficulties that GANs encounter, including regularization techniques & alternative loss functions, & difficulties with mode collapse & training instability. It also looks at how GANs have developed to produce high-fidelity text, graphics, & other data kinds, demonstrating their adaptability & potential in innovative AI applications. It provides a fundamental framework for comprehending ideas, developments, & potential paths of generative AI as seen through the prism of GANs.

Feature engineering, the process of extracting informative representations from raw data, is crucial for successful remote sensing applications. An automated feature engineering is provided by Generative Adversarial Networks (GANs), which is a type of deep learning models based on unsupervised learning. In feature engineering,

This chapter aims to explore the potential of Generative Adversarial Networks (GANs) in enhancing remote sensing applications, addressing key challenges such as data scarcity and low image resolution. The study employs a comprehensive review of existing literature and case studies to analyze the effectiveness of GANs in image generation, super-resolution, and anomaly detection. Key findings indicate that GANs significantly improve the quality of remote sensing imagery, enabling better data synthesis and interpretation. The results underscore the importance of high-quality datasets and the need for robust evaluation metrics to assess GAN performance. The implications of this research highlight the transformative role of GANs in remote sensing, paving the way for future advancements in environmental monitoring and disaster management.

Remote sensing has revolutionized Earth observation, improving environmental monitoring, agriculture, urban planning, and disaster management. However, its increasing use raises significant privacy and security concerns. This chapter examines these issues, focusing on adversarial attacks using Generative Adversarial Networks (GANs). GANs, comprising generator and discriminator networks, enhance remote sensing by improving image resolution, anomaly detection, and data augmentation. Yet, it also enable security threats, such as evasion and poisoning attacks, which compromise data integrity, privacy, and trust. Defending against these attacks involves detection, mitigation, and resilience strategies. Future research should explore robust refences, ethical frameworks, and regulations to secure remote sensing systems.

Preface

The intersection of Generative Adversarial Networks (GANs) and remote sensing technology has paved the way for revolutionary advancements in Earth observation. Remote sensing, traditionally reliant on satellite imagery and aerial data, has encountered numerous challenges such as cloud cover, low image resolution, data scarcity, and limited ground truth. The application of GANs offers a promising solution to these issues by enabling high-quality image generation, enhancing resolution, and assisting in data augmentation for more accurate analysis. This edited book, *Generative Adversarial Networks for Remote Sensing*, brings together a comprehensive collection of research and case studies to explore the transformative potential of GANs in remote sensing applications.

The book is structured to provide readers with a broad understanding of GANs' principles, architectures, and applications, particularly in remote sensing. It covers a variety of topics, from cloud removal in satellite imagery to environmental monitoring, precision agriculture, and time series analysis. The chapters delve into the practical applications of GANs, focusing on real-world challenges and how GANs can overcome them, providing insights into their future potential.

Chapter Overview

- **Chapter 1** introduces a novel approach for efficient thin cloud removal from satellite imagery. The Parallel Attention Guided Generative Adversarial Network (PACR-GAN) employs advanced spatial-attention mechanisms to reconstruct cloud-free images, demonstrating superior performance in terms of evaluation metrics.
- **Chapter 2** offers an overview of GANs in remote sensing, outlining their architecture and the various ways they can be utilized to enhance data generation, multispectral and hyperspectral imaging, and predictive modeling, while also highlighting challenges such as training instability and computational demands.

- **Chapter 3** provides an in-depth exploration of GANs, including their different types and architectures. The chapter addresses key challenges in training, such as mode collapse and instability, and discusses various methods to improve the robustness of GAN models.
- **Chapter 4** investigates the role of GANs in unsupervised feature engineering, emphasizing their ability to automatically extract informative features from remote sensing data. It also discusses the advantages and challenges of integrating GANs with existing analytical frameworks.
- **Chapter 5** further explores the architectures and types of GANs, focusing on how these models are employed in remote sensing to generate synthetic data, enhance image resolution, and improve the quality of satellite and aerial imagery.
- **Chapter 6** examines the use of GANs in object detection and segmentation in remote sensing images. The chapter discusses specialized GAN variants like cGANs and SegGANs, which improve land use analysis, urban structure detection, and environmental monitoring.
- **Chapter 7** focuses on transfer learning with GANs, highlighting the effectiveness of pretrained models and fine-tuning approaches in remote sensing. The chapter explores how these methods reduce the need for extensive training data and computational resources.
- **Chapter 8** presents case studies on the application of GANs in monitoring environmental changes, such as deforestation and climate shifts. The chapter demonstrates how GANs can generate reliable datasets that provide detailed information on environmental conditions.
- **Chapter 9** explores precision agriculture in the context of AI and remote sensing. It discusses how GANs are used to enhance satellite and drone imagery, supporting applications like crop health monitoring, soil analysis, and yield prediction.
- **Chapter 10** looks at GANs' potential in time series analysis and change detection. The chapter discusses how GANs help identify subtle temporal changes in remote sensing imagery, making them invaluable for environmental monitoring and urban development studies.
- **Chapter 11** highlights case studies of GAN applications in precision farming, demonstrating how GANs enhance the quality of remote sensing data for agricultural applications and facilitate better decision-making in areas such as crop management and weather forecasting.
- **Chapter 12** addresses the challenges and future directions of using GANs in remote sensing. The chapter emphasizes the role of GANs in improving data quality and discusses the implications of GAN-based research for environmental monitoring, disaster management, and climate change analysis.

- **Chapter 13** concludes the book by exploring the privacy and security concerns related to the use of GANs in remote sensing. The chapter discusses potential adversarial attacks and the strategies needed to defend against them, offering insights into how to maintain data integrity and privacy in future applications.

Each chapter provides critical insights and practical applications of GANs, making this book an essential resource for researchers, practitioners, and students in the fields of remote sensing, machine learning, and artificial intelligence. The collective work presented in this book highlights the growing significance of GANs in solving complex problems in Earth observation and remote sensing, offering a forward-looking perspective on this exciting area of research.

Chapter 1
A Parallel Attention-Guided Generative Adversarial Network for Efficient Thin Cloud Removal From Satellite Imagery

Jayakrishnan Anandakrishnan
https://orcid.org/0009-0006-7902-6255
National Institute of Technology, Puducherry, India

M. Venkatesan
National Institute of Technology, Puducherry, India

P. Prabhavathy
Vellore Institute of Technology, India

J. Santhana Krishnan
University College of Engineering Arni, India

Alkha Mohan
Indian Institute of Information Technology, Tiruchirappalli, India

S. Santhanakrishnan
University College of Engineering Arni, India

W. Joshua
https://orcid.org/0009-0003-2233-7763
University College of Engineering Arni, India

R. Sachin
University College of Engineering Arni, India

ABSTRACT

Earth observation depends on the spatial-temporal data from satellites. Optical observations are often affected by random thin clouds. This cloud interference impacts

DOI: 10.4018/979-8-3693-6900-5.ch001

the usefulness of satellite-based remote sensing in several application areas. This paper introduces a cloud-free reconstruction architecture based on a Generative Adversarial Network (GAN) that leverages spatial-attention mechanisms. The proposed Parallel Attention Guided Generative Adversarial Network for Efficient Thin Cloud Removal (PACR-GAN) integrates the benefits of the Convolutional Block Attention Module (CBAM) and the Coordinate Attention Module (CAM). When tested against the RICE-1 dataset, the proposed model demonstrated superior performance in terms of popular evaluation metrics when compared to existing methods. The model effectively reconstructed cloud-free images by focusing on critical features and spatial details, showing resilience to thin clouds.

INTRODUCTION

Remote sensing is an advanced technology that can transform the possibilities of observation and understandability of Earth and Earth-related phenomena. The remote sensing process gathers details about these phenomena or even ground objects without close contact with them, providing insights for various geospatial applications. This introduction section provides details for a basic understanding of remote sensing, its applications, the model of data collection, and the need and significance of atmospheric and cloud corrections for such observations. Remote sensing captures data from drones, aerial delights, or satellites. The data capture process includes using sensors to identify and quantify the emitted or reflected electromagnetic signals from the earth's surface or other bodies. This process is followed by processing measurements to find all possible information about the subjects being observed. This technology relies on the laws of physics, mainly dealing with electromagnetic waves and their properties, to determine one or several features of the surface of Earth and the atmosphere (Campbell & Wynne, 2011). Based on the energy source used in data acquisition, remote sensing is classified as active and passive sensing. Active remote sensing utilizes sensors with their energy source. For example, radar or LiDAR sends energy to the target and then measures the return signal. This method is preferable for collecting the data at night and in any weather or climate conditions, excluding the influence of energy from natural light sources such as the sun. On the other hand, Passive remote sensing uses natural energy sources and, most prominently, light energy to evaluate the reflected or radiated signals from the object. It is widely used for collecting images in near-infrared, visible, and thermal infrared regions.

Although satellite or airborne optical observations are known to provide valuable insights about ground phenomena, they are mostly affected by atmospheric distortions affecting the quality of data. Cloudiness, in particular, might be detrimental to the quality of remote sensing observations because it obscures the surface. Clouds hinder the view of the Earth's surface and, thus, mask the sensors from giving correct pictures and information. The occurrence of clouds is random, and this random nature also creates uncertainty in getting a precise observation of the ground in upcoming revisits of satellites. Since this occlusion yields missing or incomplete information, it is highly susceptible to applications that require assessing dynamic environmental processes, such as vegetation growth, development of urban areas and disasters. Further, clouds that distort and absorb electromagnetic radiation inflect the measured data with errors. Scattering reduces the quality of the images and their contrast since the cloud particles scatter the light differently from the photons, and absorption affects the changes of spectral features through which various phenomena happening on the ground surface can be identified. Much research is conducted in the direction of approximation and generative techniques to reproduce the cloud-free scene, thereby preserving valuable temporal observation. With the development of modern generative models and machine learning algorithms, generating cloud-free observations in remote sensing appears to be rather promising (Goodfellow et al., 2014). Such techniques can be useful in restoring missing or hidden information mainly because they derive patterns from large databases of cloud-free images. These generative approaches contribute to improving quality geospatial data, which facilitates the development of reliable earth observation applications.

There is a multitude of remote sensing applications that rely on quality ground data. Understanding the vast applicability can show how earth observation applications depend on quality, precise and distortion-free ground data. Environmental monitoring is one of the application areas of remote sensing that can describe deforestation, desertification in arid lands, and habitats in the hot spots for species. It also has essential applications in ecological monitoring. The use of GIS in hydrology in assessing water quality in lakes and coastal areas, early identification of algal blooms and changes in metal compositions, dissolved metals, minerals and harmful industrial effluents. In agriculture, remote sensing application helps in the early identification of crop health, production estimates, and soil moisture checks so that farmers can decide about irrigation, use of fertilizers, pest management, etc., thus increasing the efficiency of agricultural practices from maximum productivity and yield. Another important application area is estimating the intensity of the impact of natural disasters such as hurricanes, earthquakes, forest fires, landslides, and floods and making decisions that help authorities come up with appropriate measures to fight such natural disasters. Urban planners use remote sensing and mapping operations to understand the growth pattern of the city, the changing use of land, and the development of infrastructures for optimal and efficient use of resources for sustainable future cities. Furthermore, remote sensing supplies information concerning ice cover, snow extent, sea level and atmospheric conditions to forecast climate change and its consequences. Additionally, remote sensing plays a crucial role in depicting the distribution of minerals and geological structures, analyzing the surface of the earth and the ocean, and facilitating rational use of such resources and utilizing them responsibly.

Understanding these applications underlines the importance of having quality temporal to support the different remote sensing applications. The analysed spatial-temporal data contains valuable information for such earth observation activity. Continuous monitoring is also mandatory for the early detection of uncertain events and subsequent informed decision-making. The data collected through remote sensing technologies are broadly used by applying ML and DL methods to pre-process and analyze the gathered data. These sophisticated algorithms can learn and find structures within data, classify data, and predict values with much higher accuracy than simple methods. Sophisticated algorithms and techniques with the help of ML and DL must be developed to transform this data into a quality one. This is where the novelty and importance of this chapter pitch in where the advancing generating adversarial neural networks have been combined with convolutional spatial attention framework to clear the cloudy noisy observations to produce a worthy precise optical observation for every level of earth observation application. This research proposed the PACR-GAN model that utilizes parallel multi-attention operations with CBAM and CAM to generate a haze or thin cloud-free composite of ground data. These parallel attention schemes enable functional blocks that help to maintain local and global contexts during the reconstruction. The PACR-GAN implements residual connections between short-skip and long-skip additions within the parallel spatial attention frameworks. The proposed PACR-GAN was trained and evaluated on the RICE-1 dataset and compared with SOTA techniques to ensure technical validity.

The next section of this chapter provides in-depth literature on the related works performed for the cloud removal problem domain.

BACKGROUND

Remote sensing (RS) images are essential for activities such as land and sea monitoring, disaster detection and management, and agricultural oversight (Prinsa & Saritha, 2018). However, clouds often cover these satellite images, making them difficult to use (Lin et al., 2019a). Some ground data can still be recorded only when clouds are thin. Such partial ground data can provide substantial helps in reconstructing the haze affected images and there by increase the quality of data (Jiang et al., 2024a) (Wu et al., 2020). A high-resolution camera on the Sentinel-2 satellites (Sentinel-2A and Sentinel-2B) captures 13 distinct picture kinds, or spectral bands. Among these bands four are having a 10-meter resolution, six having a 20-meter resolution, and three having a 60-meter resolution. Sentinel-2's ability to take pictures of the Earth is made difficult by sporadic clouds (Khetkeeree et al., 2020). For a variety of applications, a method to eliminate thin clouds from these photos could yield far more information (Toizumi et al., 2019). Various techniques were released for cloud removal. They include multiple images' multi-spectral-based, fusion-based superposition, matching histograms, filtering with homomorphic kernels, geometric-based interpolations, and various statistical and machine learning methods (Zi et al., 2021). These primary methods use past data and process it through statistical and physical correlations to produce a clear, cloud-free image. However, these methods usually remove low-frequency data and adversely affect high-frequency data (Jiang et al., 2024a). Recently, deep-learning approaches, including multiscale interactive fusion networks, multistage self-guided separation networks, and spatial-logical aggregation networks, have gained popularity for cloud removal.

(Prinsa & Saritha, 2018) Outlined a cloud detection method based on an iterative refinement strategy. Zhan et al.'s research introduced a CNN network that effectively differentiated between clouds and snow in remote sensing images (Wu et al., 2020). These methods weren't so capable of expanding toward high-resolution data. Deep learning techniques for cloud detection and mapping became popular, and these techniques showed their superiority in moving clouds from higher-resolution satellite data (Wu et al., 2020). The abilities of CNNs made them an alternative choice for cloud removal due to their capability to perform effective nonlinear output mapping (Gao et al., 2019). (Kong et al., 2010) studies the possibilities of utilizing conventional filtering-based cloud removal. The introduction of an iterative haze optimization algorithm (IHOT) was a breakthrough in thin cloud removal, and they learned the haze trajectory for thin cloud removal (Chen et al., 2016). A Cloud Aware and Feature Extraction (CAFE) technique integrated a physical-based model that performed cloud correction, utilizing factors such as atmospheric light, levels of cloud and their reflectance, and a map of transmission (Yu et al., 2021). Methods such as Laplacian filtering in the frequency domain and advanced homomorphic filtering are effective for eliminating thin clouds (Guo et al., 2023). The incorporation of frequency domain characteristics further expanded the capabilities of cloud removal, and their integration with Deep-CNNs made the cloud removal task easy (Jiang et al., 2024b). The introduction of multi-scale feature extraction characteristics combined with deep residual structures again enhanced the ability of cloud removal algorithms (Song et al., 2021a). Further, in a multi-scale fashion (MDRN), the deep residual network extracted residuals directly from hazy and non-hazy images (Yang et al., 2020).

The capacity of GANs to model the relationship between input and output pictures for certain domains makes them very appealing (Singh & Komodakis, 2018). By using adversarial loss, they may be trained to generate cloud-free pictures that closely resemble the original, clear photos of the target domain. By learning from matched datasets of both cloudy and clear photos, Conditional Generative Adversarial Networks (cGANs) demonstrated their efficacy in producing these cloud-free images (Mirza & Osindero, 2014). With fewer parameters, PatchGAN is renowned for its efficacy and efficiency while processing photos with sparse cloud cover (Isola et al., 2018). However, a generative network based on DCGAN can effectively rebuild pictures with both dense and thin clouds (Radford et al., 2016). Recently, there has been increased interest in integrating GANs with attention mechanisms. One example is the SpA+Edges GAN, which use edge-GANs and a spatial attention map to both locate cloud-distorted regions and repair cloud-covered areas (Namboodiri et al., 2023). While the discriminator employed astute edge detectors to identify the edges impacted by the clouds, the generator determined the precise heatmap of the overcast regions. The RICE dataset series was used to assess the efficiency of the Spatial Attention GAN (SpA GAN), which used both local and global spatial attention to detect and recover obscured regions (H. Pan, 2020). In the meantime, the Haze-Aware Representation Distillation Generative Adversarial Network (HardGAN) successfully employed a normalisation layer to prevent information loss during the reconstruction process and used a multi-scale network with a specialised Haze-Aware Representation Distillation layer to achieve thin cloud removals based on a single input image (Deng et al., 2020).

Table 1. Comparative analysis of various thin cloud removal techniques with benefits and limitations

Reference	Technique	Benefits	Gaps
Zi et al. (2021)	Multi-spectral image, superposition, fusion, histogram matching, filtering, geometry interpolation	Produces clear, cloud-free images using statistical and physical correlations	Removes low-frequency data and adversely affects high-frequency data.
Jiang et al. (2024a)	Deep-learning approaches	Gains popularity due to nonlinear function mapping abilities	Limited focus on high-resolution data.
Prinsa & Saritha (2018)	Iterative refinement strategy for cloud detection	Iterative method improves cloud detection efficiency	Limited adaptation for high-resolution datasets.
Zhan et al. (Wu et al., 2020)	CNN network for differentiating clouds and snow in remote sensing images	Effective detection of clouds and snow	Ineffective for high-resolution data.
Gao et al. (2019)	CNNs for nonlinear function mapping in atmospheric correction	Broad application for addressing atmospheric correction	Limited to specific cases of cloud detection and correction.

continued on following page

Table 1. Continued

Reference	Technique	Benefits	Gaps
Kong et al. (2010)	Filtering techniques using domain transformation	Effective for correcting thin clouds	Limited effectiveness for thick cloud removal.
Chen et al. (2016)	Iterative Haze Optimization Algorithm (IHOT)	Refines cloud trajectory to generate cloud-free images	Inefficiency for random cloud distributions.
Yu et al. (2021)	Cloud Aware and Feature Extraction (CAFE) Module	Integrates physical models to address cloud distortion	Complex integration of physical parameters may reduce scalability.
Guo et al. (2023)	Laplacian filtering and homomorphic filtering	Effective for eliminating thin clouds	Limited application to non-thin clouds.
Jiang et al. (2024b)	Frequency domain dependencies in CNNs and Deep-CNNs	Generates accurate cloud-free data	High computational complexity for large datasets.
Song et al. (2021a)	Multi-scale residual network	Significant improvement in thin cloud removal	Ineffective for thick cloud cover.
Yang et al. (2020)	Multi-scale deep residual network (MDRN)	Learns residuals and directly maps cloudy to cloud-free images	Limited generalizability for different data sources.
Singh & Komodakis (2018)	Generative Adversarial Networks (GANs)	Models input-output relationships in specific domains	Requires paired datasets for training.
Mirza & Osindero (2014)	Conditional GANs (cGANs)	Generates cloud-free images by learning from paired datasets	Limited scalability for unpaired datasets.
Isola et al. (2018)	PatchGAN	Efficient for handling images with thin cloud cover	Less effective for thick cloud removal.
Radford et al. (2016)	DCGAN	Reconstructs images with both thin and thick clouds	High resource consumption during training.
Namboodiri et al. (2023)	SpA+Edges GAN	Restores cloud-covered areas and pinpoints regions distorted by clouds	Edge detection accuracy dependent on input image quality.
H. Pan (2020)	Spatial Attention GAN (SpA GAN)	Identifies and recovers clouded areas	Limited scalability across different datasets.
Deng et al. (2020)	Haze-Aware Representation Distillation GAN (HardGAN)	Multi-scale network prevents information loss during reconstruction	Focused primarily on single-image dehazing scenarios.
Han et al. (2024)	Sparse Transformer-based GAN (SpT-GAN)	Preserves ground details while removing thin clouds	Less effective for thick cloud scenarios.

continued on following page

Table 1. Continued

Reference	Technique	Benefits	Gaps
Tan et al. (2024)	GAN-UD with contrastive learning	Achieves near-supervised performance without requiring paired datasets	May underperform on highly complex cloud structures.
Baskar et al. (2024)	UNet Vision Transformer Cycle-Consistent GAN (UVCGAN)	Variational decomposition for cloud removal	Ineffective for random near-thick clouds.
Lyu et al. (2024)	Physics-based neural networks	Top-layer reflectance-based cloud removal	Complexity in integrating physical models with deep learning techniques.

Thin cloud removal techniques have advanced in recent years too. Recent studies focus on deep learning approaches focused on various GAN frameworks for this cloud removal due to their capabilities. The Sparse Transformer-based Generative Adversarial Network (SpT-GAN) integrated a globally enhanced feature extraction framework that significantly improved thin cloud removal but preserved valid ground details (Han et al., 2024). The unsupervised GAN-UD model could attain near-supervised performance by utilizing effective contrastive learning methodologies for cloud removal. They did not mandate the need for paired data for learning (Tan et al., 2024). The variational decomposition strategies for cloud removal are showcased by UNet vision transformer cycle-consistent GAN (UVCGAN); however, failed in the case of random near-thick clouds (Baskar et al., 2024). The physics-based neural networks are becoming popular and the unified physics-based models utilizing to top of layer reflectance-based cloud removal are the latest works in these regards (Lyu et al., 2024). A summary of important thin cloud removal techniques are given in Table 1.

The challenges of eliminating clouds from individual photos with a generative network design that integrates spatial attention processes are examined in the study. Despite being widely utilised for computer vision content generation, GANs typically have trouble creating realistic pictures, especially when working with real remote sensing data. It is more difficult to create genuine cloud-free pictures using GANs, and in order to accomplish a reliable reconstruction, convolutional layers must incorporate the right spatial attention cues. The following are this planned work's main contributions.

1. The elimination of thin clouds is interpreted as haze-free reconstruction by the suggested PACR-GAN model. In order to preserve local and global spatial contexts throughout reconstruction, it makes use of concurrent multi-attention processes that incorporate the well-liked CBAM and CAM designs.

2. The parallel spatial attention frameworks' use of residual connections between short-skip and long-skip additions lowers the possibility of feature loss during training and helps maintain significant features.
3. The RICE-1 dataset was used to train and assess the suggested PACR-GAN, and its results were compared to the best single-image thin cloud removal techniques.

PROPOSED METHODOLOGY

GAN Architecture

The team released a paper on Generative Adversarial Networks (GANs), a learning architecture for generative applications (Goodfellow et al., 2014). Several applications, including image synthesis, super-resolution, data augmentation, image translation, and picture generation have transformed GANs. Specifically, GANs can be used to create new images from an existing image database, or original music from a database of songs. For instance, GANs can produce pictures that appear to be photographs of actual people's faces, even though the faces are not. GANs can also be used to create standard images from week to week and across lab tech. The possibilities of GANs in the modern world are equal, positive, and negative, even though it has reached an extent to which legislations and regulations are brought in place to make GANs used in proper context only. However, GANs are proving to be highly versatile artificial intelligence tools, as evidenced by their utilization in many applications of engineering. The release of GANs has revolutionized deep learning for generative modelling. StyleGAN by NVIDIA, BigGAN by DeepMind, CycleGAN and Pix2Pix from Berkeley AI Research (BAIR), and DeepArT from DeepArt AG are a few generative models that are designed to perform several invaluable and interesting generative tasks.

GANs have two combined losses, including generator loss and discriminator loss. A GAN generator aims to create artificial samples that are realistic enough to trick the discriminator. The generator minimizes its loss function to do this. When the log probability is maximized, that is, when the discriminator has a high likelihood of classifying the produced samples as real, the loss is minimized. The discriminator decreases the negative log-likelihood of accurately categorizing both actual and manufactured samples. By this loss, the discriminator is encouraged to correctly classify produced samples as real and fraudulent. Together, the GAN problem is considered a min-max loss optimization problem. The detailed mathematical formulation and representations of generator loss, discriminator loss, and min-max representation are discussed below.

One type of deep learning architecture is a GAN. From a given training dataset, it trains two neural networks to compete with one another to produce more authentic new data. The two networks are a generator and a discriminator. An adversarial process is used to train the discriminator and generator, the two halves of a GAN, concurrently. While the discriminator seeks to discriminate between created and real data, the generator seeks to produce data that is indistinguishable from actual data. After being trained using random noise as input, the generator produces an approximate data sample. In order to distinguish between fraud and authenticity, the discriminator analyses data samples. Using a GAN, the discriminator teaches the generator to create indistinguishable false data. The following Equation (1) defines the minmax characteristics function of a GAN.

$$min_G max_D (D,G) = E_x[logD(x)] + E_y[log(1 - D(G(y)))] \tag{1}$$

In the above GAN characteristics equation, the generator is indicated by G, the discriminator is denoted by D, data from the actual distribution $P(x)$ is denoted by x, Furthermore, data from noisy distribution $p(y)$ is given by y. The terms $\max_D V(D,G)$ and $\min_G V(D,G)$ define the minimax GAN problem. $\max_D V(D,G)$ performs the training of D for maximizing the value of the mapping function $V(D,G)$ and ensuring maximum probability for correctly classifying actual samples while minimizing the probability of misclassifying fake samples. $\min_G V(D,G)$ enables the training of generator G to minimize the value of the mapping function $V(D,G)$ and create generated samples so convincing to the discriminator that it becomes incapable of telling them apart from genuine ones.

Attention Mechanism

Attention mechanisms are machine learning and artificial intelligence techniques that enhance model performance. It allows the model to selectively concentrate on the relevant information from the input data by attaching different degrees or weights of importance to different elements at hand. Attention mechanisms compute weights for elements in the input data based on their relevance and similarity to a query or context vector for their importance towards the model's output. Typically, an attention mechanism consists of three parts: value, key, and query. The query indicates the model's current context or emphasis, the key indicates the input data's elements or characteristics, and the value indicates the values connected to those elements or features. The weights are computed using the attention mechanism through a similarity measure between the query and the keys. Then, these weights are utilized to weight the values, whose summation gives the final output of the mechanism. For a number of reasons, the attention mechanism is crucial to artificial intelligence and machine learning, including better model performance, learning with variable length input, explainable AI, generative modelling and localized feature mapping. The attention mechanism is specific to the transformer architecture, which is popular for many tasks because of the robust performance for all data types. In computer vision, spatial attention helps models prioritize important regions in images. These mechanisms have become fundamental in advancing the capabilities of models, particularly in image generation. For spatial attention, understanding convolutional spatial attention is important.

Spatial Attention Framework

Spatial attention refers to the technique used by the human brain to concentrate on some spatial places in the environment or areas of sensory information. It allows the human mind to focus on some information while excluding less important or irrelevant input. A Spatial Attention Module focuses on spatial information in a convolutional neural network. The use of the inter-spatial relationship of features creates a spatial attention map. Unlike channel attention, which is concerned with where an informative element is, spatial attention is complementary. To compute spatial attention, we apply average-pooling and max-pooling along the channel axis and concatenate them to provide an efficient feature descriptor. After that, a convolution layer can be applied to the concatenated feature descriptor to create a spatial attention map, indicating which areas to highlight or suppress. Spatial attention processes in remote sensing are key to improving picture processing from satellites, aeroplanes, and drones. Advanced computational approaches can use such techniques to concentrate certain spatial areas or regions in the accumulated data. Spatial attention improves the accuracy in item recognition, land cover categorization, and change detection tasks by focusing on key characteristics such as texture patterns, spectral signatures, or geometric structures. Progress in deep learning, especially toward the attention processes within convolutional neural networks and self-attention models, further provides dynamic weighting among features regarding spatial relevance and contextual linkages across the image.

Several forms of spatial attention may be distinguished based on the direction of attention within a spatial domain. Object-based spatial attention focuses on whole objects or coherent areas within the images, mostly used for identifying larger objects such as structures, roads, patches of vegetation, etc. Feature-based spatial attention prioritizes particular visual qualities or attributes that will be used to identify analytical jobs, textural patterns, or spectral signatures. Temporal Attention operates on time series data and identifies sequences that are significant for identifying changes, anomalies, or trends throughout time that are given priority by temporal attention processes. Scale-based attention mechanisms allow zooming to specific features or out to specific sections to extract the required spatial context for the problem domain. Hierarchical and Multiscale Attention facilitate the analysis of complex spatial structures, hierarchical relationships between features, and multiscale patterns critical for tasks such as landscape ecology, habitat mapping, or urban growth monitoring.

Convolutional Neural Network (CNN)

An image-processing deep learning technique is called a convolutional neural network. CNNs are made up of layers such as convolutional, pooling, and fully connected layers. Convolutional layers use filters on the input data to identify patterns like edges, textures, and more complex forms. As a result, spatial hierarchies may be captured by the layers. These filters produce feature maps that highlight certain patterns in different input regions by performing convolution operations as they move over the picture. CNNs' ability to preserve spatial relationships within the data makes them highly successful for image processing applications. Convolutional layers are followed by pooling layers such as max pooling and average pooling. These layers reduce the spatial dimensions of the feature maps by selecting the greatest or average value from the input's sub-regions, accordingly. Max pooling, for example, minimises the overall quantity of data while preserving the most noteworthy characteristics by choosing the highest value inside each sub-region. This dimensionality reduction reduces the amount of parameters, which reduces the likelihood of overfitting and increases the network's processing performance. Furthermore, translation invariance provided by pooling reduces the network's sensitivity to the exact positions of features in the input image. Batch processing, which entails processing several data samples at once, is another crucial CNN characteristic. This approach uses parallel processing to speed up training and enhances learning stability using methods like Batch Normalisation. Batch normalisation makes ensuring that the distribution of inputs into each layer remains consistent throughout training by normalising the activations of each layer. This makes it easier to employ greater learning rates, which enhances training effectiveness and speeds up the training process' convergence.

Activation functions give the network non-linearity, which aids in the recognition of complex patterns. Standard activation functions are sigmoid, ReLU, and Tanh. ReLU is incredibly well-liked in CNNs because of its efficiency and simplicity; it mitigates the vanishing gradient issue by only activating the positive portions of the input. Deep network learning may be hampered by the vanishing gradient problem, in which gradients are minimal during backpropagation. ReLU contributes to the network's learning ability by keeping the gradient for positive values constant. Fully connected layers typically positioned at the network's end use the characteristics retrieved from the convolutional and pooling layers to perform final predictions.

These layers map the output classes or predictions to the high-level, abstracted features found by the preceding layers. For example, in an image classification task, thc fully connected layers would process the characteristics to ascertain the likelihood that each input image class exists. In particular, CNNs have transformed computer vision applications such as object identification, facial recognition, and picture categorization.

Multi-Scale Convolutional Neural Network (MSCNN)

A Multiscale Convolutional Neural Network (MSCNN) may be used to extract multiscale information from an image. Both the macro and micro features of the image are recorded by the MSCNN. The network is therefore better able to understand both small features and broader patterns, which makes it suitable for tasks such as image classification, semantic segmentation, and object recognition. An input image is subjected to multiple simultaneous convolutional layers with different filter sizes using this MSCNN module. Features of different scales will be produced by each filter set. While bigger filters seek out more general patterns, smaller filters capture finer details. These feature maps from different kernel sizes may then be concatenated to provide an integrated representation of the input. Adding multiscale properties strengthens and improves the network's recognition of objects and patterns in different sizes and forms. Applications where items of interest may appear in an image at multiple sizes, such as satellite images, medical imaging, and natural scene analysis, are particularly well-suited for MSCNNs.

Residual Network (Res-Net)

A deep network design called Residual Network (ResNet) tackles the deterioration problem that commonly occurs during massive network training. A significant advancement in ResNet is the addition of residual blocks, which enable the network to learn the residuals from layer inputs instead of directly learning unlinked functions. Shortcut connections are utilised to do this by avoiding one or more levels. In a few convolutional layers, adding an input element for a residual block directly to its output results in identity shortcut connections. Deeper networks can be trained more readily when backpropagation gradients are kept, as this helps to alleviate the vanishing gradient problem. The network must be divided into many residual blocks to integrate ResNet into convolutional layers. After feature extraction using CNNs, batch normalisation, and activation functions are applied to each of those. The input of a comparable residual block in each instance is simply added to its output to move on to the next block.

Proposed PACR-GAN Cloud Removal Architecture

The proposed GAN-based cloud removal consists of parallel spatial multi-attention operations like the Convolutional Block Attention Module (CBAM) and the Coordinate Attention Module (CAM) architectures for capturing scene details and overall structure for reconstructing areas hidden by clouds. CBAM and CAM are pivotal for the proposed PACR-GAN model, in which CBAM sequentially perform channel and spatial attention, enhancing the ability to focus on significant features and regions, while CAM refines this CBAM process by encoding positional information, which further aids in distinguishing between cloud-covered and cloud-free regions. The generator module produces high-quality cloud-free synthetic images from the input cloud-affected images, and with the integration of the CBAM and CAM module, the generator utilizes the spatial attention mechanism and ensures that superior spatial attention features are extracted for the generation of cloud-free images with fine details and overall structural integrity. The discriminator is responsible for distinguishing between generator-produced and real cloud-free images. The discriminator uses a series of convolutional layers to check the input images and determine if they are real or fake. By training in opposition to the generator, the discriminator helps improve the overall quality of the generated images, and finally, the generator fools the discriminator. The detailed architectural details of the CBAM, CAM, generator, discriminator, and loss computations with in-depth model diagrams are discussed below.

Generator Module

The generator module of the proposed cloud removal architecture consists of Spatial Feature Extractors (SFE) and a Parallel Attention Module (PAM). The SFEs include multiple convolutional and residual convolutional blocks arranged in a multiscale fashion. Meanwhile, the PAM is designed using a combination of spatial attention maps from the Convolutional Block Attention Module (CBAM) and the Channel Attention Module (CAM). The overall architecture of the proposed generator module of PACR-GAN is illustrated in Figure 1.

Figure 1. The architectural diagram of the generator module of the proposed PACR-GAN showing the generic convolutions, multi-scale residual convolutional blocks, Parallel Attention Blocks (PAM)

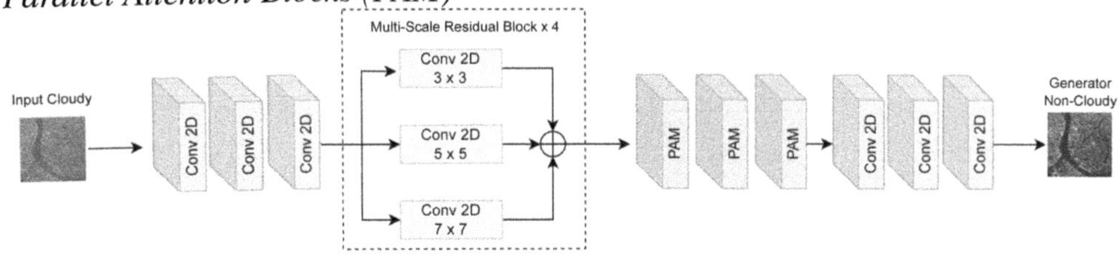

At first, a series of 2D convolutional layers handles input and gathers initial feature maps. These are then passed to the generator module as their input. This is then treated by a branched multi-scale convolutional block, that has convolutional layers of multiple branches with sizes 3×3, 5×5, and 7×7. These convolutions can capture information at different scales. A combined feature representation that can better comprehend the multi-scale features is then generated but concatenating features from each branch. These are then fed to three Parallel Attention Modules (PAM) containing residual connections. The Parallel spatial attention extractors consist of CBAM (Woo et al., 2018) and CAM (CAM) attention. This causes a feature map to reduce unnecessary characteristics while emphasising important ones. Before being sent to the parallel attention structures of CBAM and CAM, the concatenated multi-scale feature map is processed by an initial 2D convolution. To match the spatial resolution of the CBAM-extracted features, the CAM branch computes the CAM attention map, conducts two down-sampling operations, and then performs two up-sampling operations since CAM requires more computing power than CBAM. In the end, the two attention maps are joined and sent to the subsequent layer. The architecture of the proposed PAM module is shown in Figure 2. ReLU activation is applied at each convolutional layer, and the spatial resolution of the feature map is maintained by setting the stride (S) to 1 and using the same padding (P) calculated as $(k-1)/2$, where k is the filter size. A detailed understanding of CBAM and CAM is given in the following subsections for a better understanding.

Figure 2. The architectural diagram of the proposed Parallel Attention Module (PAM) of the PACR-GAN showing the Convolutional Block Attention Module (CBAM), and Coordinate Attention Module (CAM)

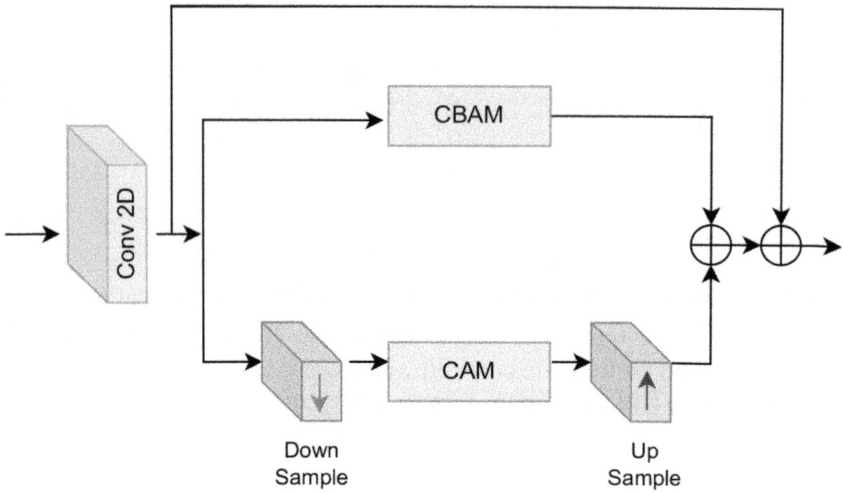

Convolutional Block Attention Module (CBAM)

The CBAM focuses on important spatial and channel areas, highlighting them to improve feature apping in traditional CNN. The Channel Attention Module (ChAM) and Spatial Attention Module (SAM) work sequentially, with ChAM identifying crucial channels and SAM pinpointing relevant spatial locations. The input feature map F of size $c \times h \times w$ undergoes both average pooling and max pooling in parallel across the channels, resulting in two independent features of size $c \times 1 \times 1$. These features are then processed by a fully connected Multi-Layer Perceptron (MLP) with ReLU activation. The outputs of these transformations are then combined and passed through a sigmoid function σ to produce a channel attention map f_c. The channel attention map f_c is multiplied with F to produce an intermediate refined feature map F' of size $c \times h \times w$. The SAM then operates on F', focusing on the significance of each spatial location across the channels. F' undergoes both average pooling and max pooling, generating two distinct feature representations. These are combined into a single feature map of size $2 \times h \times w$, which is then processed by a 7×7 convolution kernel k and a sigmoid activation function, resulting in a spatial attention map f_s with dimensions $1 \times h \times w$. The representations are combined to create a unified feature map of size $2 \times h \times w$, which is then processed with a 7×7 convolution kernel k and a sigmoid activation function to produce a spatial attention map f_s of dimension

$1{\times}h{\times}w$. Subsequently, this spatial attention map is multiplied with F' to generate a final refined feature map F'' with dimensions $c{\times}h{\times}w$, incorporating key spatial and channel information. The following equations describe the CBAM operations, where $+$, $\oplus$, $*$, and $\otimes$ represent addition, concatenation, convolution, and multiplication, respectively. The architecture of the CBAM module, including the overall diagram, ChAM and SAM, is shown in Figure 3. The mathematical derivations for finding the CBAM attention map are given in Equation (2), (3), (4), and (5).

$$f_c = \sigma[MLP(Avg(F)) + MLP(Max(F))] \quad (2)$$

$$f_s = \sigma[k^{7\times 7} * (Avg2D(F') \oplus Max2D(F'))] \quad (3)$$

$$F' = f_c \otimes F \quad (4)$$

$$F'' = f_s \otimes F' \quad (5)$$

Figure 3. The architectural diagram of the Convolutional Block Attention Module (CBAM) showing the Overall CBAM, ChAM, and SAM

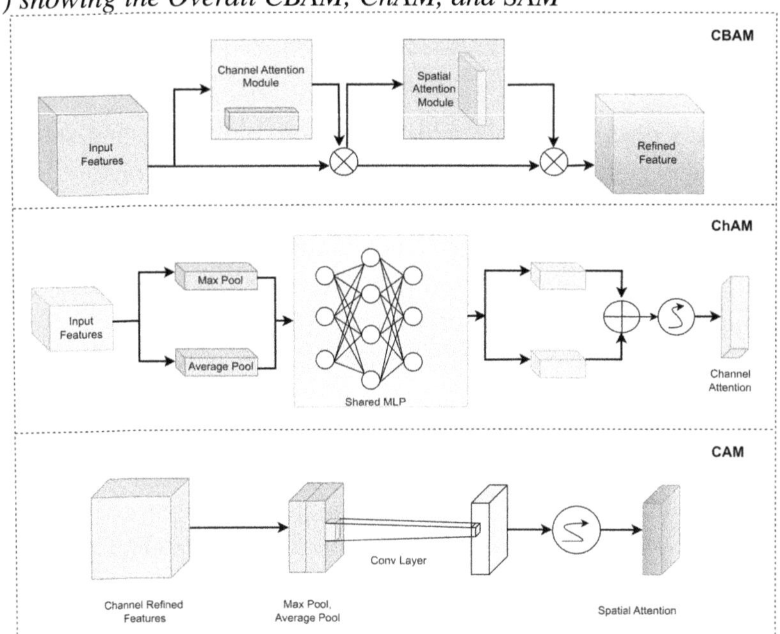

Coordinate Attention Module (CAM)

The CAM incorporates inter-channel dependencies, direction-aware features, and position-sensitive information to enhance the feature map. Two distinct 1D features are produced by the CAM approach, which combines the feature properties in both horizontal and vertical orientations. The feature maps are then independently encoded to create a position-aware attention map that precisely identifies the feature and location of the object of interest. Global average pooling applies encoding to the input I, which has dimensions $c{\times}h{\times}w$. This process is executed across both h and w axes to result in feature maps θ^h and θ^w of size $c{\times}h{\times}1$ and $c{\times}1{\times}w$. Subsequently, the generated feature map is concatenated and treated with a 1×1 convolution, followed by batch normalization and swish activation δ. The result is then divided into two unique tensors $[p^h,p^w]$, and undergo a 1×1 convolution and sigmoid activation to result in another two attention vectors q^h and q^w, with shapes $c{\times}h{\times}1$ and $c{\times}1{\times}w$, respectively. By doing vector multiplication with the original input I, an output tensor I' is generated with dimensions $c{\times}h{\times}w$. This tensor records object location and orientation in addition to inter-channel interactions. Figure 4 shows the CAM algorithm's program flow, and the mathematical formulation of the process is shown in the following Equations (6), (7), and (8).

Discriminator Module

The network's discriminator is built using both multiscale and standard convolutions. Like the generator module, the resulting cloud-free picture is put into a multiscale convolutional structure after being processed by three 2D convolutional layers. This structure consists of three convolutional branches with kernel sizes of 3×3, 5×5, and 7×7. The outputs from these convolutional branches are concatenated and further processed by a series of 2D convolutional layers. To generate a probability score between 0 and 1, this feature map is first processed by a 2D convolutional layer with an output channel size of 1. After that, it is processed by a sigmoid activation function. If the score is near to 1, it indicates the discriminator's confidence that the input is authentic; if it is near to 0, it suggests that the input is fake. The discriminator network also uses appropriate padding-stride settings and ReLU activations to maintain the spatial resolution. The architecture of the discriminator is given in Figure 5.

$$[p^h, p^w] = split\left[\delta\left[\left(f^{1\times 1} * (\theta^h \oplus \theta^w)\right)\right]\right] \tag{6}$$

$$[q^h, q^w] = \left[\sigma\left[\left(f_h^{1\times 1} * p^h\right); \sigma\left(f_w^{1\times 1} * p^w\right)\right]\right] \tag{7}$$

$$I' = I \otimes q^h \otimes q^w \quad (8)$$

Figure 4. The architectural diagram of the Coordinate Attention Module (CAM) of the proposed PACR-GAN Cloud Removal Architecture

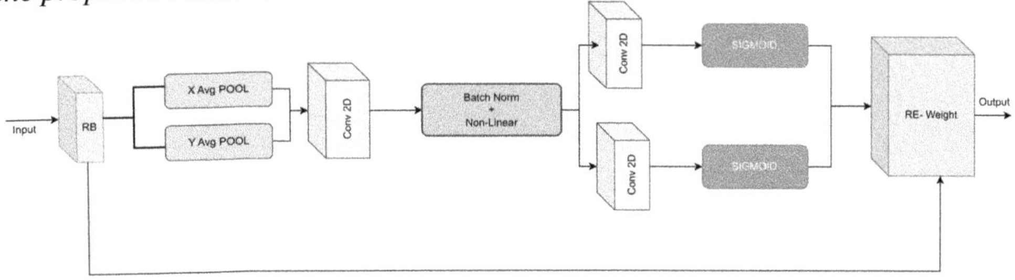

Loss Function Design

The loss function L_{total} for the proposed PACR-GAN is a combination of GAN loss L_{gan}, SStructuralSimilarity Index (SSIM) Loss L_{ssim}, and conventional L_1 loss. These combined loss functions usually result in high-quality, realistic, generative outcomes due to their combined effects in improving the training process with regularization. The overall loss L_{total} is given by the below Equation (9).

$$L_{total} = L_{gan} + L_{ssim} + L_1 \quad (9)$$

Figure 5. The architectural diagram of the discriminator block of the proposed cloud removal framework, showing the generic convolution and multiscale-convolutional blocks

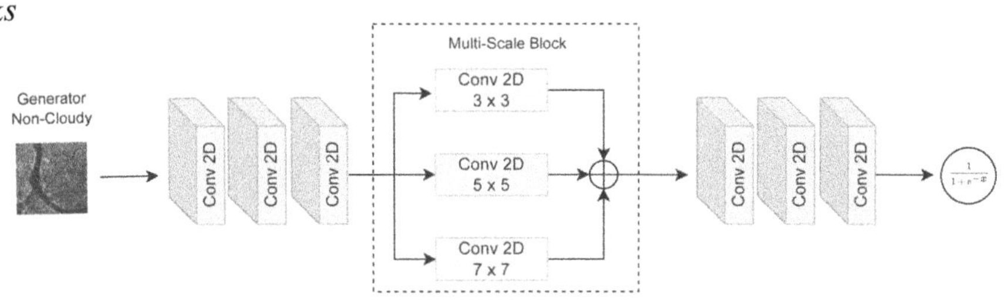

L_{gan} is the combination of generator loss L_G, and discriminator loss L_D, The L_G and L_D is given by the Equations (10) and (11).

$$L_G = -E_y\left[\log\left(1 - D(G(y))\right)\right] \tag{10}$$

$$L_D = E_x[\log(D(x))] - E_y\left[\log\left(1 - D(G(y))\right]\right] \tag{11}$$

Finally, the generator and discriminator combined loss L_{gan} is given by Equation (12)

$$L_{gan} = E_x[\log(D(x))] + E_y\left[\log\left(1 - D(G(y))\right]\right] \tag{12}$$

The L_{ssim} is defined as a metric that quantifies the quality of the generated image and original in terms of structural information, luminance, and contrast. The proposed loss function for PACR-GAN incorporates L_{ssim} to preserve key features and details that might be perceptually important, leading to realism in generated images. The structural similarity between the generated image y and actual image x is given by Equation (13), where μ, σ, σ^2, and c indicate mean, variance, covariance and stabilization constant, respectively.

$$\mathrm{SSIM}(x, y) = \frac{\left(2\mu_x\mu_y + C_1\right)\left(2\sigma_{xy} + C_2\right)}{\left(\mu_x^2 + \mu_y^2 + C_1\right)\left(\sigma_x^2 + \sigma_y^2 + C_2\right)} \tag{13}$$

The L_{ssim} the following Equation (14) defines loss between the real and generated data

$$L_{ssim} = 1 - SSIM(real, generated) \tag{14}$$

The final component of the L_{total} is the standard L_1 loss computed between the real and generated data. The L_1 loss acts as a regularization component and brings generalization to the network being trained. The L_1 loss between the actual and generated image is given in Equation (15), where N is the total number of pixels in the image.

$$L_1(\text{ real, generated }) = \frac{1}{N}\sum_{i=1}^{N} \| \text{ real }_i - \text{ generated }_i \|_1 \tag{15}$$

RESULT AND DISCUSSIONS

Experimental Setup

The RICE-1 dataset is a baseline cloud/mist dataset designed to improve satellite imagery by removing clouds and mist to reconstruct the underlying surface (Lin et al., 2019b). The proposed PACR-GAN network is trained and tested using the RICE-1 dataset, which contains 500 pairs of images, each with a spatial resolution of 512×512 pixels, both with and without clouds. The RICE-1 dataset is curated from the Google Earth Platform. Training of the PACR-GAN model is conducted on the AI computing facility at NIT Puducherry, using an NVIDIA DGX A100 GPU server with 320GB of memory. The model's performance is evaluated using quantitative metrics, including PSNR, SSIM, MSE, and RMSE. A sample from the RICE-1 dataset is shown in Figure 6. The model is trained for 100 iterations with a training and validation batch size of 1 and a learning rate set to 0.0005.

Evaluation Metrics

Evaluation metrics are quantitative measures that can be used to check up on the performance and quality of various algorithms, models, and systems. In the case of image processing and computer vision, they are very important in giving estimates on the quality with which an algorithm performs image compression, segmentation, and classification. In fact, they find wide application in the field of remote sensing to enhance the performance and effectiveness of the various algorithms and techniques applied over satellite or aerial imagery. Evaluation metrics can provide generic, quantifiable, and standardized image quality assessments, ensuring consistent evaluation among all the applied techniques. They also facilitate a fair comparison of different techniques by providing common metrics to evaluate, assess and benchmark the performance and applicability. The selection of evaluation metrics may vary according to the problem domain, and they are chosen based on their capability to identify areas for improvement in methodologies by highlighting specific aspects of quality loss or distortion, which says the user in which area to focus. These evaluation metrics are application-specific and must meet the requirements of various applications, such as medical imaging or satellite imagery, where high fidelity is crucial. The quantification of evaluation metrics further gives directions and ideas to the model designers to refine the algorithm performance and iterate until the user reaches the expected output.

Figure 6. Sample images from RICE-1 dataset columns of (a), (b), (c), and (d) represents thin cloud image and their corresponding non-cloudy counterpart

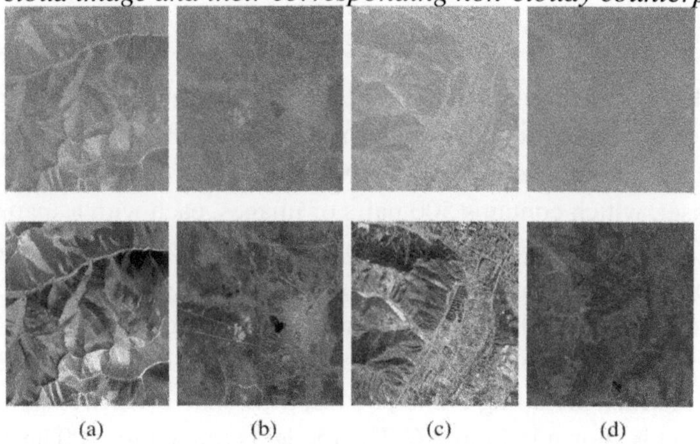

Evaluation metrics themselves do not improve image quality. They are metrics aimed at measuring and ascertaining the quality of an image. They provide insights into how far an image or processing algorithm goes, thus showing the areas for improvement. For instance, if an image has a low Peak Signal-to-Noise Ratio (PSNR) or high Root Mean Squared Error (RMSE), then the quality of the image can still be improved. These metrics thus provide real image quality enhancement by fine-tuning algorithms or processing techniques. Cloud removal or cloud-free reconstruction by GANs involves the quality assessment of the image with respect to how well GANs are located, removing the clouds and reconstructing the scene. This objective quality assessment applies automatic devices and algorithms to ensure that the reconstructed image conforms closely to the original or intended cloud-free image, without human intervention. The proposed PACR-GAN techniques are tested for superiority by comparing them with state-of-the-art techniques using popular evaluation metrics such as Mean Squared Error (MSE), RMSE, PSNR, and SSIM. The detailed mathematical derivations for the evaluation metrics are given below.

1. Mean Squared Error (MSE)

MSE is a metric used to measure the average squared difference between the values of two sets of data. This means the difference is computed between an original image and its generated image obtained by applying various algorithms and techniques. It quantifies the error between these two images by calculating how much the pixel values of one image deviate from the corresponding pixel values of the other. The similarity between the original and generated images is said to be closed if there are lower MSE values, which indicate the quality and reliable performance of the algorithm. Although they always relate to matching human perception to objective values, they are popular because of their simple implementation and low complexity in computation, MSE is the most basic but fundamental and widely used metric in image generation and quality comparison. The derivation for MSE between the real image y_i and the generated image $\tilde{y}_i$ is given in Equation (16)

$$MSE = \frac{1}{n}\sum_{i=1}^{n}(y_i - \tilde{y}_i)^2 \tag{16}$$

2. Root Mean Squared Error (RMSE)

Another popular metric for error quantification is the root mean squared error. RMSE showcase the average magnitude of errors between the predicted and the original image. This indicates that it can quantify the amount of original data that a generative algorithm can reproduce. In general, the RMSE is preferred over MSE because it can provide the measurements of error in the same units as the original image, leading to better interpretability. In contrast, MSE provides the error in squared units, while RMSE scales back to the original units, and it might be easier to understand and compare. When the RMSE is 0, the processed picture is exactly similar to the original picture. In practical cases, smaller values of RMSE are preferred because they show that the resultant image is nearer to the original image. The equation for RMSE between the actual image y_i and the generated image $\tilde{y}_i$ is given in Equation (17)

$$RMSE = \sqrt{\frac{1}{n}\sum_{i=1}^{n}(y_i - \hat{y}_i)^2} \tag{17}$$

3. Peak Signal-to-Noise Ratio (PSNR)

Peak Signal-to-Noise Ratio (PSNR) is another popular measure for determining the quality of a generated image in relation to its original image. It estimates the proportion between the maximum possible value of an image and the power of other noises affecting it. PSNR is given in decibels (dB), where more significant values mean better picture quality and lower distortion. Whereas the latter results in observable differences between the original and converted photos, the former indicates no distortions or mistakes. The many PSNR derivation variations are shown below, with MAX_I denoting the highest value that a pixel may have. For example, if it is an 8-bit color image, 255 will be the maximum possible value. Equation (18) derives PSNR.

$$
\begin{aligned}
PSNR &= 10 \cdot \log_{10}\left(\frac{MAX_I^2}{MSE}\right) \\
&= 20 \cdot \log_{10}\left(\frac{MAX_I}{\sqrt{MSE}}\right)
\end{aligned}
\tag{18}
$$

4. Structural Similarity Index Measure (SSIM)

The degree of structural similarity between a produced and reference image is evaluated by the Structural Similarity Index Measure (SSIM), a measure for image quality. In contrast to conventional metrics such as Mean Squared Error (MSE) or Peak Signal-to-Noise Ratio (PSNR), this technique takes into account variations in contrast, brightness, and structural information that correspond to human visual perception. The range of values for the SSIM is -1 to 1. When the two images have the same structure, brightness, and contrast, they have an SSIM of 1. On the other hand, a score of 0 denotes no resemblance, meaning that these two photos share no characteristics. Negative values mean dissimilarity, indicating that the two images are very structurally different. SSIM has been widely accepted so widely for image quality assessment because it has much better correspondence with human visual perception than traditional metrics like MSE. The derivation for SSIM between the generated image x and original image y is given by Equation (19), where μ, σ, σ^2, and c indicate mean, variance, covariance, and stabilization constant, respectively.

$$
\text{SSIM}(x,y) = \frac{\left(2\mu_x\mu_y + c_1\right)\left(2\sigma_{xy} + c_2\right)}{\left(\mu_x^2 + \mu_y^2 + c_1\right)\left(\sigma_x^2 + \sigma_y^2 + c_2\right)} \tag{19}
$$

Result Analysis

The efficiency of the PACR-GAN architecture is evaluated against four different GAN-based haze removal techniques: DHI (X. Pan et al., 2015), AOD-Net (Li et al., 2017), DehazeNet (Cai et al., 2016), and Cycle-Dehaze (Engin et al., 2018). Figure 7 displays the total loss recorded during both the training and validation phases of model development. Training and validation enhance the model's reliability and generalizability. As training progressed through more epochs, the validation loss decreased and better aligned with the increased training accuracy. Training and validation losses are compared at each epoch, and the model with the slightest difference between these losses is selected as the best model and saved. Table 2 presents the model's performance comparisons, and Figure 8 visually represents the PACR-GAN performance compared to SOTA techniques.

Table 2 presents a relative analysis of metrics of the performance of state-of-the-art (SOTA) cloud removal techniques using four key metrics: PSNR, SSIM, MSE, and RMSE. Peak Signal-to-Noise Ratio (PSNR) evaluates image quality, with increased values signifying improved quality. The structural Similarity Index (SSIM) extends from 0 to 1, where increased values represent better structural similarity. Mean Squared Error (MSE) and Root Mean Squared Error (RMSE) measure error, with lower values signifying higher quality for both metrics. The recommended PACR-GAN model performs better than other techniques in terms of PSNR and SSIM. The model acquired the highest PSNR value of 25.07, Showing better reconstruction quality; it also showcased the highest SSIM score of 0.865, indicating a high degree of structural similarity to the true data. The PACR-GAN model achieved the highest scores for these parameters, but DehazeNet outperformed it regarding MSE and RMSE. The 0.01 MSE difference can be ignored due to the superior PSNR and SSIM. DehazeNet excels in minimizing errors, as shown by its lowest MSE value of 0.15 and RMSE value of 0.38. These figures indicate that DehazeNet has the most minor average and root-mean-squared deviations

Table 2. PACR-GAN performance comparison against SOTA dehaze architectures with respect to popular evaluation metrics of psnr, ssim, mse, and rmse

Models	PSNR	SSIM	MSE	RMSE	CI for SSIM (95%)
DHI	15.01	0.805	0.24	0.48	[0.78, 0.83]
AOD-Net	18.76	0.814	0.36	0.6	[0.79, 0.84]
DehazeNet	23.14	0.838	0.15	0.38	[0.81, 0.86]
Cycle-Dehaze	23.73	0.851	0.22	0.46	[0.81, 0.89]
PACR-GAN (Ours)	25.07	0.865	0.16	0.40	[0.82, 0.91]

from the reference images. Its performance on the PSNR and SSIM evaluation metrics was minimal. Cycle-Dehaze and AOD-Net illustrate average performance. Cycle-Dehaze recorded higher PSNR and SSIM values than AOD-Net but also had slightly higher error values. The PACR-GAN model is excellent at creating realistic and well-structured reconstructions. It is important to note that although DehazeNet is effective at minimizing forecast errors, it does not achieve high scores for quality and structural similarity in its reconstructions.

Figure 7. Loss curves of mse, ssim, rmse, and psnr are shown in left to right top to bottom order

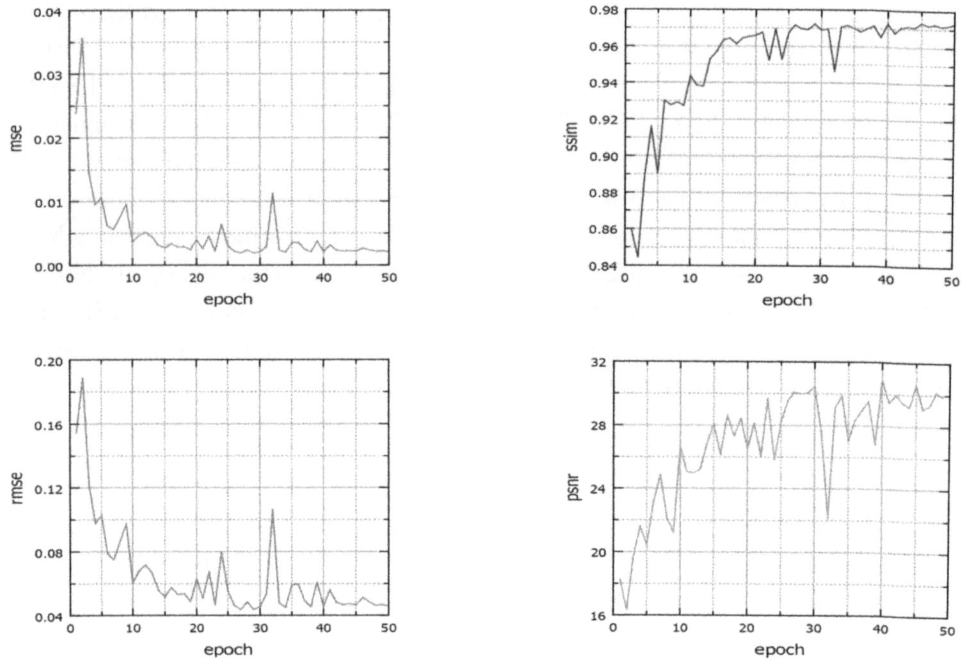

FUTURE RESEARCH DIRECTIONS

The availability of paired datasets is a significant issue in the cloud removal problem, especially in satellite-based remote sensing. This is because a cloud-free, high-quality ground observation may be available at a different revisit time of the satellite, and by that time, ground conditions and land may have changed. This disparity needs to be addressed in future researches. The future of cloud removal in remote sensing images is filled with many promising directions for research, particularly multi-sensor fusion and novel network designs. The proposed methodology is promising in the case of thin cloud removals. However, cloud occurrences are not always thin or haze-like, and there can be thick clouds that fully occlude the essence behind them and provide zero ground information. In such cases, the proposed thin cloud removal fails and may result in poor evaluation metrics. The following discusses the possibilities of incorporating recent multi-sensor fusion for effective cloud removals for all types of clouds. CNN-based multi-sensor fusion can acceptably improve cloud removal quality by integrating data from different sensor types, such as optical and SAR (synthetic aperture radar).

Figure 8. Subjective comparison of the proposed PACR-GAN against SOTA techniques, (a) Hazy Image, (b) Ground-Truth, (c) DHI, (d) AOD-Net, (e) DehazeNet, (f) Cycle-Dehaze, (g) Proposed PACR-GAN

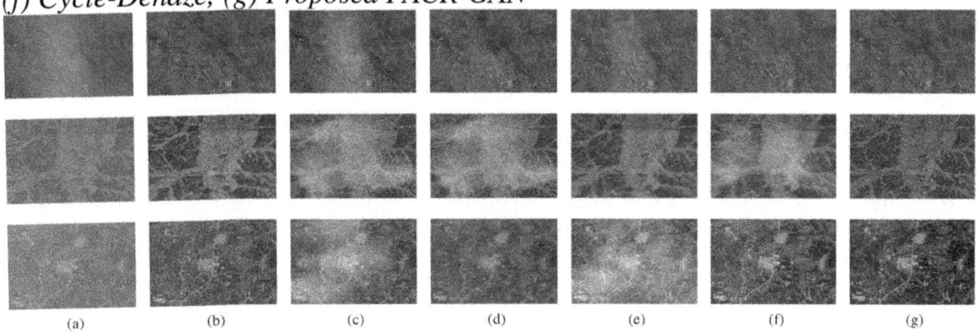

By leveraging the complementary strengths of multiple sensors, CNN models can fuse information to reconstruct clearer images even under challenging conditions. The SAR data, which is unaffected by clouds, will be able to provide structural details that assist in cloud-free reconstruction tasks. GAN-based multi-sensor fusion is another direction in which the generative capabilities of GANs are utilized to synthesize cloud-free images from combined sensor inputs. GANs can comprehend the connection between various sensor modalities, allowing them to create more accurate and visually coherent images without clouds. Aside from the multi-sensor GAN frameworks, other spatial attention frameworks, such as global-local attention, can help enhance feature learning and improve the quality of reconstructed images. One of the difficulties faced while studying cloud removal is the lack of co-registered multi-sensor data. Design for simultaneous SAR and optical sensing can be further investigated to yield a standard reference data set for generating cloud-free reconstruction models. Furthermore, developing sensors capable of capturing both SAR and optical data simultaneously could produce intrinsically aligned synchronized multimodal datasets. This can lead to more accurate and reliable cloud-free data for various applications. Recently, physics-inspired neural networks (PIN) that can model the physical behaviors of clouds or haze as mathematical functions are promising in cloud removal. Such a PIN model can provide insights into the physical principles of light scattering, cloud dynamics, and atmospheric conditions, quantify them and incorporate them into cloud removal architectures so that generative cloud removal becomes more efficient and effective. These models can achieve higher-quality reconstruction by combining data-driven learning techniques with cloud-specific physical knowledge. The accuracy and effectiveness of identifying and eliminating clouds from Earth-observation data would significantly increase by including meteorological data in cloud removal systems. Clarifying the climatic conditions when the image was taken would be much easier by understanding parameters such as temperature, humidity, wind speed, and cloud cover. Adding such data to the models is anticipated to improve the quality of cloud-free reconstruction by distinguishing clouds from other atmospheric events. Future research can examine how real-time meteorological observation and satellite imaging work together to improve cloud removal algorithms and reduce their adaptability to different environmental circumstances.

Our study further finds that cloud removal for satellite photographs can go into the future. Such techniques can provide promising applications for disaster management, responding to natural disasters, agricultural monitoring, and defense operations. These applications require the possibility and validity of real-time data and much more efficient real-time neural architectures to process them. These real-time systems for cloud removal, which enable instant access to cloud-free imagery, are particularly relevant in emergency response situations, where prompt, clear images are crucial for damage assessment, rescue operation planning, and area monitoring. Apart from conventional optical or SAR sensors, more detailed information will be obtained using hyper-spectral and multi-spectral imaging technologies, which can record data over a wide range of wavelengths in the spectrum. When cutting-edge cloud removal algorithms are combined with high-quality remote sensing imagery, the quality and depth of information collected can be greatly enhanced. This will benefit agricultural applications for in-depth soil conditions and crop health research. Further, learning models can be developed to simultaneously and mutually extract temporal and spatial dependencies present in the data, making the reliability of the applications more precise and practical. Another application is the defense and national security perspectives. Considering defense-related applications, these models will support ongoing reconnaissance and surveillance by offering images unaffected by cloud cover. Complementary and additional depth and structural information can be obtained by integrating Light Detection and Ranging (LiDAR) technology with traditional imaging sensors. This is particularly helpful for mapping topography, identifying deforestation, and conducting geological surveys in regions with persistent cloud cover. More comprehensive and precise cloud-free images may result from sophisticated data fusion from multiple sources, including satellite photography, weather models, and ground-based sensors. However, comprehending cloud dynamics and how they interact with weather patterns is crucial for atmospheric research. The capacities of space research agencies and associated organizations with regard to cloud removal will increase by researching and developing these future directions. This will provide more precise and timely data for a wide range of crucial applications that have aided in bettering our understanding of the environment, managing agriculture, conducting defense operations, and responding to natural disasters.

CONCLUSIONS

Effective cloud removal is essential for improving the quality of aerial or satellite photography, as airborne optical data is crucial in many geospatial applications, including disaster management, agricultural evaluation, and environmental monitoring. The cloud-free images are essential in disaster management for damage assessment, relief operations planning, and recovery monitoring. Crop monitoring, yield estimates, and irrigation management in agriculture are enhanced by cloud removal because it is fundamental to food security. Environmental monitoring significantly improves, as cloud-free data provides for accurate assessment of deforestation, changes in biodiversity, and management of water resources. Cloud-free imagery is essential for urban planning to monitor urban sprawl, infrastructure development, and land use patterns. Accurate, cloud-free remote sensing may greatly support the achievement of Sustainable Development Goals (SDGs) by enabling the precise monitoring of natural resources, effective land use, and improvement of climate resilience strategies. For example, cloud-free data will help develop better forestry management towards the carbon sequestered into the atmosphere necessary for mitigating climate change. Also, this was a significant step towards empowering policymakers and enabling them to leverage decisions that would meet the great challenge of balancing development and environmental conservation in a global touch.

Solid and dependable thin cloud removal techniques can enhance these applications' confidence and precision. The main goal of this research is to design and develop a robust, generalizable cloud removal model. The proposed Parallel Attention Guided Generative Adversarial Network for Efficient Thin Cloud Removal from Satellite Imagery (PACR-GAN Model) demonstrated superior performance across most metrics when tested against the RICE-1 dataset. The Convolutional Block Attention Module (CBAM) and the Coordinative Attention Module (CAM), two complex spatial attention processes, are included in the PACR-GAN model. These modules improve the model's capacity to efficiently extract better spatial context, reconstructing high-quality, cloud-free images. The maximum SSIM score of 0.865 and the highest PSNR value of 25.07 were attained by the suggested PACR-GAN model. According to these measures, it generates pictures of superior quality that are most structurally close to the reference images. The findings emphasize the importance of effectively removing thin clouds without sacrificing data integrity or quality. Future research should concentrate on improving the existing techniques for this job, such as taking advantage of newly created cloud detection technologies, physics-informed machine learning algorithms, and advances in computing power. These enhancements intend to increase the overall usability of cloud removal models. Further research may explore how well these models perform in other contexts and datasets.

REFERENCES

Baskar, D., Parambalath, N. K., & Krishnanunni, S. O. (2024). Variational mode decomposition-enhanced thin cloud removal using UNet vision transformer cycle-consistent generative adversarial network. *Journal of Applied Remote Sensing*, *18*(2), 026504–026504.

Cai, B., Xu, X., Jia, K., Qing, C., & Tao, D. (2016). DehazeNet: An End-to-End System for Single Image Haze Removal. *IEEE Transactions on Image Processing*, *25*(11), 5187–5198. DOI: 10.1109/TIP.2016.2598681 PMID: 28873058

Chen, S., Chen, X., Chen, J., Jia, P., Cao, X., & Liu, C. (2016). An Iterative Haze Optimized Transformation for Automatic Cloud/Haze Detection of Landsat Imagery. *IEEE Transactions on Geoscience and Remote Sensing*, *54*(5), 2682–2694. DOI: 10.1109/TGRS.2015.2504369

Deng, Q., Huang, Z., Tsai, C.-C., & Lin, C.-W. (2020). HardGAN: A Haze-Aware Representation Distillation GAN for Single Image Dehazing. In Vedaldi, A., Bischof, H., Brox, T., & Frahm, J.-M. (Eds.), *Computer Vision – ECCV 2020* (pp. 722–738). Springer International Publishing. DOI: 10.1007/978-3-030-58539-6_43

Engin, D., Genc, A., & Ekenel, H. K. (2018). Cycle-Dehaze: Enhanced CycleGAN for Single Image Dehazing. *2018 IEEE/CVF Conference on Computer Vision and Pattern Recognition Workshops (CVPRW)*, (pp. 938-9388). DOI: 10.1109/CVPRW.2018.00127

Gao, Y., Wang, Y., Lv, H., & Qian, J. (2019). A Revised RTM-Based Algorithm to Remove Thin Clouds within Visible Band Data of Sentinel-2A. IGARSS 2019 - 2019 IEEE International Geoscience and Remote Sensing Symposium, (pp. 1418-1421). DOI: 10.1109/IGARSS.2019.8898499

Goodfellow, I., Pouget-Abadie, J., Mirza, M., Xu, B., Warde-Farley, D., Ozair, S., . . . Bengio, Y. (2014). Generative Adversarial Nets. In Z. Ghahramani, M. Welling, C. Cortes, N. Lawrence, & K. Q. Weinberger (Ed.), Advances in Neural Information Processing Systems. 27. Curran Associates, Inc. Retrieved from https://proceedings.neurips.cc/paper_files/paper/2014/file/5ca3e9b122f61f8f06494c97b1afccf3-Paper.pdf

Guo, Y., He, W., Xia, Y., & Zhang, H. (2023). Blind single-image-based thin cloud removal using a cloud perception integrated fast Fourier convolutional network. *ISPRS Journal of Photogrammetry and Remote Sensing*, *206*, 63–86. DOI: 10.1016/j.isprsjprs.2023.10.014

He, X., Ji, W., & Xie, J. (2022). Unsupervised Haze Removal for Aerial Imagery Based on Asymmetric Contrastive CycleGAN. *IEEE Access : Practical Innovations, Open Solutions*, *10*, 67316–67328. DOI: 10.1109/ACCESS.2022.3186004

Isola, P., Zhu, J.-Y., Zhou, T., & Efros, A. A. (2018). Image-to-Image Translation with Conditional Adversarial Networks. Image-to-Image Translation with Conditional Adversarial Networks. Retrieved from https://arxiv.org/abs/1611.07004

Jiang, B., Chong, H., Tan, Z., An, H., Yin, H., Chen, S., Yin, Y., & Chen, X. (2024). FDT-Net: Deep-Learning Network for Thin-Cloud Removal in Remote Sensing Image Using Frequency-Domain Training Strategy. *IEEE Geoscience and Remote Sensing Letters*, *21*, 1–5. DOI: 10.1109/LGRS.2024.3389995

Jinqi, H. (2024). Thin Cloud Removal Generative Adversarial Network Based on Sparse Transformer in Remote Sensing Images. *Remote Sensing (Basel)*, *16*(19), 3658–3658. DOI: 10.3390/rs16193658

Khetkeeree, S., Petchthaweetham, B., Liangrocapart, S., & Srisuk, S. (2020). Sentinel-2 Image Dehazing using Correlation between Visible and Infrared Bands. 2020 8th International Electrical Engineering Congress (iEECON), (pp. 1-4). DOI: 10.1109/iEECON48109.2020.229585

Kong, J., Hu, G., & Liang, D. (2010). Thin Cloud Removing Approach of Color Remote Sensing Image Based on Support Vector Machine. *2010 Asia-Pacific Conference on Wearable Computing Systems*, (pp. 131-135). DOI: 10.1109/APWCS.2010.39

Li, B., Peng, X., Wang, Z., Xu, J., & Feng, D. (2017). AOD-Net: All-in-One Dehazing Network. *2017 IEEE International Conference on Computer Vision (ICCV)*, (pp. 4780-4788). DOI: 10.1109/ICCV.2017.511

Lin, D., Xu, G., Wang, X., Wang, Y., Sun, X., & Fu, K. (2019). A Remote Sensing Image Dataset for Cloud Removal. CoRR, abs/1901.00600. Retrieved from http://arxiv.org/abs/1901.00600

Liu, J., Hou, W., Luo, X., Su, J., Hou, Y., & Wang, Z. (2022). SI-SA GAN: A Generative Adversarial Network Combined With Spatial Information and Self-Attention for Removing Thin Cloud in Optical Remote Sensing Images. *IEEE Access : Practical Innovations, Open Solutions*, *10*, 114318–114330. DOI: 10.1109/ACCESS.2022.3213354

Lyu, H., Qian, J., & Wang, Y. (2024, July). Study on the Unified Theory of Thin Cloud Detection and Removal based on Physical Model. In IGARSS 2024-2024 IEEE International Geoscience and Remote Sensing Symposium (pp. 5549-5552). IEEE.

Mirza, M., & Osindero, S. (2014). Conditional Generative Adversarial Nets. Conditional Generative Adversarial Nets. Retrieved from https://arxiv.org/abs/1411.1784

Namboodiri, A. S., Kumar Sanodiya, R., & Arun, P. V. (2023). Remote Sensing Cloud Removal using a Combination of Spatial Attention and Edge Detection. 2023 11th International Symposium on Electronic Systems Devices and Computing (ESDC), 1, pp. 1-6. DOI: 10.1109/ESDC56251.2023.10149875

Pan, H. (2020). Cloud Removal for Remote Sensing Imagery via Spatial Attention Generative Adversarial Network. Cloud Removal for Remote Sensing Imagery via Spatial Attention Generative Adversarial Network.

Pan, X., Xie, F., Jiang, Z., & Yin, J. (2015). Haze Removal for a Single Remote Sensing Image Based on Deformed Haze Imaging Model. *IEEE Signal Processing Letters*, *22*(10), 1806–1810. DOI: 10.1109/LSP.2015.2432466

Prinsa, K., & Saritha, E. (2018). Detection and Removal of Clouds on Remote Sensing Images. 2018 4th International Conference for Convergence in Technology (I2CT), (pp. 1-5). DOI: 10.1109/I2CT42659.2018.9057879

Radford, A., Metz, L., & Chintala, S. (2016). Unsupervised Representation Learning with Deep Convolutional Generative Adversarial Networks. Unsupervised Representation Learning with Deep Convolutional Generative Adversarial Networks. Retrieved from https://arxiv.org/abs/1511.06434

Singh, P., & Komodakis, N. (2018). Cloud-Gan: Cloud Removal for Sentinel-2 Imagery Using a Cyclic Consistent Generative Adversarial Networks. IGARSS 2018 - 2018 IEEE International Geoscience and Remote Sensing Symposium, (pp. 1772-1775). DOI: 10.1109/IGARSS.2018.8519033

Song, C. H., Han, H. J., & Avrithis, Y. (2021). All the attention you need: Global-local, spatial-channel attention for image retrieval. CoRR, abs/2107.08000. Retrieved from https://arxiv.org/abs/2107.08000

Toizumi, T., Zini, S., Sagi, K., Kaneko, E., Tsukada, M., & Schettini, R. (2019). Artifact-Free Thin Cloud Removal Using Gans. *2019 IEEE International Conference on Image Processing (ICIP)*, (pp. 3596-3600). DOI: 10.1109/ICIP.2019.8803652

Woo, S., Park, J., Lee, J.-Y., & Kweon, I. S. (2018). CBAM: Convolutional Block Attention Module. CoRR, abs/1807.06521. Retrieved from http://arxiv.org/abs/1807.06521

Wu, Z., Li, J., Wang, Y., Hu, Z., & Molinier, M. (2020). Self-Attentive Generative Adversarial Network for Cloud Detection in High Resolution Remote Sensing Images. *IEEE Geoscience and Remote Sensing Letters*, *17*(10), 1792–1796. DOI: 10.1109/LGRS.2019.2955071

Yang, Q., Wang, G., Zhao, Y., Zhang, X., Dong, G., & Ren, P. (2020). Multi-Scale Deep Residual Learning for Cloud Removal. IGARSS 2020 - 2020 IEEE International Geoscience and Remote Sensing Symposium, (pp. 4967-4970). DOI: 10.1109/IGARSS39084.2020.9323261

Yu, W., Zhang, X., Pun, M.-O., & Liu, M. (2021). A Hybrid Model-Based and Data-Driven Approach for Cloud Removal in Satellite Imagery Using Multi-Scale Distortion-Aware Networks. 2021 IEEE International Geoscience and Remote Sensing Symposium IGARSS, (pp. 7160-7163). DOI: 10.1109/IGARSS47720.2021.9554963

Zhan, C. (2024, May). Tan., Wang, Man., Gui, Song, Wang., Qin, Nie. (2024). Unsupervised remote sensing image thin cloud removal method based on contrastive learning. *IET Image Processing*, *18*(7), 1844–1861. Advance online publication. DOI: 10.1049/ipr2.13067

Zhang, C., Zhang, X., Yu, Q., & Ma, C. (2022). An Improved Method for Removal of Thin Clouds in Remote Sensing Images by Generative Adversarial Network. IGARSS 2022 - 2022 IEEE International Geoscience and Remote Sensing Symposium, (pp. 6706-6709). DOI: 10.1109/IGARSS46834.2022.9884704

Zi, Y., Xie, F., Zhang, N., Jiang, Z., Zhu, W., & Zhang, H. (2021). Thin Cloud Removal for Multispectral Remote Sensing Images Using Convolutional Neural Networks Combined With an Imaging Model. *IEEE Journal of Selected Topics in Applied Earth Observations and Remote Sensing*, *14*, 3811–3823. DOI: 10.1109/JSTARS.2021.3068166

KEY TERMS AND DEFINITIONS

Active Remote Sensing: Active remote sensing makes use of sensors, like radar or LiDAR, that have their own energy source. These sensors deliver energy to the target and then measure the signal that comes back. When gathering data at night and in any type of weather or environment, this approach works best since it eliminates the impact of energy from natural light sources like the sun.

Cloud Removal: The process of removing clouds from satellite or airborne imagery that are occurring in either thin or thick format

Convolutional Neural Network: A type of deep learning model particularly effective in image processing tasks based on convolutional operations over signals.

Generative Adversarial Network: A deep learning framework consisting of two neural networks, a generator and a discriminator, that compete with each other and produce realistic images.

Geospatial Applications: Remote sensing provides invaluable data for various geospatial applications, such as environmental monitoring, urban planning, and resource management, by observing and analyzing the Earth's surface and atmosphere from a distance.

Passive Remote Sensing: Using natural energy sources, mostly sunlight, passive remote sensing measures the energy that an item radiates or reflects back. Images in the visible, near-infrared, and thermal infrared spectrums are frequently captured with it.

Remote Sensing (RS) Images: Satellite or aerial sensor images are used for surface and atmospheric monitoring and analysis. Applications for these photos include agricultural supervision, disaster identification and management, and monitoring of land and water.

Sentinel-2: A pair of satellites (Sentinel-2A and Sentinel-2B) equipped with high-resolution cameras that capture images in 13 spectral bands.

Spatial Attention: A technique in deep learning models that focuses on specific regions of an image to improve accuracy in tasks like classification, segmentation and image generation.

Thin Clouds: These clouds allow some ground data to be captured in satellite images despite partial obstruction. Removing thin clouds can significantly enhance the usability of remote-sensing images for various applications.

Chapter 2

Generative Adversarial Networks in Remote Sensing and Its Applications:

An Overview and Motivations

Rufai Yusuf Zakari
https://orcid.org/0000-0002-4645-6412
Skyline University, Nigeria

Auwal Sagir Muhammad
Xiamen University, China

ABSTRACT

In recent years, the intersection of machine learning (ML) and remote sensing has led to transformative advancements, with Generative Adversarial Networks (GANs) emerging as a powerful tool for generating synthetic images and enhancing data analysis. This chapter provides an overview of GANs' applications in remote sensing, starting with an introduction to their architecture, comprising a generator and a discriminator trained simultaneously. GANs are particularly valuable in augmenting remote sensing datasets, generating multispectral and hyperspectral images, enhancing image resolution, and enabling image-to-image translation. The chapter also explores GANs' role in data simulation for environmental monitoring and predictive modeling, especially in climate change research. Challenges such as training stability, computational demands, and ensuring geographic accuracy are discussed, along with future research directions, including integrating GANs with AI techniques like reinforcement and transfer learning. This chapter explores

DOI: 10.4018/979-8-3693-6900-5.ch002

the potential and challenges of GANs in remote sensing

1 INTRODUCTION

Generative Adversarial Networks (GANs) have been proven to be a well-known and state-of-the-art generative model in the field of computer vision. GANs aim to generate realistic or interesting samples from the distribution of the training dataset. The last decade has witnessed great interest in the development of GAN research, and it has also been widely used in remote sensing and various fields, including popular generative tools such as ChatGPT and DALL·E of OpenAI, GitHub Co-pilot, Stable Diffusion, Runway machine learning, and Mid Journey (Rui et al., 2021; Shafik, 2024a). This is mainly due to the development of deep learning, the relatively low cost of GAN model architecture in practical applications, and the birth of many improved models of GANs even if it has some notable impacts in application and consideration. Nowadays, the practical value of GANs in the field of remote sensing has been increasingly revealed, which is the result of the efforts of many researchers. The main goal is to present an overview of GANs in the field of remote sensing and discuss their applications.

In general, GAN models can help people solve two problems in the field of remote sensing, as early presented (Paola et al., 2023). The first one is the unavailability of the corresponding paired or unpaired data when training the model. GANs with an unsupervised learning structure can synthesize corresponding high-level feature or category label images. The second problem is that a small-sized labeled training set may lead to poor generalization performance of the model. GANs can generate enough training data and limit the classifier to learn some simple characteristics to reduce overfitting performance (Wu et al., 2020). Up to now, GANs can be used to solve the small labeled training set problem in the field of remote sensing. There are more than 30 GAN-based architectures that can help to produce large-scale training data in the field of remote sensing. It is expected that our findings, such as the literature on generating remote sensing images with GANs or the characteristics of the datasets/images, will be helpful for readers exploring and structuring future work or studies (Paola et al., 2023).

GANs are a system that consists of two neural networks. Starting with the generator network, it learns to generate synthetic data, for instance, images. Then, a discriminator network is trained to detect whether the received image is from the synthetic images or the real images. Both networks are trained together iteratively or simultaneously, aiming to improve the quality of the synthetic images. One major significance of GANs is to produce high-quality and realistic synthetic data, on top of generating images with high fidelity and diversity (Pengqiang et al., 2020). There have been many extensions or improved versions of GANs, gaining remarkable attention from the community to advance the GAN models. Many GAN models have been improved to solve some of the existing problems, for instance, conditional GANs for control of the features of a generated instance, to ease the incorporation of additional features or characteristics to the synthetic images, likewise, style transfer GANs that are able to perform style transfer from a base image during the synthesis phase, and many others. Several deep-learning conferences have dedicated workshops and tutorials on GANs (Zhang et al., 2020). Because of these promising results, mobile applications, including social media, computer game developers, smartphone augmented reality, and visual effect developers, have started to use GANs to generate synthetic data or add some additional characteristics or features, in other words, to modify data and perform image inpainting. For example, one can generate a synthetic face or background with ease. It can also be beneficial to create some synthetic images of remote sensing data for different types of applications and analyses that are not possible or not easy to collect ground truth data (Eltehewy et al., 2023). It is very challenging to create synthetic images that are used to perform many tasks using remote sensing data, and GANs promise to solve the problems related to the synthesis of data for many tasks. Whereas the generator has to generate realistic data samples, the discriminator assesses these created samples for the reality of the data in order to identify the difference between them. According to (Paola et al., 2023), both competitive dynamic forces keep on improvising the networks because, in this process, the generator is skilled at producing plausible data, whereas the discriminator can easily identify slight variations in real or generated data. This adversarial process results in the generation of synthetic data with high fidelity to the original dataset over time, hence making GANs a powerful tool for a wide range of applications (Wu et al., 2020).

In this regard, some of the most promising applications of GANs involve remote sensing, where they are expected to solve some of the critical challenges that researchers and practitioners face (Rui et al., 2021). Remote sensing is the acquisition of information regarding the Earth's surface and atmosphere by satellite or aerial imagery. This information is of prime importance for various applications such as environmental monitoring, urban planning, agriculture, disaster management, and climate change studies (Paola et al., 2023). Acquiring remote-sensing data is usually bound by a number of issues, including high cost, limited temporal and spatial resolution, incomplete coverage due to cloud obscuration, malfunctioning sensors, and/or restricted access to certain areas. GANs, in this way, offer a very striking balance to these challenges by offering high-quality synthesized remote sensing data that can extend the existing datasets by their resolution or create completely new data where it is lacking (Pengqiang et al., 2020; Rewhel et al., 2020). For example, in environmental monitoring, GANs generate multispectral and hyperspectral images that give critical information about vegetation health, soil moisture, and water bodies—indispensable in monitoring environmental changes over time. In urban planning, GANs are able to simulate urban growth scenarios or reconstruct missing data in satellite images, hence providing planners with more complete and accurate datasets on which to base informed decisions when needed (Zhang et al., 2020; Shafik, 2024b).

Apart from the preceding points, the potential use of GAN in disaster management, where effective and accurate timing of the required data helps, is also limitless. GAN provides whole areas that are continuations of real image data wherein places may be blocked out by clouds or other interfering surfaces of the Earth; this will further allow for finer evaluations of damages and help organize relief more systematically (Eltehewy et al., 2023). The capability to simulate different environmental eventualities, whether it be flooding, drought, or forest fire, also allows the modeling of various disaster scenarios and the prediction of their impact by researchers, leading to better preparedness and mitigating strategies. One of the key strengths of GANs lies in their capacity to learn and mimic complex data distributions (Li et al., 2020).

1.1 Significance of Remote Sensing in Various Fields

The ultimate goal of remote sensing technologies is to capture, sense, and observe the Earth and its associated processes at multiple scales. It is widely practiced for a variety of applications, including agriculture, forestry, urban planning, disaster management, and environmental monitoring. Remote sensing helps scientists, technologists, and policymakers in decision-making by providing essential information on a global long-term basis (Pan et al., 2019). In the field of agriculture, farmers in different locations can use remote sensing data to decide the scope and type of investment in agricultural activities. In environmental science, remote sensing data can monitor environmental pollution and urban sprawl. Using remote sensing data, weather-related forecasts and other applications can be successfully operated. Nowadays, land cover change, vegetation cover, and crop classification are better understood due to high internet availability from space (Zhang, et al., 2021). A foundational aspect frequently emphasized is the importance of high-quality imagery for accurate interpretation, good analysis, and successful output from any remote sensing applications. Thus, in remote sensing, accuracy and precision of data are the two main attributes that influence decision-making scenarios. In recent years, we can infer that the use of remote sensing data has been increasing in various research domains, mainly in sensor image restoration and super-resolution. Notably, research in remote sensing has seen a significant increase following the launch of various high- and medium-resolution satellite sensors (Nyberg et al., 2021; Zheng et al., 2022). Additionally, scientists integrated remote sensing with cloud environments and various data analytical frameworks, including hybrid frameworks, to extract deep insights, as a block diagram of GANs is illustrated in Figure 1.

Figure 1. Systematic Generative adversarial networks (GANs) architecture

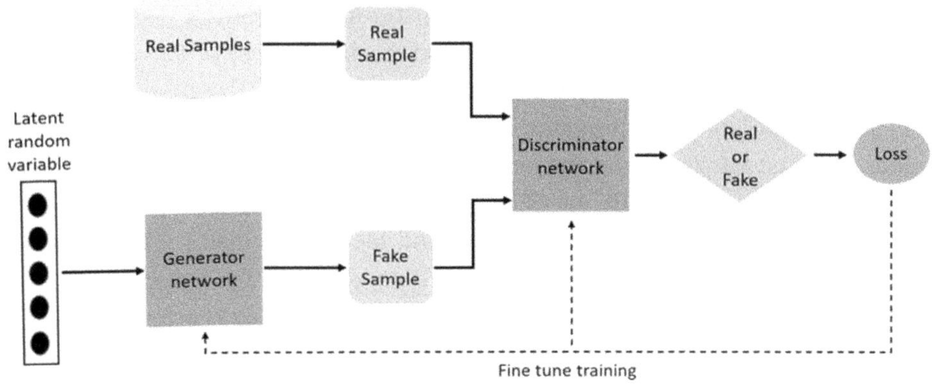

1.2 The Chapter Organization

Given these ongoing GAN developments in technologies and their applications aspect, one of the promising areas is the use of generative models for various remote sensing applications. We introduced the background of GANs, remote sensing, and their intersection, as well as the significance and contributions in Section 1. In Section 2, after introducing the basic ideas of GANs, we discuss the main reasons for introducing GANs into the remote sensing area. The research status of remote sensing with GANs in some specific application scenarios has been determined in Section 3. Our research application and research discussion are presented in Section 4. Section 5 presents the identified prospectus relating to the GANs, and finally, Section 6 presents the chapter conclusion.

2 FUNDAMENTALS OF GENERATIVE ADVERSARIAL NETWORKS

GANs are a type of generative model that allows learning from data so that after some process, it can generate data that is coherent or close to real. The model consists of two different neural networks: the generator and the discriminator. The generator neural network aims to generate data from random noise, resulting in a distribution of the data, and the output is near to the real. However, this is opposed by the neural network called the discriminator, which aims to segment the input between real data and fake data generated by the generator neural network (Gao et al., 2023; Choi et al., 2018). This results in competition between the networks, where the generator's objective is to be able to deceive the discriminator with sophisticated features and vice versa. The generated data can be trained in internal or normal training and can be optimized based on how it can provide realistic-looking results. Technically, the GAN consists of two models: the generator and the discriminator. The generator is a deep network that takes some input and then begins to synthesize data (Zheng et al., 2022). Meanwhile, the discriminator is another deep network that takes some data as input and then labels the data as either coming from a real dataset or the generator's datasets. Importantly, the model is trained using adversarial methods, which have a loss function between the generator and the discriminator. The network is then expected to be optimized based on game theory. Generally, a machine learning model is composed of supervised and unsupervised learning and currently enhanced due to the introduction of quantum computing (Zhu et al., 2018; Shafik, 2025). The GAN is included in generative models, unsupervised learning, and game theory.

2.1 Architecture and Components of GANs

The basic building block of a GAN is a pair of two components, notably the generator and the discriminator. The generator aids in creating synthetic data, while the discriminator evaluates the authenticity of that data. Both mechanisms are trained simultaneously, where the competition between the generator and the discriminator steadily improves their performance. The objective of designing a GAN is to generate realistic synthetic data, which is achieved by driving the output produced by the generator to be closer to the real data (Liu et al., 2021). The generator will aim to minimize the easy-to-get loss functions and output the data to confuse the discriminator while the discriminator is trying to distinguish real data from the fake one produced by the generator. The interplay between these two networks and the capability to learn from each other essentially gives the GAN structure. There is no one-size-fits-all in the architecture of GANs. Several variations exist that make GANs either faster or produce better samples. The basic GAN architecture can have multiple layers called the deep convolutional GAN (Miftahushudur et al., 2024). The entangled GAN is a GAN network with the capacity in the network increased such that the concatenation of the noise has a data dimension lower than the network output dimension. An alternative to this approach is the self-entangled GAN, which allows the noise vector to have a dimension equal to the generator's input data to eliminate weight misalignment. Nonetheless, training GANs has problems, such as mode collapsing or convergence, that, with extensions, have to be mitigated. All these different configurations are suited for various considerations and application environments (Pan et al., 2019). Broad knowledge of these architectures is required so that the proper type of GANs can be used for relevant remote sensing data analysis.

2.2 Training Process of GANs

Effective training of GANs is key to their performance in various applications. During the training, the generator is tasked with producing realistic-looking data, while the discriminator is trained to differentiate the generated samples from the real ones. The process is iterative and continues for a long time. During this time, the generator learns from the adversarial gradient acquired using a concept from the discriminator, while the discriminator learns based on a comparison of the real and generated data (Ma et al. 2021). Many methods, including different loss functions and techniques, which are applied to balance the generator and the discriminator, have been proposed to optimize and stabilize the training of GANs. In most methods, finding the balance between the generator and the discriminator at the convergence phase is crucial. Despite the remarkable success of GANs in various fields, they suffer from many pitfalls and issues, such as ambiguity and mode collapse, which are often encountered in the training process of GANs (Yu et al., 2022). Similar challenges and issues have been experienced in the applications of GANs in remote sensing. In most studies, GANs fulfill the same activities of photorealistic image generation and completion. Consequently, the success of a GAN model in remote sensing mostly depends on how effectively a model is trained to capture the characteristics of an aerial image. Therefore, training a GAN in remote sensing is one of the primary requirements for a useful generated image. In a broader sense, a GAN model's success in remote sensing relies on its training, thereby rendering an intense discussion of training methodology for GANs in remote sensing (Yu et al., 2022; Zhang et al., 2021). Understanding the training properties will further enhance the capability to review the models in the various applications of GANs as the taxonomy of GANs Application in Remote Sensing, as illustrated in Figure 2.

Figure 2. Taxonomy GANs Application in Remote Sensing

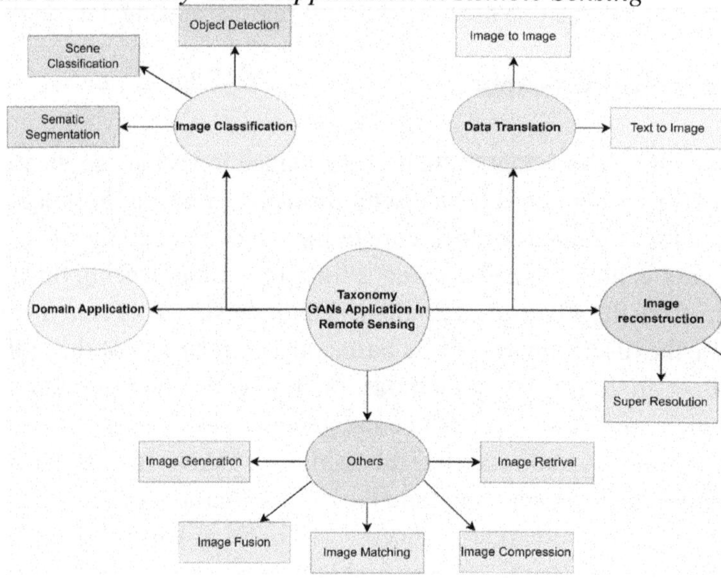

3 APPLICATIONS OF GANS IN REMOTE SENSING

GANs have, in their brief existence, provoked genuine interest in research hotspots such as energy, environment, and urban-rural informatics at the intersection of artificial intelligence, geospatial big data, and remote sensing. GAN generators can produce images of high quality, unprecedented for a machine learning system, to both synthesize more comprehensive and meaningful latent representations and create new knowledge, exploiting the huge capacity of the generator as a flexible, plausible, and smooth function fit for the multi-dimensional input distributions (Yu et al., 2022; Zhang et al., 2021). This explains GAN's increased attention and provides new possibilities in remote sensing-based imaging analysis and understanding. Addressing imaging in remote sensing, current challenges in conventional GAN applications can be broadly characterized by the fact that remote sensing imaging has additional aspects beyond visible light and the other side of the application in a confrontational dual (we propose the D2 side, the demand for data side) (Gao et al., 2023; Ali-Gombe et al., 2019). The GAN application in remote sensing and its corresponding motivation can be described from the following three perspectives. The GAN application in remote sensing not only enhances interpretation, enabling new capacity, but also extracts essential world information through more efficient strategies, improving the availability of insight-based remote sensing decision support systems and procedures. Traditionally, conventional techniques were not effective at these points in remote sensing magnetic and gravity image interpretation (Miftahushudur et al., 2024).

3.1 Image-to-Image Translation

GANs have achieved attention in the remote sensing field not only because of their powerful generation ability but also because of their virtually accessible applications. Among these, we focus on the image-to-image translation task. In remote sensing, because different kinds of sensors mounted on satellites record a different spectral range of the Earth, it is useful to convert images from one domain to another, which is exactly the form of satellite image conversion. In other cases, converting low-resolution images to high-resolution ones for accurate interpretation is necessary. Many works use GANs for satellite image reconstruction (Shamsolmoali et al., 2020). Furthermore, it is meaningful to extract useful insights from the latent data; for example, cycle-consistent GANs are often used to extract data from some new bands. To accomplish this job, GANs convert the multimodal dataset to the 2D dataset and reduce the many-to-many problem as an image-to-image translation task. For image translation using GANs, the biggest problem is how to convert the domain style from source to target while keeping the content consistent (Sara, et al., 2023). In the beginning, GAN was derived by adding a content-constrained term to the original GAN, and then eventually, cGAN was combined with GAN for condition enhancement. To alleviate the inconsistency problem and convert an input map to numerous different styles, multi-modal GAN emerged. For the GANs mentioned thus far, GAN reveals its effectiveness at many image translation tasks, including unpaired image-to-image translation, super-resolution, and de-hazing generation. GANs have been successfully used to fuse high-resolution multispectral images (MSIs) with low-resolution hyperspectral images (HSIs), thereby improving spatial and spectral resolution (Xiao et al., 2021; Su et al., 2022).

To perform large-scale image translation, and indeed, if an unsupervised approach were to be proposed in this field, many different GANs would be integrated, which can be built on arbitrarily many convolutional layers. Despite the achievement of style transfer in the domain, the hurdle in the design of GANs is the capacity to model raw Earth observation at the smallest computational budget. In practice, since the current remote sensing datasets contain various environments and data qualities, GANs for imagery transformation need to be designed to suit the tasks (Ma, et al. 2021). In general, learning the image-to-image translation pairwise is an extremely enticing direction because of the possible applications. In environmental monitoring, the transfer from one dataset to another identifies the poorly constructed greenhouses that would not normally appear. Given the AI progress in turning certain spectral configurations of the solar spectrum into expected RGB imagery, the principal areas of further application of this task in remote sensing remain under development. GANs make sense for those supervised remote sensing application cases that update the generative model from spectral data for recognizing land cover results (Liu et al., 2021). However, additional general spirit segments with labels "road" or "not a road." The value of the data can be distributed by GAN in a variety of formats, even in radiometric imagery. Large amounts of data are created, as seen in this training. Consider, for example, how pixel-based classifications would be biased by five stock images of a desert landscape, with speculative growth causing biases in many distant sensing classification assignments (Zhuet al., 2023; Zhu et al., 201). Like other applications of GAN for object segmentation, there is inadequate depth in identifying the textures of either field. Several processes are followed to enhance images that are used in GANs, as illustrated in Figure 3.

Figure 3. Process of Enhancing Imaging

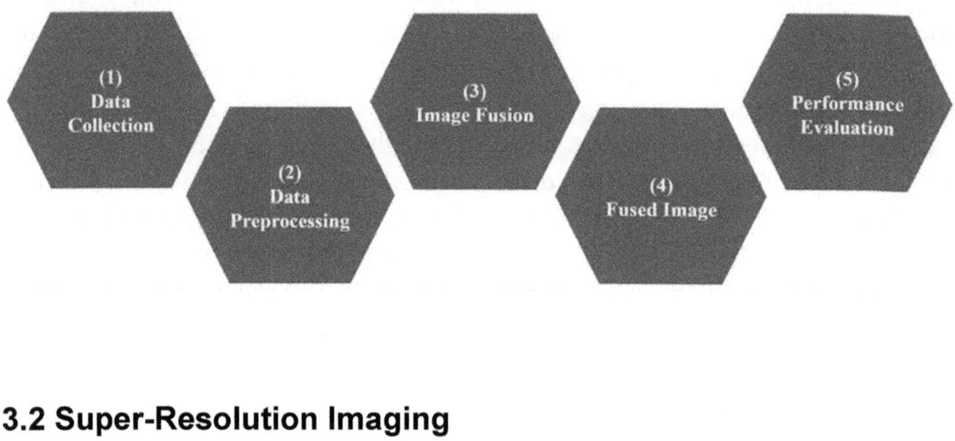

3.2 Super-Resolution Imaging

Super-resolution imaging is the process of improving or increasing the resolution of images. In the remote sensing context, this means taking lower-resolution satellite images and improving them to a higher resolution. In doing so, the GAN-enhanced images should closely follow the real distribution of images, thereby looking real. Super-resolution imaging is an important part of remote sensing imagery analytics and also acts as the handler of other remote sensing imagery-based applications (SMOTified-GAN; Ali-Gombe et al., 2019). Therefore, it is important to adopt GAN further to train the super-resolution function of GANs. The effectiveness of GAN in super-resolution is illustrated in the applications of natural scene images. A super-resolution generative adversarial network achieved good results using a combination of a per-pixel mean absolute error and a radiance per-pixel error in providing outputs that are 'convincing.' Training a generator specifically for super-resolution in the GAN framework is challenging because of two dichotomies (Ahsan, et al., 2022; Choi et al, 2021):

(1) realism against fidelity, and
(2) viability at the current scale versus long-term optimism.

With respect to the realism-fidelity balance, it refers to the idea that an image that looks real does not necessarily contain the fine details that a human can differentiate when comparing an enhanced satellite image to the ground truth image. Super-resolution GANs are typically trained using an adversarial loss function in combination with a pixel-wise loss function. The adversarial loss function aims to make the output of the generator look real, while the pixel-wise loss function measures the difference between the generated image and the corresponding ground truth (Sharma et al., 2018; Li et al., 2022; Cao et al., 2022). Although we may intend to train the SRGAN to generate high-resolution ground truth images, the adversarial loss function may bias the generator towards producing high-resolution outputs that are not necessarily faithful to the input at the current stage of training. Super-resolution GANs are typically trained using an adversarial loss function in combination with a perceptual loss function, which measures the distance between deep features extracted from the generated image and the corresponding ground truth (Kazeminia et al. 2018). This approach trains the generator to output high-resolution images that are faithful to the input over the adversarial loss while at the same time being able to extract deep features from the generated image that are similar to the deep features in the ground truth image.

4 CHALLENGES AND LIMITATIONS

The GAN framework exhibits outstanding performance when generating realistic remote sensing data. Despite the recent tremendous advancements, several fundamental challenges and limitations still exist, which hinder a series of GAN-based applications from playing a pivotal role in the field. Data scarcity: Compared with computer vision data, the amount of remote sensing data is extremely smaller, not to mention high-quality labeled data (Paolanti et al., 2023). The generalization of the GAN model can largely depend on the quantity and quality of the derived labeled training samples. If not, the GAN model would tend to generate biased results when implemented in practical situations. Transferability: Remote sensing possesses a wide range of ground-object categories and illumination landscapes. Fusing multiple remote sensing data into a model to generate diverse results remains a non-trivial task due to the scene's complexity (Crivellari et al., 2023). Data properties: Each remote sensing domain is restricted by the specific physical and environmental factors that govern the construction of the domain. This gives rise to the lack of detailed local context and results in the automatically convinced logic corresponding to loss functions and priors. The application of GAN in remote sensing furthers with more challenges and limitations. The comparative difference in the newly created diversity evaluation methods does not indicate the robustness and dependability issues that seriously hamper practical applications (Huang et al., 2023). In the coming years, we expect that generating remote scenarios based on GAN will address more work in terms of standardizing and confirming the external development and improvement of the model to ensure the performance of GAN models. In summary, despite the practical limitations, potential modifications or improvements may be considered to tackle the identified limitations in future research (Alajaji et al., 2024).

4.1 Data Scarcity and Quality Issues

While GANs are reported to be a promising tool for remote sensing, several shortcomings limit their effectiveness in remote sensing applications. GANs are inherently data-intensive and require a vast number of samples for training that are often not available in many remote sensing applications because of the high efforts needed for the acquisition of representative training data, such as samples from different remote sensing sensors, platforms, spatial scales, and so on. The nonexistence of large-scale labeled datasets in remote sensing makes the training of generative models more difficult (Jozdani et al., 2022; Ahsan, et al., 2022). The quality of the collected data is also indisputable in the success of deep learning-based models and GANs. However, in many remote sensing applications, acquiring high-quality ground truth for training is quite challenging due to the unavailability of a wealth of information, expert knowledge, and highly accurate operational data available to the public. One of the methods developed to work around this problem is to generate synthetic data through GANs. Several low-cost tools attempt to synthesize more data for model training when they are somewhat prone to overfitting. The main motivation of GAN-driven synthesis is that perturbations of input data will result in improved generalization performance (Wang et al., 2023). There are a variety of factors that make ensuring high-quality datasets needed for GANs in remote sensing models difficult.

4.2 Transferability to Different Remote Sensing Domains

While these recent improvements facilitated the wider adoption of GANs, one should note that there is still a disparity between the application of many newly proposed architectures on only a few datasets and the development of fundamentally generic GANs, which could be used systematically in a wide range of remote sensing research and application scenarios. The obvious challenge related to this issue is the performance of a trained GAN model on a different dataset than it was trained on (Liu et al., 2023). The notion of adapting model performance to a different dataset is a subject of research in the domain adaptation and transfer learning fields, respectively. In domain adaptation or transfer learning scenarios, a well-performing GAN trained to generate images from a given dataset domain is expected to function effectively by employing its learned discriminative feature detection capabilities when generating images based on another dataset (Sara, et al., 2023). This is a challenge mainly because there exist domain-specific variations resulting in different data characteristics between those two domains. Several existing works propose the use of cyclical adversaries, specialized loss functions, ensembling, or rearrangement of data for training to deal with the issue of domain adaptation for remote sensing-related tasks. These strategies rather allow for the effective transfer of models while focusing on the development of more advanced neural network architectures (Nyberg et al., 2021). This contributes to the overview presented by noting challenges pertaining to the transferability of GANs to different remote sensing domains. In comparison with the overfitting to a specific test set, it focuses on the operational phase, as it is concerned with working with GANs in real-world scenarios. In particular, we emphasize the importance of model versatility, especially in the context of latent space-based data augmentation techniques for a fused classifier. In contrast, the present section examines the nuanced challenges associated with transferring GANs not to different but to a wider range of remote sensing datasets (Wang et al., 2023). Table 1 presents sampled contrasts of image fusion approaches in HS and MS images.

Table 1. Comparison of Image Fusion Techniques for multispectral images (MSIs) with low-resolution hyperspectral images (HSIs)

Method	Specific Technique	Description	Advantages	Disadvantages	Application
Component Substitution (CS) Methods	Principal Component Analysis (PCA)	Transforms HS image into principal components, replaces first with MS image.	Simple; enhances spatial resolution	May distort spectral information	Pansharpening, urban planning
	Intensity-Hue-Saturation (IHS)	Transforms HS image into IHS components; replaces intensity with MS image	Preserves spectral info	Can introduce color distortions	Land use classification, vegetation mapping
	Brovey Transformation	Normalizes HS bands, fuses with MS by adjusting brightness	Sharpens images; simple	Alters original spectral relationships	High-resolution mapping, remote sensing
Multiresolution Analysis (MRA) Methods	Wavelet Transform	Decomposes images into frequency components; fuses high and low-frequency details	Balances spatial and spectral fidelity	Computationally intensive	Image restoration, environmental monitoring
	Contourlet Transform	Captures geometric structures in images, preserving edges during fusion	Effective edge preservation	Complex; requires more computation	Urban feature extraction, texture analysis
	Pyramid Decomposition	Decomposes into multi-resolution layers, then fuses and reconstructs	Enhances spatial detail	Loss of spectral accuracy at finer scales	Remote sensing, multispectral imaging

continued on following page

Table 1. Continued

Method	Specific Technique	Description	Advantages	Disadvantages	Application
Model-based Fusion Techniques	Bayesian Approaches	Models fusion as a probabilistic problem, estimating the most probable image	Can incorporate prior knowledge	Complex implementation	Remote sensing, environmental monitoring
	Sparse Representation	Represents HS and MS images sparsely, combined to generate a fused image	High-quality fusion; handles noise	Requires learned dictionary; computational cost.	Image denoising, object detection
Hybrid Methods	Guided Filtering-based Fusion	Uses guided filter to combine HS and MS, preserving spatial details	Effective spatial detail preservation	May introduce artifacts	Object detection, change detection
	Deep Learning-based Fusion	Uses CNNs or GANs to learn optimal fusion, balancing spatial and spectral fidelity	Captures complex patterns; high accuracy	Requires large datasets and high computational power.	Advanced pansharpening, high-resolution imaging
Spectral Unmixing-based Methods	Linear Spectral Unmixing	Decomposes HS into endmembers; refines with MS for enhanced resolution	Good for mixed pixels	Assumes linearity, which may not always hold.	Land cover classification, resource management
	Non-linear Spectral Unmixing	Accounts for non-linear interactions, improving accuracy in heterogeneous environments	More accurate in complex scenes	More computationally intensive than linear methods.	Complex land cover analysis, environmental studies

continued on following page

Table 1. Continued

Method	Specific Technique	Description	Advantages	Disadvantages	Application
Optimization-based Methods	Variational Methods	Frames fusion as an optimization problem, balancing spatial and spectral consistency	Balances multiple aspects of fusion	It can be computationally intensive.	Image enhancement, data fusion
	Low-rank Approximation	Models images as low-rank matrices, fuses by solving rank minimization problem	Effective in handling noise; preserves structure	It may require complex regularization.	Image restoration, noise reduction
Tensor-based Fusion	Coupled Tensor Decomposition	Treats images as tensors; applies decomposition to fuse, preserving multi-dimensional structure.	Captures complex multi-dimensional data	High computational cost	Multi-modal data fusion, advanced remote sensing
	Tucker Decomposition	Decomposes into core tensors and factor matrices, then fuses for final output.	Preserves spatial and spectral details	Computationally demanding, complex	Multispectral and hyperspectral image fusion
Pansharpening	Intensity-Hue-Saturation (IHS)	Replaces intensity component of MS image with PAN image to increase resolution	Enhances spatial resolution while maintaining color balance	It may introduce color distortions, spectral distortion	High-resolution satellite imaging, urban planning
	Principal Component Analysis (PCA)	Replaces the first principal component with PAN image to enhance spatial detail	Improves spatial resolution effectively	Risk of altering spectral content	High-resolution mapping, remote sensing
	Brovey Transformation	Fuses MS and PAN images by adjusting brightness to match spatial details.	Simple; enhances spatial resolution	Alters original spectral relationships	Remote sensing, image sharpening

5 FUTURE DIRECTIONS AND RESEARCH OPPORTUNITIES

Considering the emerging trends, some possible research directions and problems that are likely to be at the forefront in the future are discussed and summarized as follows.

(1) Hybrid models based on GAN. GAN has been effective in synergy with other aspects of advanced machine learning.
(2) Adapting GAN and PriAdv to real-time applications. In some practical systems, especially in rapidly changing environments or with rapidly generated large-scale data streams, GAN may be worth adapting to better address the need for change.

In summary, in addition to extensive research on new advanced model optimization algorithms, there are new ideas for research in model and data expansion. To understand the potential benefits that GANs can bring to the remote sensing field, it is important to integrate GANs with advanced machine learning or multi-source data fusion (Xiao et al., 2021). In addition, the training of a generative model for a large dataset typically requires a tremendous amount of computation. Therefore, developing a low-power system to address this problem is important. In the future research process, to obtain practical results in the field of remote sensing, it is necessary to propose a novel research method for the specific issues of the remote sensing of GANs. Especially in recent years, the rapid development of sensors and satellite platforms has led to an explosion of high-resolution remote sensing data with both geometric and radiometric resolutions (Rui et al., 2021).

5.1 Incorporating GANs with Other Machine Learning Techniques

Hybrid models developed by fusing GANs and other machine learning approaches can be conceived as a promising future direction for emerging GANs in remote sensing and its applications. The main framework of this direction is to unite the advantages of GANs and another complementary expressive method that is suitable for a particular application scenario (Rewhel et al., 2020). The complementary methods can be either a direct extension of supervised or semi-supervised GANs with certain labels, such as refinement of the outputs of the corresponding existing GAN models, or the combination of GANs with other machine learning methods, such as reinforcement learning, transfer learning, or supervised/semi-supervised learning, which can provide additional guidance to enhance their performances in terms of robustness, generalization, adaptation, etc., orthogonal to the effects of GANs. Each method might have its strengths and play different roles in various challenging remote sensing tasks. Usually, the effectiveness and synergy of combining GANs and other machine learning approaches for addressing a particular task should be systematically explored and analyzed (Pan et al., 2019).

At this stage, one of the key challenges in practice is that the complementarity of combined methods should be guaranteed, or the experimental results could even be worse than single models without less performance gain. We described the insights into developing strong hybrid models by mutually enriching methodologies. We have elaborated the analysis on which aspects of challenges the hybridization can address when combined and the considerations for designing a strong hybrid model (Li et al., 2020). The restricted set of case studies proposes solutions involving GANs among the wide range of possibilities focusing on supervised/semi-supervised learning strategies. These case studies represent different types of land covers, from urban to rural, with different challenges, and with a variety of resolution input data from coarse to very high resolution. The simplification or enhancement of GAN by other methods at the moment is that fewer datasets are being made available; we believe that the number of GANs combined with techniques will still increase (Zhang, et al., 2021).

In recent years, great emphasis has been placed on hybrid models that combine GANs with other machine learning paradigms to benefit from them and produce more effective outputs jointly. This can be seen as a natural progression of the effort to create robust GAN models with state-of-the-art performance across a range of tasks. Hybrid models combine the strengths of different models, resulting in higher performance, data efficiency, and model robustness as they perform feature learning or inference more efficiently by leveraging the strengths of different models. This can be exploited for generating domain-specific data that is more stable. In particular, hybrid GAN models can be achieved by combining a GAN with other machine learning techniques, such as reinforcement learning, transfer learning, clustering, and traditional supervised learning, as well as developing modified versions of GAN that also use labeled data in the generative process (Nyberg et al., 2021). Hybrid models have a promising future, as they can yield powerful models with improved generative capability.

They have been widely used in various fields and can be extended to remote sensing for improved model performance. Modeling has received a great deal of attention in discovering its potential, and transferable learning has been shown to have the potential to evolve in the context of deep learning. Various methods have been proposed and utilized the premise of domain-invariant representations within a variety of applications on transfer learning models (Zheng et al., 2022). Overall, GANs, specifically those related to deep, adversarial, and conditional variants, in conjunction with other techniques, may serve as a crucial stepping stone in harnessing the promise of remote sensing data in scenarios such as few-shot learning, transfer learning, domain adaptation, cross-view geomorphological translation, and data augmentation in the presence of limited and unbalanced data (Zhu et al., 2018).

5.2 Enhancing GANs for Real-Time Applications

Enhancing GANs for Real-Time Applications in Remote Sensing Time is money; time is data; time is everything. Rapid decision-making is essential, especially in the field of Earth imagery, which is basically required to operate from 'minutes' to a week timeframe. Around-the-clock surveillance, outbreak analysis, disaster notifications, and urban planning may benefit from real-time remote sensing data. This makes it imperative to cope with real-time field operations. Consequently, a reasonable adaptation of the GAN architecture and training algorithms is envisaged (Liu et al., 2021). Moreover, possible GAN optimization techniques for increasing performance with maximum speed are also explored in this review. Maintaining good performance is crucial in making GANs achievable in real-time applications. In general, the recent GAN advancements aiming at increasing training speed can be classified into two approaches: one is modeling-based methods focusing on optimizing the network architecture, while the other is data-based strategies oriented towards improving data processing. They may sometimes collaborate in practice (Miftahushudur et al., 2024). Furthermore, GANs that work seamlessly in real-time operations are suitable for end-users striving for on-the-go intelligence, such as paramedics considering urban growth or decline. In time-sensitive applications, near real-time monitoring of dynamic changes may augment profit, safety, and national security. Given that such alerts and warnings are sufficient, GANs can significantly benefit your life in the future (Yu et al., 2022).

5.3 Ethical Concerns in Remote Sensing with GANs

While there are ethical concerns associated with how remote sensing is conducted in general, this project is most concerned with the creation of remote sensing imagery itself. The advent of models has allowed for a wide range of applications that capitalize on their generation capabilities. The ability to generate high-resolution, realistic images is a double-edged sword. It brings about a host of privacy issues, ranging from the trivial to the morally egregious (Paolanti et al., 2023). These technologies have pushed fake images into mainstream media, attracting attention to deepfakes, which are created using similar techniques with identical underlying technologies. It has become impossible to guarantee the credibility of any video or image. These technologies have also allowed for the creation of detailed, high-resolution images of the Earth. In an era where everyone can look up satellite imagery of any place on

Earth, how should we handle malicious actors who generate worthwhile imitations (Crivellari et al., 2023)? If the field of remote sensing with GANs continues to grow at the speed it has, without an ethical reporting system in place, it will be impossible to control the actions of the networks. The ability to generate 'real-looking' remote sensing images could lead to unethical actions being carried out by anyone who aims to create these images. The importance of creating ethical guidelines surrounding the generation, use, and distribution of remotely sensed images using GANs. Not only should these guidelines cover the creators but also those who believe they may be selected (or not selected) for the images and all involved in these newly created acts, e.g., private and public organizational units (Huang et al., 2023). In the creation of remote sensing GAN images, confusion, deception, and disorientation are innovations that must be understood properly. Trying to find a simple definition of the purpose of this project feels almost impossible, but as there are eight principles created in consultation with different sectors affected by the increase of information in media, and with the intent to understand the consequences of the modern digital world, it enables not only an ethical guide for the network creators but also an ethical concept that mainly affects the sector involved (Alajaji et al., 2024).

6 CONCLUSION

In this chapter, we reviewed the current state of the art of recent research focusing on the development of GANs in the remote sensing domain and investigated the motivation that attracted influential lines of research during the last five years. Among the conclusions drawn in this paper, we highlight the increasing level of interest and constant appeal that GANs have witnessed from scientists and researchers. The motivation behind GANs in remote sensing has shown great promise in addressing the limitations. We envisage that GAN utility can be applied in several remote sensing domains, mainly to enhance data and improve the highly automated exploitation of diverse sensing techniques. We identified three challenges for future researchers to address: (1) The state of the art in applications is promising as long as high-quality data is used for training; (2) transferability is a current limitation for deployment; and (3) further research is needed to integrate GAN with other machine learning algorithms, deep learning, and operational assessment. In spite of the efforts invested in this line of research, we strongly believe that the conducted work outlines an extensive class of problems and contributes to creating opportunities for the development of innovative applications in generating and translating remote sensing features despite the public opinion cast by the proponents of GANs in remote sensing to a well-developed research or application domain.

REFERENCES

Ahsan, M., Ali, M., & Siddique, Z. (2022). Imbalanced Class Data Performance Evaluation and Improvement using Novel Generative Adversarial Network-based Approach: SSG and GBO. *ArXiv*, abs/2210.12870. https://doi.org//arXiv.2210.12870.DOI: 10.48550

Alajaji, S. A., Khoury, Z. H., Elgharib, M., Saeed, M., Ahmed, A. R., Khan, M. B., Tavares, T., Jessri, M., Puche, A. C., Hoorfar, H., Stojanov, I., Sciubba, J. J., & Sultan, A. S. (2024). Generative adversarial networks in digital histopathology: Current applications, limitations, ethical considerations, and future directions. *Modern Pathology*, *37*(1), 100369. DOI: 10.1016/j.modpat.2023.100369 PMID: 37890670

Ali-Gombe, A., & Elyan, E. (2019). MFC-GAN: Class-imbalanced dataset classification using Multiple Fake Class Generative Adversarial Network. *Neurocomputing*, *361*, 212–221. DOI: 10.1016/j.neucom.2019.06.043

Cao, J., Luo, M., Yu, J., Yang, M., & He, R. (2022). ScoreMix: A Scalable Augmentation Strategy for Training GANs With Limited Data. *IEEE Transactions on Pattern Analysis and Machine Intelligence*, *45*, 8920–8935. DOI: 10.1109/TPAMI.2022.3231649 PMID: 37015400

Choi, H., Jung, D., Kim, S., & Yoon, S. (2021). Imbalanced Data Classification via Cooperative Interaction Between Classifier and Generator. *IEEE Transactions on Neural Networks and Learning Systems*, *33*(8), 3343–3356. DOI: 10.1109/TNNLS.2021.3052243 PMID: 33531305

Choi, Y., Choi, M., Kim, M., Ha, J. W., Kim, S., & Choo, J. (2018). Stargan: Unified generative adversarial networks for multi-domain image-to-image translation. In *Proceedings of the IEEE conference on computer vision and pattern recognition* (pp. 8789-8797). DOI: 10.1109/CVPR.2018.00916

Crivellari, A., Wei, H., Wei, C., & Shi, Y. (2023). Super-resolution GANs for upscaling unplanned urban settlements from remote sensing satellite imagery–the case of Chinese urban village detection. *International Journal of Digital Earth*, *16*(1), 2623–2643. DOI: 10.1080/17538947.2023.2230956

Eltehewy, R., Abouelfarag, A., & Saleh, S. N. (2023). Efficient classification of imbalanced natural disasters data using generative adversarial networks for data augmentation. *ISPRS International Journal of Geo-Information*, *12*(6), 245. DOI: 10.3390/ijgi12060245

Gao, G., Lai, H., & Jia, Z. (2023). Unsupervised image dedusting via a cycle-consistent generative adversarial network. *Remote Sensing (Basel), 15*(5), 1311. DOI: 10.3390/rs15051311

Huang, Y., & Cao, L. (2023). Privacy-Preserving Remote Sensing Image Generation and Classification With Differentially Private GANs. *IEEE Sensors Journal, 23*(18), 20805–20816. DOI: 10.1109/JSEN.2023.3267001

Jozdani, S., Chen, D., Pouliot, D., & Johnson, B. A. (2022). A review and meta-analysis of generative adversarial networks and their applications in remote sensing. *International Journal of Applied Earth Observation and Geoinformation, 108*, 102734. DOI: 10.1016/j.jag.2022.102734

Kazeminia, S., Baur, C., Kuijper, A., Ginneken, B., Navab, N., Albarqouni, S., & Mukhopadhyay, A. (2018). GANs for Medical Image Analysis. *Artificial Intelligence in Medicine, 109*, 101938. DOI: 10.1016/j.artmed.2020.101938 PMID: 34756215

Li, J., Shen, Y., & Yang, C. (2020). An adversarial generative network for crop classification from remote sensing timeseries images. *Remote Sensing (Basel), 13*(1), 65. DOI: 10.3390/rs13010065

Li, Z., Wu, X., Xia, B., Zhang, J., Wang, C., & Li, B. (2022). A Comprehensive Survey on Data-Efficient GANs in Image Generation. *ArXiv*, abs/2204.08329. https://doi.org//arXiv.2204.08329.DOI: 10.48550

Liu, L., Lei, S., Shi, Z., Zhang, N., & Zhu, X. (2021). Hyperspectral Remote Sensing Imagery Generation From RGB Images Based on Joint Discrimination. *IEEE Journal of Selected Topics in Applied Earth Observations and Remote Sensing, 14*, 7624–7636. DOI: 10.1109/JSTARS.2021.3099242

Liu, P. (2023). A review on remote sensing data fusion with generative adversarial networks (GAN). *Authorea Preprints*.

Ma, T., Ma, J., Yu, K., Zhang, J., & Fu, W. (2021). Multispectral Remote Sensing Image Matching via Image Transfer by Regularized Conditional Generative Adversarial Networks and Local Feature. *IEEE Geoscience and Remote Sensing Letters, 18*(2), 351–355. DOI: 10.1109/LGRS.2020.2972361

Miftahushudur, T., Grieve, B., & Yin, H. (2024). Permuted KPCA and SMOTE to Guide GAN-Based Oversampling for Imbalanced HSI Classification. *IEEE Journal of Selected Topics in Applied Earth Observations and Remote Sensing, 17*, 489–505. DOI: 10.1109/JSTARS.2023.3326963

Nyberg, D. (2021). *Exploring the Capabilities of Generative Adversarial Networks in Remote Sensing Applications*. Linkoping University Electronic Press.

Oh, J., Hong, J., & Baek, J. (2019). Oversampling method using outlier detectable generative adversarial network. *Expert Systems with Applications*, *133*, 1–8. DOI: 10.1016/j.eswa.2019.05.006

Pan, X., Yang, F., Gao, L., Chen, Z., Zhang, B., Fan, H., & Ren, J. (2019). Building extraction from high-resolution aerial imagery using a generative adversarial network with spatial and channel attention mechanisms. *Remote Sensing (Basel)*, *11*(8), 917. DOI: 10.3390/rs11080917

Paola, Z. L., Jesús, L. S., Christian, A. H., & Sonia, R. U. (2023). Correction of Banding Errors in Satellite Images With Generative Adversarial Networks (GAN). *IEEE Access : Practical Innovations, Open Solutions*, *11*, 51960–51970. DOI: 10.1109/ACCESS.2023.3279265

Paolanti, M., Tiribelli, S., Giovanola, B., Mancini, A., Frontoni, E., & Pierdicca, R. (2024). Ethical framework to assess and quantify the trustworthiness of artificial intelligence techniques: Application case in remote sensing. *Remote Sensing (Basel)*, *16*(23), 4529. DOI: 10.3390/rs16234529

Pengqiang, Z. H. A. N. G., Bing, L. I. U., Xuchu, Y. U., Xiong, T. A. N., Fan, Y. A. N. G., & Zenghua, Z. H. O. U. (2020). Generative adversarial networks for hyperspectral image classification. *Bulletin of Surveying and Mapping*, (3), 29. http://tb.chinasmp.com/EN/10.13474/j.cnki.11-2246.2020.0073

Rewhel, E. M., Li, J., Hamed, A. A., Keshk, H. M., Mahmoud, A. S., Sayed, S. A., Samir, E., Zeyada, H. H., Mohamed, S. A., Moustafa, M. S., Nasr, A. H., & Helmy, A. K. (2023). Deep Learning Methods Used in Remote Sensing Images: A Review. *Journal of Environment and Earth Science*, *5*(1), 33–64. DOI: 10.30564/jees.v5i1.5232

Rui, X., Cao, Y., Yuan, X., Kang, Y., & Song, W. (2021). Disastergan: Generative adversarial networks for remote sensing disaster image generation. *Remote Sensing (Basel)*, *13*(21), 4284. DOI: 10.3390/rs13214284

Sara, D., Mandava, A. K., Kumar, A., Duela, S., & Jude, A. (2021). Hyperspectral and multispectral image fusion techniques for high resolution applications: A review. *Earth Science Informatics*, *14*(4), 1685–1705. DOI: 10.1007/s12145-021-00621-6

Shafik, W. (2024a). Deep Learning Impacts in the Field of Artificial Intelligence. In *Deep Learning Concepts in Operations Research* (pp. 9–26). Auerbach Publications., DOI: 10.1201/9781003433309-2

Shafik, W. (2024b). Introduction to ChatGPT. In *Advanced applications of generative AI and natural language processing models* (pp. 1–25). IGI Global., DOI: 10.4018/979-8-3693-0502-7.ch001

Shafik, W. (2025). Quantum Computing and Generative Adversarial Networks (GANs): Ethics, Privacy, and Security. In *Quantum AI and its Applications in Blockchain Technology* (pp. 111-156). IGI Global Scientific Publishing. https://doi.org/DOI: 10.4018/979-8-3373-1657-4.ch007

Shamsolmoali, P., Zareapoor, M., Shen, L., Sadka, A., & Yang, J. (2020). Imbalanced Data Learning by Minority Class Augmentation using Capsule Adversarial Networks. *ArXiv*, abs/2004.02182. https://doi.org/.DOI: 10.1016/j.neucom.2020.01.119

Sharma, A., Singh, P., & Chandra, R. (2021). SMOTified-GAN for Class Imbalanced Pattern Classification Problems. *IEEE Access : Practical Innovations, Open Solutions*, *10*, 30655–30665. DOI: 10.1109/ACCESS.2022.3158977

Sharma, S., & Namboodiri, V. (2018). No Modes left behind: Capturing the data distribution effectively using GANs. *ArXiv*, abs/1802.00771. https://doi.org/.DOI: 10.1609/aaai.v32i1.11790

Su, L., Sui, Y., & Yuan, Y. (2023). An Unmixing-Based Multi-Attention GAN for Unsupervised Hyperspectral and Multispectral Image Fusion. *Remote Sensing (Basel)*, *15*(4), 936. DOI: 10.3390/rs15040936

Suh, S., Lee, H., Lukowicz, P., & Lee, Y. (2020). CEGAN: Classification Enhancement Generative Adversarial Networks for unraveling data imbalance problems. *Neural networks: the official journal of the International Neural Network Society*, 133, 69-86 . https://doi.org/.DOI: 10.1016/j.neunet.2020.10.004

Wang, C., Chen, B., Zou, Z., & Shi, Z. (2023). Remote sensing image synthesis via semantic embedding generative adversarial networks. *IEEE Transactions on Geoscience and Remote Sensing*, *61*, 1–11. DOI: 10.1109/TGRS.2023.3335484

Wang, X., Sun, L., Chehri, A., & Song, Y. (2023). A review of GAN-based super-resolution reconstruction for optical remote sensing images. *Remote Sensing (Basel)*, *15*(20), 5062. DOI: 10.3390/rs15205062

Wu, A. N., Stouffs, R., & Biljecki, F. (2022). Generative Adversarial Networks in the built environment: A comprehensive review of the application of GANs across data types and scales. *Building and Environment*, *223*, 109477. DOI: 10.1016/j.buildenv.2022.109477

Xiao, J., Li, J., Yuan, Q., Jiang, M., & Zhang, L. (2021). Physics-Based GAN With Iterative Refinement Unit for Hyperspectral and Multispectral Image Fusion. *IEEE Journal of Selected Topics in Applied Earth Observations and Remote Sensing*, *14*, 6827–6841. DOI: 10.1109/JSTARS.2021.3075727

Yu, M., Zhang, W., Chen, X., Liu, Y., & Niu, J. (2022). An End-to-End Atrous Spatial Pyramid Pooling and Skip-Connections Generative Adversarial Segmentation Network for Building Extraction from High-Resolution Aerial Images. *Applied Sciences (Basel, Switzerland)*, *12*(10), 5151. DOI: 10.3390/app12105151

Zhang, H., Song, Y., Han, C., & Zhang, L. (2020). Remote sensing image spatiotemporal fusion using a generative adversarial network. *IEEE Transactions on Geoscience and Remote Sensing*, *59*(5), 4273–4286. DOI: 10.1109/TGRS.2020.3010530

Zhang, N., Wang, Y., Zhang, X., Xu, D., & Wang, X. (2020). An unsupervised remote sensing single-image super-resolution method based on generative adversarial network. *IEEE Access : Practical Innovations, Open Solutions*, *8*, 29027–29039. DOI: 10.1109/ACCESS.2020.2972300

Zheng, Y., Su, J., Zhang, S., Tao, M., & Wang, L. (2022). Dehaze-AGGAN: Unpaired remote sensing image dehazing using enhanced attention-guide generative adversarial networks. *IEEE Transactions on Geoscience and Remote Sensing*, *60*, 1–13. DOI: 10.1109/TGRS.2022.3225843

Zhu, C., Deng, S., Li, J., Zhang, Y., Gong, L., Gao, L., Ta, N., Chen, S., & Wu, Q. (2023). Hyperspectral and multispectral remote sensing image fusion using SwinGAN with joint adaptive spatial-spectral gradient loss function. *International Journal of Digital Earth*, *16*(1), 3580–3600. DOI: 10.1080/17538947.2023.2253206

Zhu, C., Deng, S., Zhou, Y., Deng, L., & Wu, Q. (2023). QIS-GAN: A Lightweight Adversarial Network With Quadtree Implicit Sampling for Multispectral and Hyperspectral Image Fusion. *IEEE Transactions on Geoscience and Remote Sensing*, *61*, 1–15. DOI: 10.1109/TGRS.2023.3332176

Zhu, L., Chen, Y., Ghamisi, P., & Benediktsson, J. (2018). Generative Adversarial Networks for Hyperspectral Image Classification. *IEEE Transactions on Geoscience and Remote Sensing*, *56*(9), 5046–5063. DOI: 10.1109/TGRS.2018.2805286

Zhu, L., Chen, Y., Ghamisi, P., & Benediktsson, J. A. (2018). Generative adversarial networks for hyperspectral image classification. *IEEE Transactions on Geoscience and Remote Sensing*, *56*(9), 5046–5063. DOI: 10.1109/TGRS.2018.2805286

Chapter 3
An Exploration of Generative Adversarial Networks, Architectures, Types, and Training Challenges

Pooja Dehankar
Ajeenkya D.Y. Patil School of Engineering, India

Rohini Tarade
https://orcid.org/0009-0002-4682-0250
Ajeenkya D.Y. Patil University, India

Susanta Das
https://orcid.org/0000-0002-9314-3988
Ajeenkya D.Y. Patil University, India

ABSTRACT

The advent of GANs by Generative AI has completely transformed ML profession. In addition to outlining the fundamental ideas & significant turning points that have influenced the development of GANs, this chapter offers a thorough overview of the technology. In addition to discussing different varieties of GANs, such as progressive, Wasserstein, and conditional GANs, & highlighting their special qualities & uses, the chapter delves into the basic architecture of GANs, which includes generator & discriminator networks. Further, it addresses methods to overcome training difficulties that GANs encounter, including regularization techniques & alternative loss functions, & difficulties with mode collapse & training instability. It also looks

DOI: 10.4018/979-8-3693-6900-5.ch003

at how GANs have developed to produce high-fidelity text, graphics, & other data kinds, demonstrating their adaptability & potential in innovative AI applications. It provides a fundamental framework for comprehending ideas, developments, & potential paths of generative AI as seen through the prism of GANs.

INTRODUCTION

In the realm of artificial intelligence, deep learning has proven very successful, and numerous deep learning models have been created. One deep learning model that was introduced based on zero-sum game theory is called Generative Adversarial Networks (GAN), and it has emerged as a new area of focus for research. The purpose of the model variation is to produce more actual and realistic data by using unsupervised learning to obtain the data distribution. Due to its vast array of potential applications, such as image and vision processing, video and language processing, etc., GANs have now been the subject of extensive research (Gonog & Zhou, 2019).

To create fresh instances of text, music, images, and videos, generative machine learning techniques like GANs are applied.

Although GANs are already widely used on social media, are ingrained in children's lives, and carry significant ethical issues, generative AI is not covered in the K–12 AI education curricula that are now in place (Ali et al., 2021)

The machine learning community has focused a lot of emphasis on GAN because of its ability to learn high-dimensional, complex real data distribution. In particular, they are independent of any distributional assumptions and are able to easily produce real-like samples from latent space. Because of this strong characteristic, GANs are used in a wide range of academic domains and applications, including image synthesis, image translation, picture attribute editing, and domain adaption (Hong et al., 2019).

Deep learning has seen many revolutions in the previous few years, with GANs having the biggest impact. In addition to offering a distinctive model architecture, GANs provide amazing outcomes that have a direct bearing on society. It is nearly hard to stay up to date with the community's ongoing research due to the tremendous advancements and new fields of study that GANs have brought about (Iglesias et al., 2023).

An artificial intelligence algorithm called generative adversarial networks was created to address the generative modeling issue. A generative model looks at a set of training examples and aims to identify the probability distribution that produced each one. The estimated probability distribution can then be used to create more instances using GANs. Deep learning-based generative models are popular, but GANs are among the most effective generative models (particularly when it comes to producing realistic, high-resolution images). Because GANs are based on game theory, whereas the majority of other generative modeling techniques are based on optimization, they continue to present special problems and research opportunities despite their successful application to a wide range of tasks (Goodfellow et al., 2020).

Training GANs has been found to have poor generalization qualities; for example, while training may seem effective, the learned distribution may deviate significantly from the target distribution according to standard metrics. On the other hand, generalization is possible for neural net distance, a weaker measure. Additionally, it is demonstrated that for modest generator capacity and training set sizes, there is an approximate pure equilibrium in the discriminator/generator game given a natural training aim (Wasserstein). The MIX+GAN methodology, which can be used with any current GAN training and has been empirically proved to improve certain of them, is inspired by the existence of this equilibrium (Arora et al., 2017).

As a successful means of avoiding discriminator overfitting, recent efforts in GANs are actively exploring different data augmentation strategies. Still unknown, though, is which augmentations might really make GANs better—specifically, how to use a larger variety of augmentations during training. By integrating the recently developed ContraD contrastive representation learning method into the GAN discriminator, the authors suggested a fresh approach to these problems. By allowing the discriminators to operate with significantly stronger augmentations without making their training unstable, this "fusion" helps to better prevent the discriminator overfitting problem in GANs. Better yet, the authors note that GAN training actually enhances contrastive learning itself (Jeong & Shin, 2021).

Deep representations can be learned using GANs, which do not require heavily annotated training material. These networks learn by using a competitive process between two networks to derive back propagation signals. There are numerous uses for the representations that GANs can learn. GANs have performed incredibly well in many applications and made great strides. Semantic picture editing, style transfer, image synthesis, super-resolution, and categorization are among the crucial uses (Alqahtani et al., 2021).

By merging two neural networks that use zero-sum game theory to compete with one another, GANs produce outputs that are far more distinct and crisp. GANs are useful for a variety of computer vision tasks, including image processing, video production, and prediction. In addition, GANs can be used for several science-related tasks like as protein engineering, crystal structure synthesis, remote sensing image dehazing, and astronomical data processing. Sports, music, fashion design, marketing, banking, and sports are other noteworthy industries where GANs have achieved success (Dash et al., 2023).

A significant advancement in artificial intelligence is Generative Adversarial Networks (GANs), which employ two neural networks—a discriminator and a generator—that operate in an adversarial environment. The discriminator assesses the data's legitimacy, while the generator creates artificial data. This dynamic interaction creates a minimax game that generates synthetic data of superior quality. Since its debut by Ian Goodfellow in 2014, GAN has developed through a number of novel designs, such as BigGAN, CycleGAN, StyleGAN, Deep Convolutional GAN (DCGAN), Conditional GAN (cGAN), Vanilla GAN, and Wasserstein GAN (WGAN). To solve technological issues like training stability, data diversification, and result quality, each of these systems offers a unique strategy. GANs have been widely used in many different fields. GANs are utilized in the medical field to create artificial medical images that aid in the development of diagnostics while protecting patient privacy. In the media and entertainment sector, GANs help to improve the quality of images and videos and provide realistic material. However, issues including mode collapse, training instability, and insufficient quality evaluation hinder the advancement of GANs. GANs have ethical problems in addition to technological ones, such as the misuse of the technology to create deepfakes. Public education, detecting systems, and legal requirements are crucial mitigating strategies. In order to facilitate cross-disciplinary innovation, future trends indicate that GANs will be utilized more and more in text-to-image synthesis, realistic video production, and integration with multimodal systems (Purwono et al., 2025).

In computer vision, image restoration is a difficult task that aims to recover high-quality images from damaged or distorted ones. Three well-known image restoration techniques—GFP GAN, DeOldify, and MIRNet—are thoroughly compared. GFP GAN presents an AI-centric method with a customized GAN architecture made for image restoration challenges. While MIRNet provides a lightweight network designed especially for image restoration within an AI framework, DeOldify, a deep learning-based approach, concentrates on colorizing and restoring aged photographs using cutting-edge AI approaches. Training and testing each approach on a varied dataset that includes both degraded and ground truth photos is part of the comparative analysis. To fully evaluate the effectiveness of these AI-based techniques, precision, accuracy, recall, and other evaluation metrics are calculated using a confusion matrix. The matrix offers a deeper knowledge of each AI-driven approach's performance by revealing its advantages and disadvantages. Experimental analyses reveal how well GFP GAN, DeOldify, and MIRNet perform in resolving picture restoration problems, including noise, blur, compression artifacts, and other degradations, in the framework of AI techniques. The findings not only highlight the benefits and drawbacks of each AI-enhanced technique, but they also provide researchers and practitioners with a useful tool for choosing the best AI-driven strategy for their particular image restoration needs. Using a confusion matrix to examine precision, accuracy, and other relevant metrics, it provides a comprehensive comparison of GFP GAN, DeOldify, and MIRNet in the field of AI-driven picture restoration. By carefully examining these AI-powered assessment criteria, it provides priceless information about the complex performance of different approaches, enabling well-informed choices across a range of AI-driven image restoration applications (Rastogi et al., 2025).

MOTIVATION, OBJECTIVES AND CONTRIBUTIONS

GAN's distinct adversarial training mechanism is the driving force for the decision to use it as the main algorithm component rather than more conventional techniques or other generative models. Through this approach, GANs can handle non-linear, high-dimensional mental health data and capture more implicit information, increasing the model's accuracy and dependability. Furthermore, GAN can potentially get better on its own. As training data increases, the generated samples' quality and diversity will also improve, better representing college students' true mental health conditions. GAN offers excellent technical support for creating a more transparent, equitable, and scientific evaluation system, which makes it a promising tool for assessing the quality of mental health (Fowlis et al., 2024).

Our heavy reliance on the Internet for activities like social networking, e-learning, mobile banking, entertainment, and digital commerce is a reflection of the widespread use of digital transactions in today's digital era. Although these technologies are convenient, there is increasing fear that they could be abused for fraudulent purposes, surpassing the efficacy of conventional fraud detection techniques. Using variational autoencoders (VAEs) and generative adversarial networks (GANs), generative artificial intelligence (AI) can identify minor irregularities in financial information that are frequently missed by traditional methods. Its ability to produce synthetic data safeguards the confidentiality of the actual data while also allowing for thorough testing of detection systems. Latent frauds can be predicted with the aid of generative AI's predictive powers. Generative models are supported by the examination of historical data in order to deduce trends and spot potential flaws in the current financial systems. Thus, generative AI proactively makes it possible to foresee con artists and prevent losses. With its capacity to generate synthetic data, identify irregularities, and predict possible fraud, generative AI has become increasingly important in preventing fraud as a result of our increasing reliance on digital transactions and the shortcomings of conventional fraud detection (Kulkarni et al., 2025).

GANs have become a popular and active area of study in machine learning and artificial intelligence in recent years. By producing realistic synthetic samples, GANs offer a sufficient data augmentation technique that enhances model accuracy, particularly for underrepresented classes (Swadi & Haider, 2025).

GANs, which produce realistic and high-quality images, have transformed image processing tasks in recent years. However, there are particular difficulties that need for specialized methods in order to achieve real-time image processing capabilities on neuromorphic systems that are modeled after the functioning of the human brain (Bindu & Satyanarayana, 2025).

GAN ARCHITECTURES

Supervised learning using convolutional networks (CNNs) has been increasingly popular in computer vision applications in recent years. In contrast, there has been less research done on unsupervised learning using CNNs. The goal of the authors' work is to close the gap between supervised and unsupervised learning with CNNs. The authors present a type of CNNs known as deep convolutional generative adversarial networks (DCGANs), show that they are a good fit for unsupervised learning, and outline some architectural restrictions. The authors provide strong evidence that our deep convolutional adversarial pair learns a hierarchy of representations in both the generator and discriminator, ranging from object pieces to scenes, through training on a variety of picture datasets. Furthermore, the learnt features are applied to new tasks by the authors, proving that they are applicable as general image representations (Radford, 2015).

With its convolutional architecture, DCGAN plays a major role in maintaining the balance of GAN training, which makes it important. Natural DCGAN and GAN networks are unsupervised by nature. The network is not provided with the category label information, even though it is aware whether the samples it receives as input are authentic or not.a technique that makes it possible to use multi-category data sets with DCGAN structure to create supervised network structures. The technique makes sure that noise can be assigned a category label, and that the output layer may use the created category label information. This technique is called Supervised DCGAN (SDCGAN), and it is very adaptable and efficient (Öcal & Özbakır, 2021).

Deep Convolutional GAN (DCGAN) with Attention Module is a novel Unsupervised Deep Feature Learning Method for Remote Scene Classification that was proposed by Patil (2021). DCGAN and the Attention module are linked to maximize the potential of feature extraction. A suggested feature fusion architecture is coupled with Discriminator to extract contextual information. The discriminator and generator losses are optimized by the suggested module. SVM receives the retrieved features as input for classification. Publicly available remote sensing scene datasets, the UC-Merced dataset with 21 distinct scene classes and the RSSCN7 dataset with 7 different scene classes, are used to develop the suggested module DCGAN with Attention. The experimental outcomes produced by the suggested DCGAN model with Attention module outperform those of the most advanced machines (Patil, 2021).

In numerous application sectors, Conditional Generative Adversarial Networks, or cGANs, are becoming more and more popular. Even with remarkable advancements, quantitative assessment of these models frequently requires several separate measures to evaluate various desirable attributes, including intra-conditioning variety, conditional consistency, and picture quality. Model benchmarking becomes difficult in this context since several metrics could point to distinct "best" models. In order to implicitly incorporate the aforementioned qualities in a single metric, DeVries et al. (2019) propose the Frechet Joint Distance (FJD), which is defined as the Frechet distance between joint distributions of pictures and conditioning. Proof-of-concept trials on a controlled synthetic dataset were carried out by DeVries et al. (2019), and the results consistently show that FJD is superior than currently used metrics.

Complex data distributions can be successfully approximated by Generative Adversarial Networks (GANs), as demonstrated lately. Conditional GANs (cGANs), which decide certain representations of the generated images by introducing external information, are a pertinent extension of this approach. In order to translate a real image into a latent space and a conditional representation, Perarnau et al. (2016) assess encoders for inverse mapping of a cGAN. This enables the reconstruction and alteration of authentic facial images based on random characteristics, for instance. Perarnau et al. (2016) assess the cGANs' design as well. Real images can be regenerated with predictable complicated alterations thanks to the Invertible cGAN (IcGAN) combination of an encoder and a cGAN.

GAN TYPES

Theoretical research has gained more attention as a result of the empirical success of GANs. The Wasserstein GANs and their generalizations, which particularly permit good dimension reduction properties, were the subject of Kunkel & Trabs' (2024) study. The initial optimization problem, known as Vanilla GANs, has rather narrow statistical results that rely on assumptions like equal latent and ambient space dimensions and smooth activation functions. The authors make a link between Vanilla GANs and the Wasserstein distance in order to close this gap. This allows for the extension of Wasserstein GAN results to Vanilla GANs. Specifically, an oracle inequality for Vanilla GANs in Wasserstein distance is obtained by the authors.

A family of CNNs known as deep convolutional generative adversarial networks (DCGANs), which show promise for unsupervised learning despite significant architectural limitations. Deep convolutional adversarial pair learns a hierarchy of representations from object pieces to scenes in both the generator and discriminator after training on a variety of image datasets (Radford et al., 2015).

GAN, is a popular and very effective tool in both academia and industry. It draws inspiration from game theory, in which two models—a generator and a critic—compete, share knowledge, and strengthen one another over time. The majority of advancements concentrated on stability, and numerous models, like Wassertein GAN (WGAN) and deep convolutional GAN (DCGAN), have already shown success. Nonetheless, another crucial skill that must be improved during training is the rate of convergence. To speed up the training process, a decreasing noise is added to the MNIST data and sent to the vanilla GAN's discriminator. The outcome demonstrates that this modification might quicken convergence without significantly lowering the caliber of the output. Also, this enhancement ought to be sufficiently universal to encompass the other models (Si, 2023).

When it comes to data-driven unconditional generative picture modeling, the style-based GAN architecture (StyleGAN) produces cutting-edge outcomes. The capacity to manipulate the generated images by style mixing—that is, by feeding distinct latent w to various layers at inference time—is one of StyleGAN's primary advantages. Certain feature maps may be amplified by an order of magnitude or more in practice due to style modulation. In order for style mixing to function, this amplification must be specifically counteracted on a per-sample basis; otherwise, the data cannot be meaningfully processed by the next layers.

Carneiro (2020) defines and describes the Vanilla GAN and its key modifications (e.g., DCGAN, WGAN, and SAGAN) as well as illustrates many applications (e.g., style-transfer and image-to-image translation) that make GAN an exceptionally fascinating technology for the entertainment sector. Lastly, the Frechet Inception Distance (FID) and Inception Search (IS) metrics are offered to evaluate the quality of the generated images (Carneiro, 2020).

Realistic samples from complex image distributions can be produced using Wasserstein Generative Adversarial Networks (WGANs). Based on the idea of distance between individual images, the Wasserstein metric utilized in WGANs generates an idea of distance between image probability distributions (Adler & Lunz, 2018).

Because Wasserstein GANs are easy to train, they are being employed in computer vision applications more and more (Liu et al., 2019). An alternative to conventional GAN training is WGAN. This new model demonstrated that it is possible to eliminate issues such as mode collapse, increase learning stability, and produce meaningful learning curves that are helpful for debugging and hyperparameter searches (Arjovsky et al., 2017).

Style transfer technology has advanced significantly along with the field of computer vision's rapid growth of deep neural networks. Cycle-GAN is capable of picture augmentation, style transfer, and object modification without requiring a one-to-one mapping between source and target domains. The painting style transfer task demonstrates Cycle-GAN's performance. The generator model is a critical decision in Cycle-GAN, and ResNet and U-Net are popular backbones. By using the jump connection of the residual structure, the ResNet generator preserves a portion of the original characteristics, preventing the image from losing crucial information and possibly preserving the image's authenticity. With its intricate and sophisticated network design, the U-Net generator recovers more features and details, making it a great choice for tasks involving a large amount of feature extraction (Ma, 2024).

There are two ways in which LSGANs are advantageous. Firstly, LSGANs penalize samples that are far on the correct side of the decision boundary, even though they are correctly categorized, in contrast to ordinary GANs that incur essentially little loss for those samples. The decision boundary, or the discriminator's parameters, are fixed when we update the generator.

Consequently, the generator will produce samples closer to the decision boundary as a result of the penalization. However, for a GAN to learn effectively, the decision boundary needs to span the manifold of real data. The learning process will get saturated otherwise. Thus, generating samples that are closer to the manifold of real data can be achieved by bringing them toward the decision boundary. Second, when updating the generator, punishing the samples that are far from the decision boundary can produce more gradients, which solves the issue of disappearing gradients. As a result, LSGAN performance is more stable during the learning phase. Another way to look at it is this: the sigmoid cross entropy loss function will saturate at reasonably big x, but the least squares loss function is flat only at one point (Mao et al., 2017).

When employing gradient descent learning techniques to train both the generator and discriminator artificial neural networks, the vanishing gradient problem arises. Back propagation algorithms are used to find the gradients of neural networks. These algorithms compute the gradient of the loss function with respect to the network weights layer by layer using a chain rule that starts at the final layer and works its way down to the first one to compute the derivatives of the first layers (Kim & Seo, 2015).

Nevertheless, the issue arises when 'n' tiny derivatives are multiplied collectively when 'n' hidden layers use activation similar to the sigmoid function. Thus, when the network propagates down toward the starting layers, the gradient causes a decrease exponentially. Networks cannot properly update parameters like weights and biases at the first layers when they are trained with modest gradient values for every session.

Given that the first few layers are crucial for identifying the fundamental components of the incoming data. This may result in a network that is difficult to train overall and inaccurately trained. Sometimes, the gradient will begin to efficiently evaporate, preventing the weight from changing its value and, as a result, totally halting further training of the neural network (Arjovsky et al., 2017).

Although GANs have demonstrated success in predicting data distributions, training them is still a difficult undertaking. The non-intuitive structure of GAN loss curves makes it necessary to subjectively assess the produced output in order to determine training progress, which adds to the challenge. Recently, duality gap has been proposed as a domain agnostic measure to track GAN training, inspired by game theory. It is limited to the situation where the GAN converges to a Nash equilibrium, though. However, GANs can represent the data distribution without necessarily convergent to a Nash equilibrium. The proximal duality gap useful for training GANs in general contexts where Nash equilibria might not exist (Sidheekh et al., 2021).

Mode breakdown in GANs often has four causes. First, when a GAN's goal is misrepresented, the solitary generator produces output in a predictive manner, but the discriminator in the model assesses the output according to authenticity rather than output diversity. In that case, the discriminator's capacity would be very limited, but the generator would produce a large number of samples. Thus, the generator cannot be instructed to approach the distribution by the discriminator. Second, the generator takes into account individual examples without taking into account forward or backward samples when it ignores the goal function. In this instance, the generator is not appropriately motivated, leading to an erratic output. Third, a gradient exploding when there is an imbalance or inconsistency between the discriminator and generator can be used to explain the mode collapse problem. Four: the mechanism for separating the output fails because the discriminator procedure is inefficient and processes each sample separately and one at a time. The real issue with this entire process, to put it briefly, is with mode discovery and mode retention. The concept pertains to generator optimization, enabling the identification of potential data clusters while adjusting the focus in order to trick the discriminator (Hong et al., 2019).

TRAINING CHALLENGES

Numerous real-world artificial intelligence applications have demonstrated the effectiveness of the GAN. But the majority of the time, its network architecture is created by a laborious, slow, and less-than-ideal human trial-and-error method. To automate the entire GAN design process, an evolutionary architectural search (EAS) technique is devised. Specifically, as variation operators, many goal functions are employed in the GAN generator. By playing adversarial training against the GAN discriminator, this aids in training the generator with a variety of candidate designs and the weights that go along with them. Superior candidate generators that pass the discriminator's evaluations move on to the next generation, where they concurrently evolve for possibly better designs and their weights, producing more stabilized GANs with higher performance (Lin et al., 2022).

GANs aims to learn a target distribution in an unsupervised way. Even though GANs have been successfully used to solve a wide range of issues, training them is a well-known difficult task that calls for a substantial amount of neural network engineering, hyperparameter tuning, and non-trivial amounts of "tricks." Numerous suggested losses, regularization and normalization algorithms, and neural architectures were produced as a result of the success of GANs in numerous practical applications and the lack of a metric to evaluate the failure mechanisms of GANs (Kurach et al., 2018).

GANs are prone to instability and can be challenging to train. Among the difficulties in training GANs are: Mode collapse, Non-convergence, Instability, Data requirements, Initialization and Configuration (Saad et al., 2024).

Training issues with GANs include mode collapse, non-convergence, and instability. Because of these drawbacks, GANs may produce images that are unrealistic, hazy, and less varied. The mode collapse problem arises when the generator takes varied input information and generates similar output images. Minibatch discrimination (Xue et al. 2019), skip connections (Segato et al. 2020), VAEGAN (Kwon et al. 2019), varying layers of generator and discriminator (Qin et al. 2020), spectral normalization (Xu et al. 2020), perceptual image hashing (Neff et al. 2017), Gaussian mixture model as generator (Wu et al., 2018), discriminator with conditional information vector (Modanwal et al. 2021), self-attention mechanism (Abdelhalim et al. 2021), and adaptive input-image normalization (Saad et al. 2022b).

In order for GANs to function well during training, a delicate balance between the discriminator and the generator must be maintained. In order to address this problem, a novel regularization method called progressive augmentation of GANs (PA-GAN) was presented. The main concept is to gradually augment the discriminator's input or feature space, increasing the task difficulty and allowing for continual learning of the generator (Zhang & Khoreva, 2019).

VERIFICATION AND VALIDATION

It is crucial that we employ high quality evaluation measures since deep generative models are strong tools that have yielded remarkable outcomes in recent years, most of which have been empirically driven. Several measures are used to evaluate the quality of the data that is created, such as a pre-trained image classifier's ability to classify generated images is measured by the Inception Score. Inception Score is the most popular metric for assessing image-generating models. Although this metric is significant in and of itself, it also provides a framework that highlights many of the challenges in creating a successful approach to evaluating blackbox generative models (Barratt & Sharma, 2018).

One kind of generative modeling framework is generative adversarial networks, or GANs. In GANs, two neural networks compete iteratively to produce fictitious data that is identical to the actual thing. Because the GAN framework is based on deep learning algorithms, one prominent use of GANs is creating "deep fakes," or fake human faces. Although evaluating the quality of the generated images is subjective by nature, efforts have been made to use defined measures in order to objectively measure quality. The Frechet Inception Distance (FID), which calculates the difference between feature vector distributions for two distinct picture datasets, is an example of an objective metric. Images with poor perceptual properties are sometimes not given the proper FID scores. The distance between the feature distributions of produced and real images is determined by the Frechet Inception Distance (FID); a lower FID denotes higher quality (Nunn et al., 2021).

One of the most important things manufacturing organizations do to make sure their products meet the standards and specifications is quality control. Artificial intelligence models have made it possible to automate visual quality assessment, which expedites the procedure and guarantees that all items are assessed using the same standards. Data augmentation strategies to address data imbalance in the context of automated visual inspection, as well as supervised and unsupervised fault detection systems is investigated. Augmentation of data using Generative Adversarial Networks to improve the discriminative performance of the classifiers. The results indicate that while state-of-the-art unsupervised defect detection can lessen the labeling effort provided some labeling errors are tolerated, it does not perform as well as supervised models. Checking for glaring differences by manually comparing generated samples to actual data (Rožanec et al., 2023).

FUTURE SCOPE

The production of audiovisual media is undergoing a rapid transformation due to deep learning algorithms. Deep learning-generated synthetic audiovisual media, commonly referred to as "deepfakes," have several remarkable features. They are becoming easier to create and can be mistaken for actual sounds and images captured by a sensor. The ethical issues brought up by this technical advancement have received a lot of attention. a number of questions about what is meant by synthetic audiovisual media, where it fits into a larger taxonomy of audiovisual media, and how deep learning methods vary from more conventional methods of media synthesis. After going over key etiological aspects of deep learning pipelines for media generation and manipulation, it is clear that "deepfakes" and related synthetic media created with these pipelines do more than just provide slight enhancements over earlier techniques; they also question established taxonomical divisions and open the door for truly innovative types of audiovisual media (Millière, 2022).

In order to facilitate creative decision-making, research focus is moving toward more regulated and steerable generative processes as symbolic music creation advances. Prior approaches emphasize fine-grained control or global conditioning through input sequences, but they frequently restrict the flexibility for real-time interventions and necessitate changes to the model's design. a Transformer encoder-decoder and probe models to create a new framework for the generation of symbolic music. This allows us to use pre-defined non-differentiable musical features to interpret the encoder's hidden state and then manipulate the hidden state to produce a set of desired attributes in the generated music. Without changing the fundamental model architecture, this technique enables fine-grained control over particular musical characteristics. In order to achieve flexible control without retraining the model, probes can be used post-training or taught in conjunction with the generative model (Hachana & Rasheed, 2025).

Achieving training stability in generative modeling is a significant challenge, particularly for adversarial models like GANs. Models will become more stable and scalable as more research is done on stronger architectures, better loss functions, and training techniques. Despite their superiority in producing high-quality samples, models like as GANs still struggle to produce varied and high-resolution data, especially in complicated fields. Improvements in the future might concentrate on increasing the variety of generated samples without compromising quality. The ability of generative models to provide not just generated samples but also an understanding of the uncertainty or confidence in the generated data has led to an increase in interest in uncertainty quantification. This can be vital in applications like as medical diagnostics, where it's critical to understand the generated data's degree of confidence. The usefulness of DGMs will continue to grow with the creation of models that can produce various kinds of data. For multi-modal systems, the capacity to generate several data formats from a shared latent space offers enormous promise. As DGMs gain authority, significant ethical concerns are also brought forward. In order to address deepfakes, privacy concerns, and the exploitation of generated content, ethical standards and protections for the usage of generative models must be developed (Cate, 2025).

Evolutionary computation (EC) (Fogel, 1995) is a developing algorithm influenced by genetics and the survival of the fittest as an evolutionary approach. Selection, crossover, mutation, and initialization are frequently included in EC algorithms. Every iteration is an evolutionary process that leads to a constant improvement in population quality. Iterations and spontaneous development lead to a gradual approach to the ideal answer (Eiben and Smith 2015). It has been extensively utilized in the field of deep learning and has demonstrated acceptable performance in neural network architecture search and parameter tuning. The integration of these two systems, which have demonstrated a great potential for problem resolution in the real world (Bharti et al. 2020; Bernard & Leprévost 2019).

Recently, there has been a growing body of work proposing to use machine learning models to drive evolutionary algorithms. Usually, the training characteristics of the chosen models have a significant impact on how well these model-based evolutionary algorithms perform. Due to the curse of dimensionality, performance rapidly deteriorates as issue scales increase because model training often requires a specific quantity of data (i.e., the candidate solutions created by the algorithms). A multiobjective evolutionary approach powered by GANs solve this problem. The parent solutions are initially divided into real and fake samples to train the GANs at each generation of the suggested method; the trained GANs then sample the offspring solutions (He et al., 2020).

CONCLUSION

A novel class of deep generative models that has received a lot of interest lately are GANs. Across images, audio, and data, GANs implicitly learn complicated and high-dimensional distributions. However, due to improper network architecture design, usage of goal function, and optimization algorithm selection, there are significant obstacles in training GANs, namely mode collapse, non-convergence, and instability. In response to these difficulties, a number of strategies for improved GAN design and optimization have recently been studied. These strategies draw on approaches such as redesigned network designs, novel objective functions, and alternative optimization algorithms.

Deep representations can be learned without the need for heavily annotated training material thanks to generative adversarial networks (GANs). They accomplish this by using two networks in a competitive process to derive backpropagation signals. Applications for the representations that GANs can learn include picture synthesis, semantic image manipulation, style transfer, image superresolution, and classification. This review chapter aims to give the signal processing community an overview of GANs by utilizing well-known ideas and analogies whenever feasible.

Learning generative models for real data has proven to be successful with GANs. Yet, as generative tasks become more difficult, current GANs (GAN and its variations) frequently experience various training issues like instability and mode collapse. This chapter presented evolutionary GANs (E-GANs), a unique GAN framework for better generative performance and stable GAN training. Various adversarial training objectives are utilized as mutation operations, and the updates of each individual (i.e., generator candidature) are made according to these mutations.

One of the most widely used deep learning techniques for creating synthetic data is Generative Adversarial Networks, or GANs. Other application fields where GANs are being used include text-to-image translation, policy learning in deep reinforcement learning, object recognition and segmentation, voice and audio synthesis, and many more. Generative Adversarial Networks (GANs) and its derivatives have been increasingly popular in academia and industry in the past few years. The architecture and goal functions of various cutting-edge GAN-based techniques and their applications varies. After introducing generative models, the chapter discussed common training challenges for GANs, including mode collapse, convergence, and vanishing gradients.

In the area of artificial intelligence, deep learning has shown remarkable promise, leading to the development of numerous deep learning models. GAN was introduced using zero-sum game theory and has since emerged as a new area of interest for research. The model variation is important since it allows for the distribution of data to be obtained through unsupervised learning and produces more accurate and realistic data. GANs have been extensively researched recently because of their vast array of potential applications, such as language and video processing, image and vision computing, etc. This chapter introduced the background of GANs, theoretical models, and extensional variations of GANs, which can modify the fundamental structures or further optimize the original GAN.

REFERENCES

Abdelhalim, I. S. A., Mohamed, M. F., & Mahdy, Y. B. (2021). Data augmentation for skin lesion using self-attention based progressive generative adversarial network. *Expert Systems with Applications, 165*, 113922. DOI: 10.1016/j.eswa.2020.113922

Adler, J., & Lunz, S. (2018). Banach Wasserstein GAN. *Advances in Neural Information Processing Systems*, 31. https://doi.org/10.48550/arXiv.1806.06621

Ali, S., DiPaola, D., & Breazeal, C. (2021, May). What are GANs?: Introducing generative adversarial networks to middle school students. *Proceedings of the AAAI Conference on Artificial Intelligence, 35*(17), 15472–15479. DOI: 10.1609/aaai.v35i17.17821

Alqahtani, H., Kavakli-Thorne, M., & Kumar, G. (2021). Applications of generative adversarial networks (gans): An updated review. *Archives of Computational Methods in Engineering, 28*(2), 525–552. DOI: 10.1007/s11831-019-09388-y

Arjovsky, M., Chintala, S., & Bottou, L. (2017, July). Wasserstein generative adversarial networks. In *International conference on machine learning* (pp. 214-223). PMLR.

Arora, S., Ge, R., Liang, Y., Ma, T., & Zhang, Y. (2017, July). Generalization and equilibrium in generative adversarial nets (gans). In *International conference on machine learning* (pp. 224-232). PMLR.

Barratt, S., & Sharma, R. (2018). A note on the inception score. *arXiv preprint arXiv:1801.01973.*

Bernard, N., & Leprévost, F. (2019). Evolutionary algorithms for convolutional neural network visualisation. In *High Performance Computing:5th Latin American Conference, CARLA 2018,Bucaramanga, Colombia,September 26–28, 2018, Revised Selected Papers 5* (pp. 18-32). Springer International Publishing DOI: 10.1007/978-3-030-16205-4_2

Bharti, V., Biswas, B., & Shukla, K. K. (2020, January). Recent trends in nature inspired computation with applications to deep learning. In *2020 10th International Conference on Cloud Computing, Data Science & Engineering (Confluence)* (pp. 294-299). IEEE. DOI: 10.1109/Confluence47617.2020.9057841

Bindu, N. M., & Satyanarayana, S. (2025). Designing of GAN for a real-time image processing in neuromorphic system. In *Primer to Neuromorphic Computing* (pp. 21–44). Academic Press. DOI: 10.1016/B978-0-443-21480-6.00007-9

Carneiro, G. (2020, November). Why are generative adversarial networks so fascinating and annoying? In 2020 33rd SIBGRAPI Conference on Graphics, Patterns and Images (SIBGRAPI) (pp. 1-8). IEEE.

Cate, M. Fundamentals of Deep Generative Models.

Dash, A., Ye, J., & Wang, G. (2023). A review of generative adversarial networks (GANs) and its applications in a wide variety of disciplines: From medical to remote sensing. *IEEE Access : Practical Innovations, Open Solutions.*

DeVries, T., Romero, A., Pineda, L., Taylor, G. W., & Drozdzal, M. (2019). On the evaluation of conditional GANs. *arXiv preprint arXiv:1907.08175.*

Eiben, A. E., & Smith, J. (2015). From evolutionary computation to the evolution of things. *Nature, 521*(7553), 476–482. DOI: 10.1038/nature14544 PMID: 26017447

Fogel, D. B. (1995, December). Phenotypes, genotypes, and operators in evolutionary computation. In *Proc. 1995 IEEE Int. Conf. Evolutionary Computation (ICEC'95)* (pp. 193-198). DOI: 10.1109/ICEC.1995.489143

Fowlis, N., Barnett, N., Banks, S., & Jubraj, B. (2024). A qualitative evaluation of weekly reflective practice sessions for the intensive care unit pharmacy team during the COVID-19 pandemic. *European Journal of Hospital Pharmacy. Science and Practice, 31*(1), 57–62. DOI: 10.1136/ejhpharm-2021-003164 PMID: 35428696

Gonog, L., & Zhou, Y. (2019, June). A review: generative adversarial networks. In *2019 14th IEEE conference on industrial electronics and applications (ICIEA)* (pp. 505-510). IEEE. DOI: 10.1109/ICIEA.2019.8833686

Goodfellow, I., Pouget-Abadie, J., Mirza, M., Xu, B., Warde-Farley, D., Ozair, S., Courville, A., & Bengio, Y. (2020). Generative adversarial networks. *Communications of the ACM, 63*(11), 139–144. DOI: 10.1145/3422622

Hachana, R., & Rasheed, B. (2025). Probe-Assisted Fine-Grained Control for Non-Differentiable Features in Symbolic Music Generation. *IEEE Access : Practical Innovations, Open Solutions, 13*, 28059–28070. DOI: 10.1109/ACCESS.2025.3540543

He, C., Huang, S., Cheng, R., Tan, K. C., & Jin, Y. (2020). Evolutionary multi-objective optimization driven by generative adversarial networks (GANs). *IEEE Transactions on Cybernetics, 51*(6), 3129–3142. DOI: 10.1109/TCYB.2020.2985081 PMID: 32365041

Hong, Y., Hwang, U., Yoo, J., & Yoon, S. (2019). How generative adversarial networks and their variants work: An overview. *ACM Computing Surveys, 52*(1), 1–43. DOI: 10.1145/3301282

Iglesias, G., Talavera, E., & Díaz-Álvarez, A. (2023). A survey on GANs for computer vision: Recent research, analysis and taxonomy. *Computer Science Review, 48*, 100553. DOI: 10.1016/j.cosrev.2023.100553

Jeong, J., & Shin, J. (2021). Training gans with stronger augmentations via contrastive discriminator. *arXiv preprint arXiv:2103.09742.*

Karras, T., Laine, S., Aittala, M., Hellsten, J., Lehtinen, J., & Aila, T. (2020). Analyzing and improving the image quality of stylegan. In *Proceedings of the IEEE/CVF conference on computer vision and pattern recognition* (pp. 8110-8119). DOI: 10.1109/CVPR42600.2020.00813

Kim, S. E., & Seo, I. W. (2015). Artificial neural network ensemble modeling with exploratory factor analysis for streamflow forecasting. *Journal of Hydroinformatics, 17*(4), 614–639. DOI: 10.2166/hydro.2015.033

Kulkarni, P., Pathak, P., Pillai, S., & Tigga, V. (2025). Role of Generative AI for Fraud Detection and Prevention. *Generative Artificial Intelligence in Finance: Large Language Models, Interfaces, and Industry Use Cases to Transform Accounting and Finance Processes*, 175-198.

Kunkel, L., & Trabs, M. (2024). A Wasserstein perspective of Vanilla GANs. *arXiv preprint arXiv:2403.15312.*

Kurach, K., Lucic, M., Zhai, X., Michalski, M., & Gelly, S. (2018). The gan landscape: Losses, architectures, regularization, and normalization.

Kwon, G., Han, C., & Kim, D. S. (2019, October). Generation of 3D brain MRI using auto-encoding generative adversarial networks. In *International Conference on Medical Image Computing and Computer-Assisted Intervention* (pp. 118-126). Cham: Springer International Publishing. DOI: 10.1007/978-3-030-32248-9_14

Lin, Q., Fang, Z., Chen, Y., Tan, K. C., & Li, Y. (2022). Evolutionary architectural search for generative adversarial networks. *IEEE Transactions on Emerging Topics in Computational Intelligence, 6*(4), 783–794. DOI: 10.1109/TETCI.2021.3137377

Liu, H., Gu, X., & Samaras, D. (2019). Wasserstein gan with quadratic transport cost. In *Proceedings of the IEEE/CVF international conference on computer vision* (pp. 4832-4841).

Ma, X. (2024, February). A comparison of art style transfer in Cycle-GAN based on different generators. [). IOP Publishing.]. *Journal of Physics: Conference Series, 2711*(1), 012006. DOI: 10.1088/1742-6596/2711/1/012006

Mao, X., Li, Q., Xie, H., Lau, R. Y., Wang, Z., & Paul Smolley, S. (2017). Least squares generative adversarial networks. In *Proceedings of the IEEE international conference on computer vision* (pp. 2794-2802).

Millière, R. (2022). Deep learning and synthetic media. *Synthese, 200*(3), 231. DOI: 10.1007/s11229-022-03739-2

Modanwal, G., Vellal, A., & Mazurowski, M. A. (2021). Normalization of breast MRIs using cycle-consistent generative adversarial networks. *Computer Methods and Programs in Biomedicine, 208*, 106225. DOI: 10.1016/j.cmpb.2021.106225 PMID: 34198016

Neff, T., Payer, C., Stern, D., & Urschler, M. (2017, May). Generative adversarial network based synthesis for supervised medical image segmentation. In *Proc. OAGM and ARW joint Workshop* (Vol. 3, p. 4).

Nunn, E. J., Khadivi, P., & Samavi, S. (2021). Compound frechet inception distance for quality assessment of gan created images. *arXiv preprint arXiv:2106.08575.*

Öcal, A., & Özbakır, L. (2021). Supervised deep convolutional generative adversarial networks. *Neurocomputing, 449*, 389–398. DOI: 10.1016/j.neucom.2021.03.125

Patil, A. (2021, December). DCGAN: Deep convolutional GAN with attention module for remote view classification. In *2021 International Conference on Forensics, Analytics, Big Data, Security (FABS)* (Vol. 1, pp. 1-10). IEEE. DOI: 10.1109/FABS52071.2021.9702655

Perarnau, G., Van De Weijer, J., Raducanu, B., & Álvarez, J. M. (2016). Invertible conditional gans for image editing. *arXiv preprint arXiv:1611.06355.*

Purwono, P., Wulandari, A. N. E., Ma'arif, A., & Salah, W. A. (2025). Understanding Generative Adversarial Networks (GANs): A Review. *Control Systems and Optimization Letters, 3*(1), 36–45.

Qin, Z., Liu, Z., Zhu, P., & Xue, Y. (2020). A GAN-based image synthesis method for skin lesion classification. *Computer Methods and Programs in Biomedicine, 195*, 105568. DOI: 10.1016/j.cmpb.2020.105568 PMID: 32526536

Radford, A. (2015). Unsupervised representation learning with deep convolutional generative adversarial networks. *arXiv preprint arXiv:1511.06434.*

Radford, A., Metz, L., & Chintala, S. (2015). Unsupervised representation learning with deep convolutional generative adversarial networks. *arXiv preprint arXiv:1511.06434.*

Rastogi, R., Rawat, V., & Kaushal, S. (2025). Advancements in Image Restoration Techniques: A Comprehensive Review and Analysis through GAN. *Generative Artificial Intelligence and Ethics: Standards, Guidelines, and Best Practices*, 53-90.

Rožanec, J. M., Zajec, P., Theodoropoulos, S., Koehorst, E., Fortuna, B., & Mladenić, D. (2023). Synthetic data augmentation using GAN for improved automated visual inspection. *IFAC-PapersOnLine*, *56*(2), 11094–11099. DOI: 10.1016/j.ifacol.2023.10.817

Saad, M. M., O'Reilly, R., & Rehmani, M. H. (2024). A survey on training challenges in generative adversarial networks for biomedical image analysis. *Artificial Intelligence Review*, *57*(2), 19. DOI: 10.1007/s10462-023-10624-y

Saad, M. M., Rehmani, M. H., & O'Reilly, R. (2022a, December). A self-attention guided multi-scale gradient GAN for diversified x-ray image synthesis. In *Irish Conference on Artificial Intelligence and Cognitive Science* (pp. 18-31). Cham: Springer Nature Switzerland.

Saad, M. M., Rehmani, M. H., & O'Reilly, R. (2022b, July). Addressing the intra-class mode collapse problem using adaptive input image normalization in gan-based x-ray images. In *2022 44th annual international conference of the IEEE Engineering in Medicine & Biology Society (EMBC)* (pp. 2049-2052). IEEE. DOI: 10.1109/EMBC48229.2022.9871260

Salehi, P., Chalechale, A., & Taghizadeh, M. (2020). Generative adversarial networks (GANs): An overview of theoretical model, evaluation metrics, and recent developments. *arXiv preprint arXiv:2005.13178.*

Segato, A., Corbetta, V., Di Marzo, M., Pozzi, L., & De Momi, E. (2020). Data augmentation of 3D brain environment using deep convolutional refined auto-encoding alpha GAN. *IEEE Transactions on Medical Robotics and Bionics*, *3*(1), 269–272. DOI: 10.1109/TMRB.2020.3045230

Si, H. (2023, November). On the convergence acceleration of vanilla Generative Adversarial Network. [). IOP Publishing.]. *Journal of Physics: Conference Series*, *2634*(1), 012047. DOI: 10.1088/1742-6596/2634/1/012047

Sidheekh, S., Aimen, A., & Krishnan, N. C. (2021, July). On characterizing GAN convergence through proximal duality gap. In *International Conference on Machine Learning* (pp. 9660-9670). PMLR.

Swadi, L. A., & Haider, A. M. (2025). A Review of Generative Adversarial Networks for Addressing Data Imbalance and Enhancing Model Performance. [IJICI]. *Iraqi Journal of Intelligent Computing and Informatics*, *4*(1), 18–32.

Wu, Y., Yue, Y., Tan, X., Wang, W., & Lu, T. (2018, October). End-to-end chromosome Karyotyping with data augmentation using GAN. In *2018 25th IEEE international conference on image processing (ICIP)* (pp. 2456-2460). IEEE. DOI: 10.1109/ICIP.2018.8451041

Xu, L., Zeng, X., Huang, Z., Li, W., & Zhang, H. (2020). Low-dose chest X-ray image super-resolution using generative adversarial nets with spectral normalization. *Biomedical Signal Processing and Control*, *55*, 101600. DOI: 10.1016/j.bspc.2019.101600

Xue, Y., Zhou, Q., Ye, J., Long, L. R., Antani, S., Cornwell, C., Xue, Z., & Huang, X. (2019). Synthetic augmentation and feature-based filtering for improved cervical histopathology image classification. In *Medical Image Computing and Computer Assisted Intervention–MICCAI 2019: 22nd International Conference, Shenzhen, China, October 13–17, 2019* [Springer International Publishing.]. *Proceedings*, *22*(Part I), 387–396.

Zhang, D., & Khoreva, A. (2019). Progressive augmentation of gans. *Advances in Neural Information Processing Systems*, •••, 32.

ADDITIONAL READING

Dalmasso, M., Civera, M., De Biagi, V., Surace, C., & Chiaia, B. (2025). Conditional generative adversarial networks for the data generation and seismic analysis of above and underground infrastructures. *Tunnelling and Underground Space Technology*, *157*, 106285. DOI: 10.1016/j.tust.2024.106285

Hayajneh, A., Serpedin, E., Shaqfeh, M., Glass, G., & Stotland, M. A. (2025). Adapting a style based generative adversarial network to create images depicting cleft lip deformity. *Scientific Reports*, *15*(1), 3614. DOI: 10.1038/s41598-025-86588-6 PMID: 39875471

Khalid, M. N. A., Ullah, A., Numan, M., & Majid, A. (2025). Words shaping worlds: A comprehensive exploration of text-driven image and video generation with generative adversarial networks. *Neurocomputing*, •••, 129767.

Lee, J., Jung, D., Moon, J., & Rho, S. (2025). Advanced R-GAN: Generating anomaly data for improved detection in imbalanced datasets using regularized generative adversarial networks. *Alexandria Engineering Journal*, *111*, 491–510. DOI: 10.1016/j.aej.2024.10.084

Ma, C., Mi, J., Gao, W., & Tao, S. (2025). DESRGAN: Detail-enhanced generative adversarial networks for small sample single image super-resolution. *Neurocomputing, 617*, 129121. DOI: 10.1016/j.neucom.2024.129121

Zhang, J., Nie, R., Li, Y., & Tan, Y. (2025). Random generation of three-dimensional realistic ballast particles using generative adversarial networks. *Computers and Geotechnics, 178*, 106923. DOI: 10.1016/j.compgeo.2024.106923

Zhang, Z., Li, X., Cheng, Y., Chen, Z., & Liu, Q. (2025). Credit Risk Identification in Supply Chains Using Generative Adversarial Networks. *arXiv preprint arXiv:2501.10348.*

KEY TERMS AND DEFINITIONS

Conditional GAN (cGAN): Generative Adversarial Networks (GANs) have recently demonstrated to successfully approximate complex data distributions. A relevant extension of this model is conditional GANs (cGANs), where the introduction of external information allows to determine specific representations of the generated images.

Cycle-GAN: CycleGAN, is an approach to training a deep convolutional neural network for image-to-image translation tasks. The Network learns mapping between input and output images using unpaired dataset.

Deep convolutional GAN (DCGAN): DCGAN is one of the popular and successful network design for GAN. It mainly composes of convolution layers without max pooling or fully connected layers. It uses convolutional stride and transposed convolution for the downsampling and the upsampling.

Evolutionary GANs (E-GANs): Evolutionary generative adversarial networks (E-GANs) are a type of generative adversarial network (GAN) that uses evolutionary strategies to improve training and performance. E-GANs can help address issues like mode collapse and vanishing gradients that often plague GANs.

Least Squares Generative Adversarial Networks (LSGANs): LSGAN: is a type of generative adversarial network that adopts the least squares loss function for the discriminator.

StyleGAN: StyleGAN is an advanced generative model developed by NVIDIA designed to generate highly realistic images such as faces of people who don't exist. Unlike traditional image generation methods StyleGAN provides a unique advantage by giving you control over various levels of image features. This includes broad shapes and structures as well as finer details like texture and lighting. In this article we will learn more about them.

Vanilla GAN's: A vanilla GAN is a basic generative adversarial network (GAN) that uses a generator and discriminator to create and classify images. It's the simplest type of GAN and is often used as a starting point for more advanced GANs.

Wassertein GAN (WGAN): It is a modification of Deep Learning GAN with few changes in the algorithm. GAN is a way to build an accurate generative model. This network was introduced by Martin Arjovsky, Soumith Chintala, and Léon Bottou in 2017. It is widely used to generate realistic images.The key difference between GANs and WGANs is the loss function and the gradient penalty. WGANs were introduced as the solution to mode collapse issues. The network uses the Wasserstein distance, which provides a meaningful and smoother measure of distance between distributions.

Chapter 4 Generative Adversarial Networks in Feature Engineering: An Unsupervised Approach

Kavita Khobragade
https://orcid.org/0009-0002-5776-9147
Fergusson College, India

Poonam Ponde
Nowrosjee Wadia College, India

ABSTRACT

Feature engineering, the process of extracting informative representations from raw data, is crucial for successful remote sensing applications. An automated feature engineering is provided by Generative Adversarial Networks (GANs), which is a type of deep learning models based on unsupervised learning. In feature engineering, training strategies for GANs are also reviewed, with an emphasis on methods to enhance convergence, stabilize training, and reduce problems like mode collapse and overfitting. In addition, the chapter addresses issues like computing efficiency, scalability, and integration with current analytical frameworks, when implementing GAN-based feature engineering techniques in practical remote sensing applications. This chapter concludes with a strong argument that using GANs is an effective tool for unsupervised feature engineering. Through the utilization of GANs' capacity to acquire insightful latent representations from unlabeled data, scholars and professionals might open up novel avenues for the extraction of significant insights from remote sensing data.

DOI: 10.4018/979-8-3693-6900-5.ch004

1. INTRODUCTION

Artificial Intelligence (AI) became a very common technology used in our lives. The machine sense lots of complex data and handle it easily by providing fast and accurate decisions. For example –remote sensing data, facial recognition, etc. Machine learns through the images, videos, and text data extracted and then solves problems. This process of learning is known as Machine Learning (ML). Deep Learning is a part of AI, ML & Data Science (55) (Figure 1). The complex problems can be solved by using artificial neural networks (ANN) is a process referred as Deep Learning. (56) The DNN [Deep Neural Networks] is a machine learning algorithm, comprises multiple layer of neurons between the layers which accepts input and layer produce the output (Figure 2). (Katy War, (2019), Nikhil Buduma, (2019), Shahab Jozdani et.al.,(2022), 55, 56)

Figure 1. Understanding Artificial Intelligence, Machine Learning, Deep Learning (57)

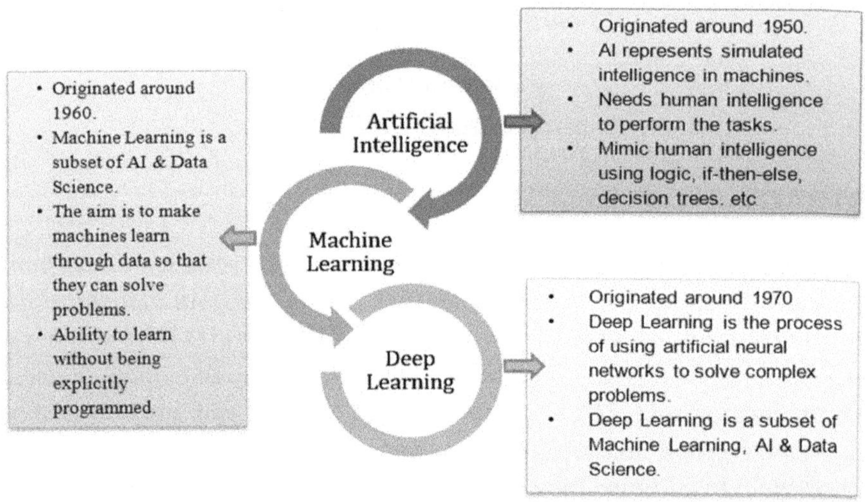

Today, the computers were already trained with available data and the computers can make better estimates than individuals in most of the situations. But learning a pattern from the input like image, voice, text etc., is quite difficult. The machine learning algorithms (Figure 3) are used to acquire such patterns using computers. After acquiring such patterns, the algorithms learn and extract few features from the pattern and make predictions about the image or voice or text by classifying the features. This is helpful to identify the image, voice or text. The deep learning algorithms, automatically learns and train itself for the recognition without asking for the features separately.

Figure 2. A Simplified DNN implementation

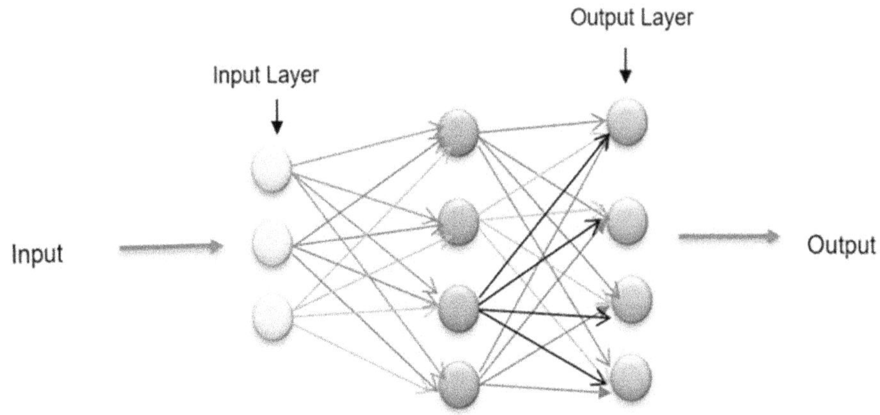

To understand this in a simple word, let's see the following concept. An individual learns to multiply numbers or solve any equations by listening to the set of instructions in the classroom. This is machine learning. But the things we learn at an extremely early age are learnt through real example. Our parents did not teach us how to recognize a cat or cow by measuring shape of its nose, tail or contours of its body. But we learnt it by multiple examples and being corrected when we make mistakes. It means our concepts are reinforced. Deep learning algorithms (Figure 4) learns exactly the same way. (Nikhil Buduma, (2019), Seth Weidman, (2019), Tanya Tiwari et.al., (2018), Mohsen Soori,(2023), Rene Y. Choi et. Al., (2024), Ayushi Chahal, & Preeti Gulia, (2019))

Figure 3. Machine Learning Algorithms

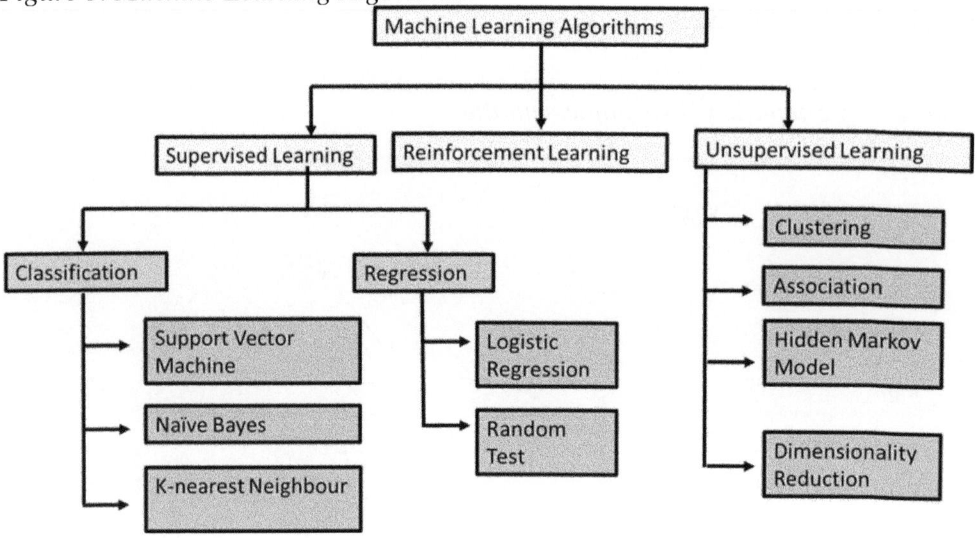

Figure 4. Deep Learning Algorithms

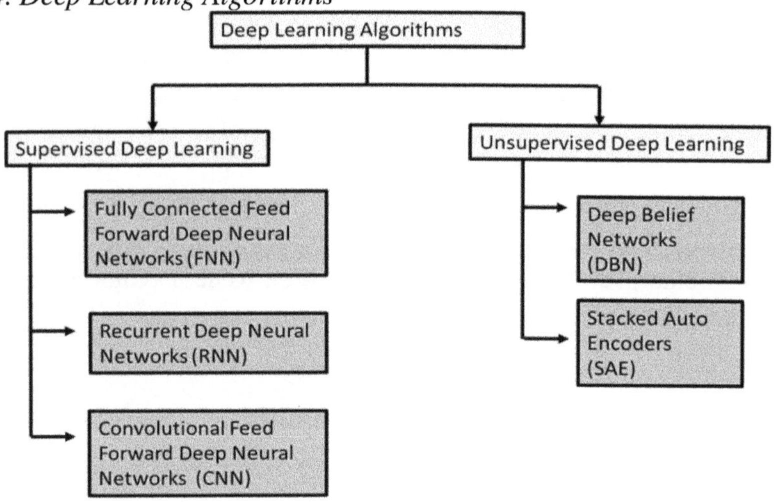

The major objective of this chapter are:

1. To understand the Generative Adversarial Networks
2. To understand the use of GANs in unsupervised classification
3. To open-up the new opportunities of Feature Engineering using GANs

2. UNSUPERVISED LEARNING

When the data is unlabeled and a mathematical model needs to build from the available input then it is called as unsupervised learning. The expected output labels may not be available in this type of learning. The algorithms are unsupervised because the patterns may or may not exist in available dataset The main goal of unsupervised learning is to find the hidden patterns within a dataset and the other important goal is to allow the computational machine to discover the feature representations automatically i.e. feature learning; that the classification of raw data becomes easy. Some of the most common unsupervised learning tasks are clustering, association rules, K-means and anomaly detection, etc. For example, an unsupervised learning is frequently used to identify anomalies, such as credit card fraud and recommender systems that suggest what to buy next.

Transactional data is a popular application for unsupervised learning. Even with a sizable dataset of consumers and their purchases, a human being would probably be unable to deduce what similar characteristics may be inferred from the profiles of the customers and the kinds of purchases they make.

By utilizing an unsupervised learning algorithm to process this data, it is possible to ascertain that pregnant women within a specific age group who purchase fragrance-free soaps are inclined that they might be pregnant. Consequently, a promotional drive featuring pregnancy related products and baby products can be directed towards this demographic data to enhance the quantity of purchases made by them. (Yongpeng Luo, (2023), Zhongwei Li,(2022), Endang Suryawati et.al.,(2021), Wang, J.,et.al,(2020))

3. REMOTE SENSING

By using sensors the different forms of energy can be measured or detected, such as electromagnetic wave and sonic or sound signals. These signals can be discharge, bounce back, or disperse by the object under inquiry. The remote sensing technology allows data gathering to be done without having to come into close touch with the topic. Many platforms and sensors have been created for remote sensing. The amount of data collected by remote sensing has increased to astounding levels as sensors continue to evolve.

In order to integrate remote sensing techniques with artificial intelligence, it is necessary to grasp the basic concepts of the field and its many methodologies and applications. Systems for remote sensing are designed to view diverse targets by utilizing atmospheric windows and different regions of the electromagnetic spectrum (Figure 5).

Figure 5. The electromagnetic spectrum

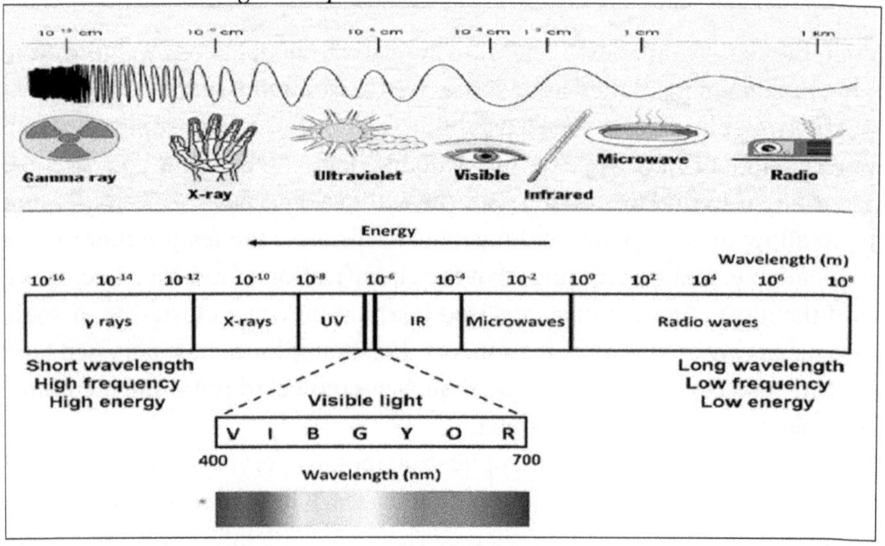

Image source: https://pyimagesearch.com/2020/10/19/adversarial-images-and-attacks-with-keras-and -tensorflow/, 43https://kaushiklade27.medium.com-image-generation-using-generative-advarsarial-networks-gans-cd82afd71597

Optical sensors that record sunlight reflection are examples of passive sensors that pick up natural energy released or reflected by the Earth (Figure 6), while active sensors are those which emits an energy and measures the reflected or backscattered signals. A vast array of sensors makes it possible to obtain remote sensing data using different tools namely aircraft can be used for enhancing image resolution and drones can be used for limited (modest) data collection, and satellites for worldwide coverage. Because of this broad spectrum of sensors, remote sensing data can be collected using drones for small-scale data gathering, aircraft for higher spatial resolution, and satellites for global coverage. (Zhongwei Li, et al., (2022), Janga, B., Asamani et. Al., (2023), Zhang, L. et al., (2016), Shahab Jozdani et. Al., (2022))

After obtaining remote sensing data, deciphering digital and picture data is essential to obtaining relevant information. The extraction of important insights is made possible by digital image processing methods. These techniques comprise image cleaning, blending the images, extracting the features, and classifying the images, etc.

There are various remote sensing techniques such as optical remote sensing, radar remote sensing, LiDAR remote sensing, Thermal remote sensing, Multi-spectral and Hyper-spectral Imaging techniques, etc. Most of the remote sensing community is using traditional machine learning techniques for performing a variety of tasks like detecting an object, classifying it and for estimating geophysical parameters. When managing materialistic and multi-sensorial remote sensing data, these techniques have shown to be effective, yielding useful information for environmental monitoring.

Figure 6. The Process of Remote Sensing

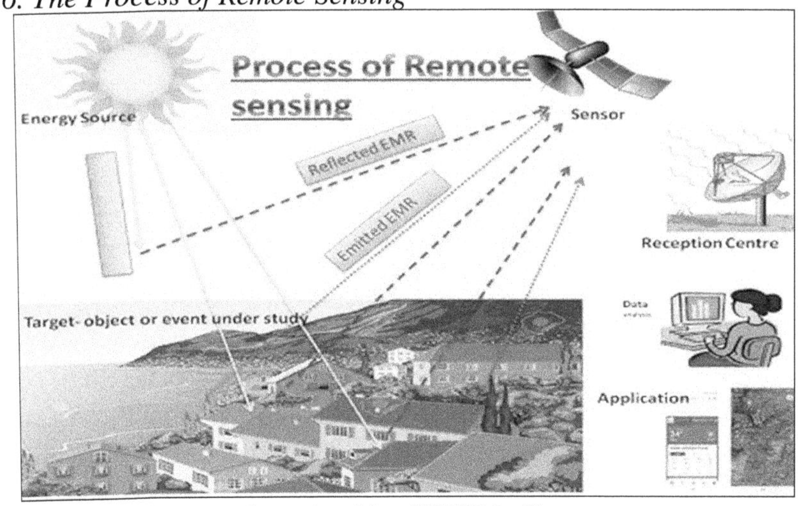

Image source- https://www.youtube.com/watch?v=rSUMJP9Oz_M

One area of machine learning called deep learning has become very useful for remote sensing applications, providing answers to hitherto unmet problems and opening up new avenues for research. Deep learning extracts significant characteristics from big and complicated datasets by using hierarchical artificial neural networks. Using ANN patterns are extracted from the data. (Zhongwei Li et. Al., (2022), Janga, B., Asamani et. Al., (2023))

In order to improve its capacity to identify patterns and its relationships as it analyzes additional data, the network uses a technique called backpropagation to modify weights and biases during training. Image preprocessing, object recognition, and pixel-based categorization are only a few of the tasks for which deep learning networks progressively turn the input into representations. Some popular deep learning algorithms are briefly introduced in this chapter.

4. GENERATIVE ADVERSARIAL NETWORKS: AN OVERVIEW

In 2014, the use of Generative Adversarial Networks was first initiated by Ian Goodfellow and his colleagues. The GAN consist of generator and discriminator. These are the two neural networks (Figure 7). The generator is for creating fake data, whereas the discriminator is for evaluating its authenticity against real data. (Katy War, (2019))

The generator network takes randomly generated noise as input and generates the training data samples. (Figure 8). The objective of generator is to produce outputs which is identical to real data.

Figure 7. Generative Adversarial Networks

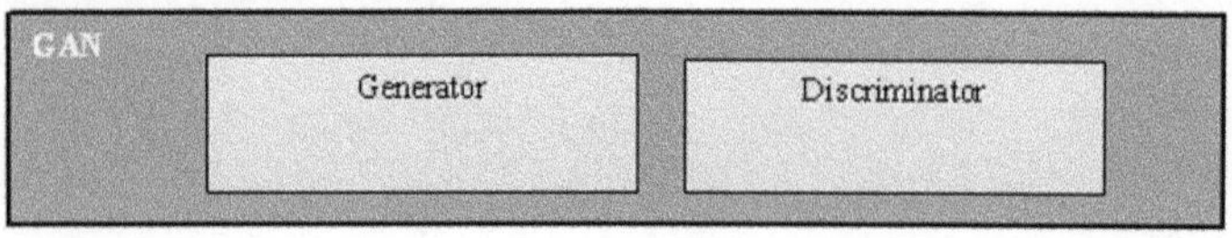

The discriminator network receives both real and fake data and generates the outputs as a probability score which indicates whether the data is real or generated. Its goal is to correctly classify data samples as real or fake (Figure 9). (Aggarwal, A., Mittal et. Al., (2021), Bendaoud et.al., (2021), Alankrita Aggarwal et. Al., (2021), Goodfellow. et.al. (2020). Gui, J. et. Al., (2021), Jieren Cheng et. Al. (2020))

Figure 8. Generator

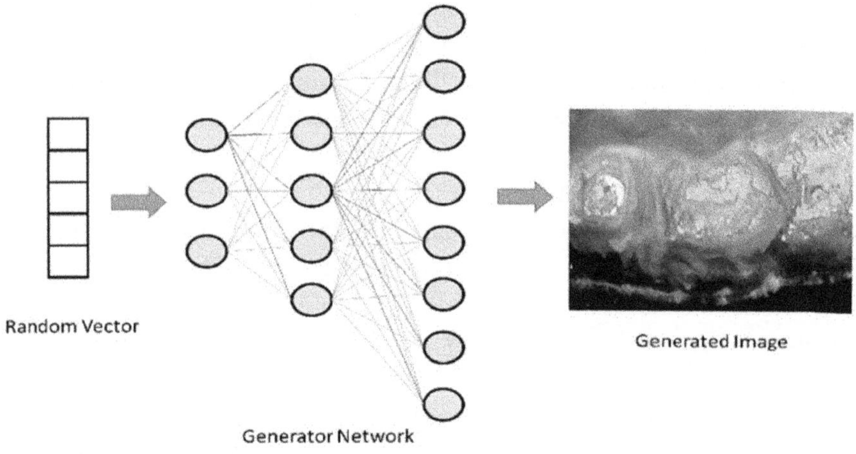

Image source: https://www.google.com/search?q=images+of+ocean+floor+on+remote+sensing&sca_esv =d620e9c68447bd3d&sxsrf=ADLYWII

Figure 9. Discriminator

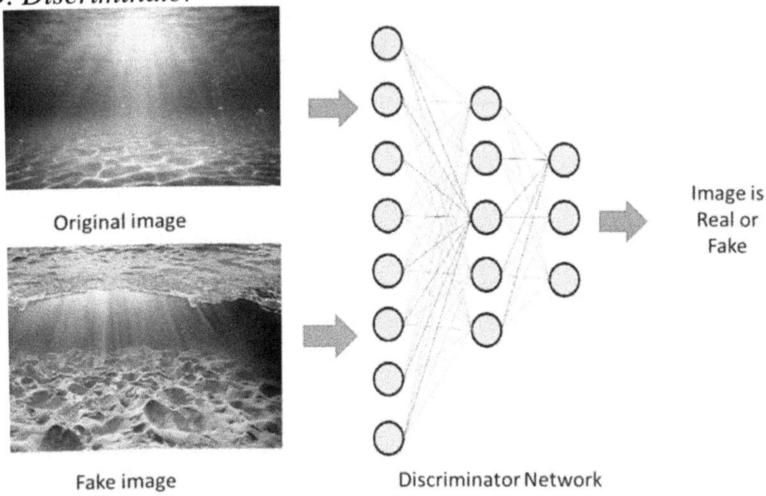

Image source: https://www.google.com/search?q=images+of+ocean+floor+on+remote+sensing&sca_esv=d620e9c68447bd3d&sxsrf=ADLYWII

5. ADVERSARIAL INPUT

The image processing domain uses the concept of adversarial input for creating hallucination (i.e. optical illusions) to which Artificial Intelligence is responsive. The adversarial input might be generated by sprinkling apparently unimportant pixels across the particular image which is classified as another image in artificial intelligence. (Katy War, (2019))

For example, sprinkle pixels to a cat image so it will be classified as an image of a dog. This is done without introducing notifiable features which a person figures out as a dog. Another example, sprinkle pixels to a rice crop image so it will be classified as an image of a wheat crop. This is done without introducing notifiable features which a person figures out as a wheat plant. (Figure 10)

Basically an adversarial input is generated with the focus where a model is returning an incorrect result, so that the network can be fooled successfully and produce the predictions that humans would consider incorrect. The adversarial input exploits the reality that deep learning models handle many possible input variants based on a small proportion of learning examples. But the learned models must be flexible, so they can handle the complexity. (War, (2019))

When we make some calculated changes to the original image and generate an adversarial image by modifying each pixel by a small amount, it is called **Perturbation attack**. This attack may not be used for image but it also works on the voice frequencies. (Katy War, (2019))

Figure 10. Fooling Image Classifier with Adversarial Perturbation (a,b,c,d, are fool images) (Shahab Jozdani et.al., 2022)

6. GAN's IN FEATURE ENGINEERING

Feature engineering, the process of extracting informative representations from raw data. Feature engineering is a very important step in machine learning algorithm which improves the performance of predictive models. Traditionally, feature engineering is based on the domain knowledge and manual processes. However, Generative Adversarial Networks has introduced a novel, innovative and unsupervised approach to feature engineering through Adversarial Frameworks which offers a way to create, refine, and transform features without explicit supervision.

The above two networks Generator and Discriminator are used in a continuous adversarial process. The generator improves its capability to produce real samples and the discriminator improves its ability to differentiate between real and fake data.

The adversarial training allows the generator to create high quality synthetic data which shows like a copy of real data distribution. The generative capabilities of GANs for enhancing, transforming and creating features, needs to be optimized in feature engineering. This process has many key areas like feature extraction, data augmentation, data transformation, etc.

7. FEATURE EXTRACTION

The feature extraction is used to obtain the most appropriate information from the raw data and understands the complex patterns from it and represent them. The remote sensing data is high dimensional, GANs are used for feature extraction.

Feature Extraction (Images): For extracting features from the remote sensing images, GANs works with deep convolutional architectures (DCGANs or Style-GANs). These architectures handles the hierarchical features as inputs, send them to other machine learning models or sometimes used these hierarchical features for classifying images and detecting objects. (Karras, T., Laine, S., & Aila, T. (2019).

Feature Representation (Text): while handling text data, dense vector of words or sentences needs to be generated. This dense vector captures the semantic similarities and fine distributions. GANs can be used to learn this in Natural Language Processing (NLP) applications.

8. DATA AUGMENTATION

Data augmentation is a process of using the available data to create new samples of data for training any machine learning model. GANs can generate synthetic data which looks like a copy of real data, thus it augments the dataset with different examples. For example, the images are captured from satellite for Ocean floors. The ocean floors shows various features like flat plains, sharp mountains, rugged canyons, etc. Sometimes in few areas, the features are deep sea or coral reef. Many images are available for the same. (Veenu Rani et.al.,(2023), 47) The GANs can generate variation in available images while classifying these images. It improves the robustness of a system as the additional training images have been added. For instance, augmenting an ocean floor images with images of different lighting conditions, orientations, or occlusions, can assist the model to generalize it better.

This is important when the original dataset is small or imbalanced. It follows two steps in it as shown below:

- **Synthetic Data Generation:** The new features can be created by combining and transforming existing features is called Feature synthesis. GANs can be used to synthesize features by learning complex distributions and relationships within the data. GANs can create realistic synthetic samples which preserves the statistical properties of the original data.

For example, in medical imaging, GANs can generate additional images of rare conditions, enhancing the dataset and improving model performance. Same is applicable for the remote sensing images. (Sadeena Sabnam & Sivakumar Rajagopal, (2024))

Application: In tabular data, GANs can be used to generate new feature combinations or interactions that were not initially present. This can be particularly useful in high-dimensional datasets where manual feature engineering is challenging.

- **Balancing Datasets:** GANs can be used to generate samples for underrepresented classes in classification tasks, which will address the class imbalance issues and improves the performance of the machine learning models.

9. DATA TRANSFORMATION

GANs can transform data into different representations which makes it more suitable for various tasks as shown below:

- **Domain Adaptation:** CycleGANs and other domain adaptation techniques can transform data from one domain to another, such as converting images from day to night or changing styles in art. This transformation can help in training models that are robust across different domains.

For example, in remote sensing images, domain adaptation helps a lot for images as they might have taken into day time or night time.

- **Feature Manipulation:** GANs can be used to modify specific features of the data while keeping other aspects unchanged.

For example, in image editing, GANs can alter attributes like age or facial expressions while preserving other features. In remote sensing images, GANs can alter colours of the plants in the forest to recognize which different plants are available in the forest.

For example, a GAN trained on text data might generate embedding's that capture semantic relationships between words, which can then be used for progressing the tasks such as topic modeling or sentiment analysis.

GANs can be employed to attain meaningful features from an unlabeled data. The generator's latent space can be explored to find hidden patterns or structures within the data. For example, in unsupervised learning scenarios, the generator's output can be used as features for clustering or dimensionality reduction tasks.

From all above processes, the new image will be generated (figure 11), which can be used for further processing.

Other than this, there are many advanced techniques for features such as conditional GANs, Autoencoders and GANs, progressive growing GANs etc. are discussed below:

10. CONDITIONAL GANs FOR FEATURE SYNTHESIS

GAN is basically a deep learning framework. The generator and discriminator can be conditioned with some additional information such as class labels or specific attributes. Then it is called as Conditional GAN (cGAN). The synthetic images can be generated by using cGAN's. For example, in the images of ocean floors, some additional information such as different angles, different lightening conditions can be added and synthetic images can be generated. By using these synthetic images, classifiers can be trained, and different features can also be detected. This allows to generate more controlled features and their manipulations. The following things are possible:

- cGANs can generate class conditioned features based on synthetic images.
- cGANs can allow to have detailed control over feature engineering process and manipulate specific attributes.

Figure 11. Feature Extraction with GANs

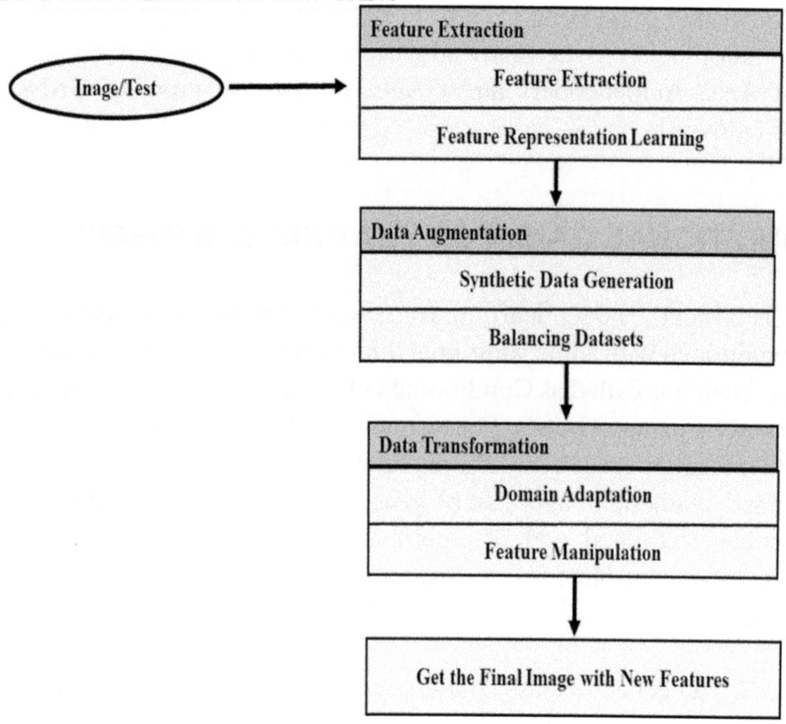

11. VARIATIONAL AUTOENCODERS (VAES) AND GANS

The GANs used for generating high-quality samples and translation of them whereas VAEs (Variational Autoencoders) provides attention on the learning of unused representation of data. VAEs are basically used for image compression and image de-noising. Both VAE and GAN can be combined together to provide strengths to the models. (Shahab Jozdani et. Al.,(2023))

For example, in the images of ocean floors, as these images are in high-dimensional space, VAE and GAN can be used together (called VAE-GAN Hybrid Model) to capture essential patterns from the features like the ocean trench and hydrothermal vents. These patterns (representations) can be used to identify meaningful data or to associate with missing data in the dataset.

12. PROGRESSIVE GROWING GANS

There are GANs which gradually increasing its complexity using generator and discriminator, those are called as Progressive Growing GANs (PGGANs). These PGGANs improves the training and produce the higher quality samples. For example, generating synthetic ocean floor data, PGGANs can be used to imitate the day night time and complex lightening conditions.

The generated data can enhance feature sets used for risk modeling and financial forecasting.

13. PRACTICAL APPROACH

For practical considerations we need to know 2 major aspects:

1. Data Preprocessing
2. Model Selection and Training
3. Evaluation metrics
4. Addressing Mode Collapse
5. Ethical Considerations

13.1. Data Preprocessing

Pre-processing is critical for ensuring that GAN-generated features are of high quality. The performance of GANs can be significantly affected by proper normalization, scaling, and augmentation of input. For example, in a dataset of satellite images, pre-processing steps such as normalization and augmentation can help the GAN learn better representations of various land features, leading to more accurate feature extraction and analysis.

13.2. Model Selection and Training

For effective feature engineering, choosing the right GAN architecture and training process is crucial. This includes the following two aspects:

- **Architecture Choice:** Select GAN architectures based on the type of data used and the feature engineering task.

For example, DCGANs for image data or GANs with self-attention mechanisms for complex data.

- **Training Stability:** GANs can be challenging to train as it requires vigilant tuning of hyper-parameters and optimization techniques. Regularization, loss function adjustments, and careful monitoring of training progress can help ensure stable and effective training.

13.3. Evaluation Metrics

The quality of the GAN generated features is assessed by considering both i.e. the synthetic data and the extracted features. This also requires robust metrics and validation methods. The common approaches include:

- **Qualitative Assessment**: Visual inspection and domain expert reviews can help evaluate the realism and usefulness of generated features.
- **Quality Metrics:** The Inception Score (IS) is a quality metrics which is used to estimates the quality of collected synthetic images. This scores shows quality and diversity amongst the collected synthetic images. Fréchet Inception Distance (FID) is used to quantify the reality in the synthetic data. The FID score is lower reflects that the synthetic images looks real. (49)
- **Feature Utility:** Feature utility is basically for capturing meaningful patterns from the data. The generated features can be evaluated and their impact on the performance can also be evaluated.

13.4. Addressing Mode Collapse

In GAN training, mode collapse is used by the generator to produces very less variety of outputs. The Mini-batch Discrimination is a discriminative method which discriminate between whole mini-batches of samples rather than between individual samples for generative adversarial networks. This technique is for avoiding collapse of the generator. For example, in ocean floor synthesis, mode collapse can result in repetitive or non-diverse generated ocean floor samples. (50)

13.5. Ethical Considerations and Bias Mitigation

When GANs are used for feature engineering, it is necessary that the ethical implications and potential biases must be considered.

Ensuring fairness and transparency in the use of GAN-generated data is crucial for responsible AI development. For example, in ocean floor recognition systems, it is necessay to ensure that the synthetic data generated by GANs does not perpetuate existing biases or inaccuracies, leading to fairer and more accurate models.

14. TYPES OF GANs

Therc arc many GANs used in the world. The selection of the GAN is as per the functionality of a researcher. The following are the basic types of GANs used:

Vanilla GANs - This original GAN architecture supports a basic generator and discriminator.

Conditional GANs (eGANs) - It also supports generator and discriminator and they both receive additional information (example labels) to produce the samples which are conditioned on the additional information.

Deep Convolutional GANs (DCGANs) - It improves performance and stability in generator and discriminator by using Convolutional layers.

CycleGANs – The CycleGANs are designed for the tasks like image-to-image translation were paired data is not available. An image in one domain can be converted to another domain and again retranslated back to its original domain by using CycleGANs. For this process, CycleGANs uses the Cycle consistency loss.

Wasserstein GANs (WGANs) - This GAN focuses on improving training stability and quality by using the Wasserstein distance as a loss metric which can reduce issues with mode collapse and vanishing gradients.

Progressive Growing GANs – The Progressive Growing GANs are used during training to strengthen the training process and produce high resolution images. For this, PGGANs has to gradually increase the complexity and size of the generator and discriminator networks.

StyleGANs – This is an advanced variant of GANs, are designed to generate high-quality, photorealistic images with fine control over various stylistic attributes.

15. GANS IN PRIVACY PRESERVING FEATURE ENGINEERING

While using GANs to generate synthetic data, privacy must be preserved. For example, in in remote sensing analysis, GANs can generate synthetic ocean floor records which mimic the real data without revealing sensitive information about ocean floors. This allows researchers to develop models and conduct analysis while adhering to privacy regulations.

16. UNSUPERVISED APPROACHES TO FEATURE ENGINEERING WITH GANS

1. Discovering Latent Structures

The latent structures are the models which attempts to capture the understanding the causality. That is why they are called as casual models. In unsupervised learning GANs captures the variation in data by understanding the latent structure in data. For example, for ocean data, GANs can generate new samples using the latent space by revealing hidden attributes such as ridges or vents. Using this the anomalies like deviations can be detected if there are any in the ocean floor data. (51)

2. Unsupervised Feature Learning with GANs

In unsupervised learning GANs captures the variation in data and can learn useful features from the raw data. But it does not require labeled data. The adversarial training is applied to get these labels where generator learns to produce synthetic data samples which looks like real. For example, for ocean data, GANs can apply adversarial training and generate new samples. The clustering or dimensionality reduction technique applied on these learned features.

3. Self-Training with GANs

Self-training represent self-supervised learning. GANs can use synthetic data for self-training to enhance the feature extraction process. For example, for ocean data, GANs can apply self-training on synthetic samples and can generate variations in the existing samples.

4. Feature Transfer and Domain Adaptation

GANs can propelled the feature transfer and domain adaptation. This is done by learning the features in one domain and applying them in other domain. This is useful in heterogeneous data sources. For example, for ocean data, GAN can transfer features learned from ocean trench for finding the effect of water quakes on ocean trench.

5. Multi-Modal Feature Learning

GANs are capable of handling multi-modal data also. The features are extracted from multiple sources and different types of data. GAN can provide a rich set of features by applying joint representations with different modalities. For example, Ocean floor and Diving in ocean go hand in hand as multi-model.

17. ADVANCED TECHNIQUES IN UNSUPERVISED GAN-BASED FEATURE ENGINEERING

1. Deep Convolutional GANs (DCGANs)

The generator and discriminator uses CNN i.e. convolutional neural networks in Deep Convolutional GANs (DCGANs). This is very effective for image data. DCGANs can capture hierarchical features and generate high-quality images. For example, in image analysis of medical images, DCGANs can be used to generate synthetic images of specific pathologies, enriching the dataset and aiding in the development of diagnostic models. This approach can help address issues related to limited availability of labeled medical images.

2. InfoGANs for Mutual Information Maximization

InfoGANs are the extension of traditional GANs which escalate the mutual information between the generated data and the latent variables. Due to this, the model can learn interpretable and disentangled representations. For example, the underwater acoustic target audio data is converted into time-frequency image data by applying Short-Time Fourier transform method. Then this time-frequency image data is used as an input to InfoGAN to generate underwater acoustic target data. (Yang et. Al., (2023))

3. CycleGANs for Unsupervised Domain Translation

CycleGAN model is used to carry out image translation over the pairs of input images. This model provides new regularization constraints while training the model. Due to this, the model prevent itself from creating the real scenarios i.e. artifacts which may hamper the subsequent change in classification. CycleGANs are designed especially for unsupervised domain translation. (Pedro Juan et.al., (2021))

4. Self-Supervised Learning with GANs

Self-supervised learning creates supervisory signals from the data itself, for example predicting some parts of the data from other parts. Self-supervised learning (SSL) is nothing but a sort of unsupervised learning which helps to improve the performance of computer vision tasks such as detecting objects and images, segmenting images, etc. The generic artificial intelligence systems can be developed at a low price by using un-structured and un-labeled data. (Veenu Rani et. Al., (2023))

GANs can be used to generate such signals, enabling effective learning from unlabeled data. For example, in video analysis, self-supervised GANs can generate predictions of future frames based on past frames, learning temporal features that can be used for tasks like action recognition or video summarization.

18. PRACTICAL APPROACH EXAMPLE

Adversarial attacks are small, intentional perturbations i.e. changes applied to input data that can fool a deep learning model to make incorrect predictions. Integrating adversarial inputs into the workflow can help to improve the strength of remote sensing image classifier by training it to be more resistant to such attacks. (Nicholas Carlini, & David Wagner, (2018))

In the practical approach, the following things are explained:

1. Create adversarial inputs using the FGSM i.e. Fast Gradient Sign Method.
2. Train the model using both clean and adversarial examples.
3. Build a deep learning model capable of handling both types of inputs.

For the experiment, ocean floor images are used which are stored in two directories i.e. catagory1 (figure 12) and category2 (figure 13).

Figure 12. Catagory1 images

Image source: https://www.google.com/search?q=fake+ocean+floor+images&sca_esv=d620e9c68447bd3d &sxsrf=ADLYWIIVshKLztfwUpIaZx2NSJ459s

Figure 13. Catagory2 images

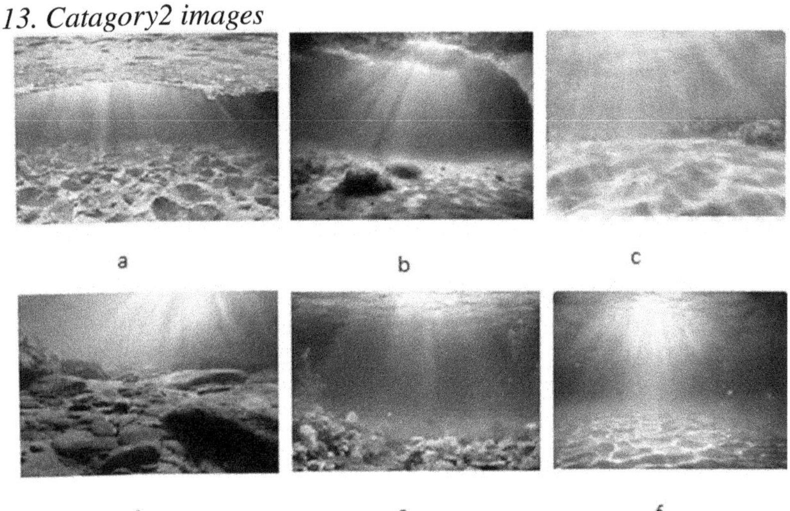

Image source: https://www.google.com/search?q=fake+ocean+floor+images&sca_esv=d620e9c68447bd3d &sxsrf=ADLYWIIVshKLztfwUpIaZx2NSJ459s

In this practical approach, implementation is done by building a deep learning model for classifying remote sensing images, incorporating adversarial inputs to improve robustness. For this a Python Code is written for Remote Sensing Image Classification with Adversarial Input (FGSM).

1. Import the necessary libraries like numpy, tensorflow with Sequential, Conv2D and sklearn model for training and testing data.
2. Use keras for optimization of the code.
3. Import the os to find the directories.
4. Mount the drive, so the images can be used, for example-
   from google.colab import drive
   drive.mount('/content/drive')
5. Set up paths of the folder where remote sensing images are stored and modify the categories based on the classes, using following statement
   Data_dir = '/content/drive/MyDrive/Colab Notebooks/image'
6. Provide Image dimensions for width and height
7. Load images from folder and load labels
8. check the category to which image belongs by using a for loop
9. Define images and labels and then check the category to which image belongs by using a for loop
10. Once images and labels are recognized, normalize the images
11. Split data into training and testing sets
12. Build the CNN model on the data by applying relu and softmax function
13. Compile the model after building it
14. Apply the FGSM (Fast Gradient Sign Method) Adversarial Attack Function
15. Find the adversarial image and ensure entered pixel values are valid.
16. Train the model with both clean and adversarial examples.
17. Convert the list of adversarial images to numpy array using following statement
    adversarial_examples = np.array(adversarial_examples).squeeze()
18. Combine both clean and adversarial examples
19. Train the model on both clean and adversarial examples using following statement
    model.train_on_batch(combined_x, combined_y)
20. Fit the model on clean data, use the following statement
    model.fit(X_train, y_train, epochs=5, batch_size=32, validation_split=0.1)
21. Perform adversarial training by deciding epochs number, this can be changed as per the users requirement for datasets.
22. Evaluate model on clean test data and find test loss and accuracy using the statement
    test_loss, test_acc = model.evaluate(X_test, y_test)

23. Display test loss and accuracy values.
24. Evaluate model on adversarial test data using a for loop, use the following code

```
X_test_adv = []
for i in range(len(X_test)):
X_test_adv.append(create_adversarial_pattern(model, X_test[i:i+1],
y_test[i:i+1], epsilon=0.01))
X_test_adv = np.array(X_test_adv).squeeze()
test_loss_adv, test_acc_adv = model.evaluate(X_test_adv, y_test)
print(f 'Test accuracy on adversarial data: {test_acc_adv}')
```

The result is shown below (figure 14). 5 Epoch were considered for the execution of the code. The result shows the accuracy and loss in each epoch. The result also reflects that the accuracy and loss on clean data and adversarial data. As the size of the dataset increases, the result may show variation in the result.

Figure 14. Result on Catagory1 and Catagory2 images

```
Drive already mounted at /content/drive; to attempt to forcibly remount, call drive.mount("/content/drive", force_remount=True).
/usr/local/lib/python3.11/dist-packages/keras/src/layers/convolutional/base_conv.py:107: UserWarning: Do not pass an `input_shape`/`input_dim` argument to a layer. When using Sequential models, prefer using an `Input(shape)` object a
  super().__init__(activity_regularizer=activity_regularizer, **kwargs)
Epoch 1/5
1/1 ──────────── 4s 4s/step - accuracy: 0.1429 - loss: 0.7550 - val_accuracy: 0.0000e+00 - val_loss: 1.0513
Epoch 2/5
1/1 ──────────── 1s 544ms/step - accuracy: 0.7143 - loss: 0.6302 - val_accuracy: 1.0000 - val_loss: 0.6146
Epoch 3/5
1/1 ──────────── 1s 611ms/step - accuracy: 0.5714 - loss: 0.6221 - val_accuracy: 0.0000e+00 - val_loss: 0.7709
Epoch 4/5
1/1 ──────────── 1s 582ms/step - accuracy: 0.7143 - loss: 0.4736 - val_accuracy: 0.0000e+00 - val_loss: 1.1015
Epoch 5/5
1/1 ──────────── 1s 628ms/step - accuracy: 0.7143 - loss: 0.7230 - val_accuracy: 1.0000 - val_loss: 0.6205
Epoch 1/5
Epoch 2/5
Epoch 3/5
WARNING:tensorflow:5 out of the last 13 calls to <function TensorFlowTrainer.make_train_function.<locals>.one_step_on_iterator at 0x7ff006662340> triggered tf.function retracing. Tracing is expensive and the excessive number of traci
Epoch 4/5
WARNING:tensorflow:6 out of the last 14 calls to <function TensorFlowTrainer.make_train_function.<locals>.one_step_on_iterator at 0x7ff006662340> triggered tf.function retracing. Tracing is expensive and the excessive number of traci
Epoch 5/5
1/1 ──────────── 0s 51ms/step - accuracy: 0.3333 - loss: 1.1605
Test accuracy on clean data: 0.3333333432674408
1/1 ──────────── 0s 51ms/step - accuracy: 0.3333 - loss: 1.3686
Test accuracy on adversarial data: 0.3333333432674408
```

19. DESCRIPTION OF THE ABOVE PROGRAM

1. **Create Adversarial Inputs**: The create_adversarial_pattern function generates adversarial examples using the Fast Gradient Sign Method (FGSM). It calculates the slope of the loss for the input and perturbs the input in the slops direction.
2. **Adversarial Training**: The adversarial_training function trains the model on both clean and adversarial examples. It alternates between clean inputs and their adversarial counterparts to make the model robust.

3. **Model Architecture:** The CNN architecture remains the same as before but now includes adversarial inputs during training. In this case two functions such as Relu and Softmax were used.

The Rectified Linear Unit (ReLU) function is a activation function in neural networks which is used for image classification, especially in convolutional neural networks (CNNs). It introduces irregularities into the model while maintaining simplicity, which helps in learning complex patterns. Because ReLU is used to improve the network efficiency and regularize the data. RELU provides the zero as an outputs for negative values which can lead to sparse activations.

The Softmax function is used to transform raw output scores (logits) into the probabilities which sum up to 1. Due to this, they can interpret probabilities of different classes. Each output value is between 0 and 1, and all output values sum to 1. Softmax function can also be used for multi-class classification problems to determine the likelihood of each class. The class with the highest probability is considered as the predicted class.

4. **Evaluation:** The model is trained and evaluated on clean test data and adversarial perturbed test data to measure its robustness.

Fast Gradient Sign Method (FGSM) is a popular technique for generating adversarial examples. It is used in machine learning. The inputs are the adversarial examples to a model that intentionally perturbed to root the model to create incorrect predictions. FGSM is a straightforward, computationally efficient and especially designed method to create such perturbations. It uses the loss function gradient with respect to the input to create perturbations; that maximize the loss that is how it fools the model. (Weikang Yu et.al., (2024))

The FGSM can be implemented using the following 3 steps:

1. **Compute the Gradient**: FSGM is developed in such a way that it manipulates the input data or image which calculate the gradient of the loss function. (Weikang Yu et.al., (2024))
2. **Generate Perturbation**: To create a perturbation, Use the sign of the gradient.
3. **Apply Perturbation**: Add the perturbation to the original image to create the adversarial example.

We need to remember following points when using FGSM:

Epsilon: Epsilon is nothing but the size of the adversarial perturbation. In practical usage, a larger epsilon will create stronger adversarial examples, then it may become unrealistic or too noticeable.
Training: The model is pre-trained first by using clean data and then adjusted with both clean and adversarial examples.
Robustness: Adversarial training generally makes the model more resilient to adversarial attacks but can slightly reduce performance on clean data.

20. FUTURE DIRECTIONS

Today, the use of GANs in feature engineering is an evolving field. This is used by many researchers for ongoing research and development. Future directions of GAN may include the following:

Integration with Other Techniques: GANs can be combined with other machine learning techniques to enhance the feature extraction and interpretation processes, such as reinforcement learning or self-supervised learning. (Ngoc-Trung Tran et. Al. (2019))

Improved Training Methods: A new training techniques and architectures may be developed to address the challenges of GANs, such as mode collapse and non-convergence.

Self-Supervised Learning with GANs: In this case, the model learns from unlabeled data and creates its own supervision signal. These signals could be combined with GANs to improve feature learning and extraction process. (Ngoc-Trung Tran et. Al. (2019))

Explainability and Interpretability: In various applications, transparency is crucial. GANs can be combined with developing techniques to interpret and explain the features generated by GANs for gaining trust and understanding in applications.

Real-Time Feature Engineering: The use of new advanced changes in computational efficiency and GAN architecture may enabling dynamic and adaptive systems which will respond to changing data patterns. It will lead to real-time feature engineering,

Ethical Considerations: Ethical issues related to the generation and use of synthetic data must be addressed correctly as there are many problems happening in today's world for data privacy, misuse, and fairness.

21. CONCLUSION

Generative Adversarial Networks offers a powerful, effective and innovative approach to feature engineering. GAN enables the creation, refinement, and transformation of features in an unsupervised manner for remote sensing applications. The practitioners can enhance data representation, augment datasets, and transform features by using the support from GANs. This will help to improve machine learning models. As the use of GANs increasing day-by-day, their role in feature engineering is likely to expand which will offer new possibilities and challenges in the field of GAN.

In this chapter, we have explored the fundamentals of GANs, their application in feature engineering, and advanced techniques for leveraging these networks effectively. Understanding and utilizing GANs for feature engineering can open up the avenues to make the machine learning models more robust and effective, particularly in scenarios where traditional methods may fall short. This chapter includes the practical approach which shows that through the utilization of GANs' capacity to acquire insightful latent representations from unlabeled data, scholars and professionals might open up novel avenues for the extraction of significant insights from the continuously expanding amount of remote sensing data.

REFERENCES

Aggarwal, A., Mittal, M., & Battineni, G. (2021). Generative adversarial network: An overview of theory and applications. *International Journal of Information Management Data Insights*, *1*(1), 100004. DOI: 10.1016/j.jjimei.2020.100004

Aggarwal, A., Mittal, M., & Battineni, G. (2021). Generative adversarial network: An overview of theory and applications. *International Journal of Information Management Data Insights*, *1*(1), 00004. DOI: 10.1016/j.jjimei.2020.100004

https://www.coursera.org/in/articles/ai-vs-deep-learning-vs-machine-learning-beginners-guide on 15/1/2025 at 10.06 pm

Bendaoud, N. M. M., Farah, N., & Ahmed, S. B. (2021). Comparing generative adversarial networks architectures for electricity demand forecasting. *Energy and Building*, *247*, 111152. DOI: 10.1016/j.enbuild.2021.111152

Brophy, E., Wang, Z., She, Q., & Ward, T. (2023). Generative adversarial networks in time series: A systematic literature review. *ACM Computing Surveys*, *55*(10), 1–31.

Buduma, N. (2019), Fundamentals of Deep Learning, O'Reilly publications, November 2019, ISBN:978-93-5213-560-8.

Carlini, N., & Wagner, D. (2018), Audio Adversarial Examples: Targeted Attacks on Speech-to-Text, IEEE Deep Learning and Security Workshop, http://bit.ly/21FXTIW

Chahal, A., & Gulia, P. (2019). Machine learning and deep learning. *International Journal of Innovative Technology and Exploring Engineering*, *8*(12), 4910–4914.

Cheng, J., Yang, Y., Tang, X., Xiong, N., Zhang, Y., & Lei, F. (2020), Generative Adversarial Networks: A Literature Review, Ksii Transactions On Internet And Information Systems Vol. 14, No, , ISSN: 1976-7277.DOI: 10.3837/tiis.2020.12.001

Choi, R. Y., Coyner, A. S., Kalpathy-Cramer, J., Chiang, M. F., & Peter Campbell, J. (2024), Introduction to Machine Learning, Neural Networks, and Deep Learning, TVST | Special Issue | Vol. 9 | No. 2 |, Article 14.

Goodfellow, I., Bengio, Y., Courville, A., & Bengio, Y. (2016). Deep learning: Vol. 1. *No. 2*. MIT press.

Goodfellow, I., Pouget-Abadie, J., Mirza, M., Xu, B., Warde-Farley, D., Ozair, S., & Bengio, Y. (2020). Generative adversarial networks. *Communications of the ACM*, *63*(11), 139–144. DOI: 10.1145/3422622

Gui, J., Sun, Z., Wen, Y., Tao, D., & Ye, J. (2021). A review on generative adversarial networks: Algorithms, theory, and applications. *IEEE Transactions on Knowledge and Data Engineering*, *35*(4), 3313–3332. DOI: 10.1109/TKDE.2021.3130191

Hadid, A., Chakraborty, T., & Busby, D. (2024). When geoscience meets generative AI and large language models: Foundations, trends, and future challenges. *Expert Systems: International Journal of Knowledge Engineering and Neural Networks*, *41*(10), e13654.

Hughes, R. T., Zhu, L., & Bednarz, T. (2021). Generative Adversarial Networks–Enabled Human–Artificial Intelligence Collaborative Applications for Creative and Design Industries: A Systematic Review of Current Approaches and Trends. *Frontiers in Artificial Intelligence*, *4*, 604234. DOI: 10.3389/frai.2021.604234 PMID: 33997773

Janga, B., Asamani, G. P., Sun, Z., & Cristea, N. (2023). A Review of Practical AI for Remote Sensing in Earth Sciences. *Remote Sensing (Basel)*, *2023*(15), 4112. DOI: 10.3390/rs15164112

Jozdani, S., Chen, D., Pouliot, D., & Johnson, B. A. (2022). A review and meta-analysis of Generative Adversarial Networks and their applications in remote sensing. *International Journal of Applied Earth Observation and Geoinformation*, *108*, 102734. DOI: 10.1016/j.jag.2022.102734

Karras, T., Laine, S., & Aila, T. (2019), A style-based generator architecture for generative adversarial networks. In *Proceedings of the IEEE/CVF conference on computer vision and pattern recognition* (pp. 4401-4410). DOI: 10.1109/CVPR.2019.00453

Li, Z., Shi, S., Wang, L., Xu, M., & Li, L. (2022). Unsupervised Generative Adversarial Network with Background Enhancement and Irredundant Pooling for Hyperspectral Anomaly Detection. *Remote Sensing (Basel)*, *14*(5), 1265. Advance online publication. DOI: 10.3390/rs14051265

Lim, W., Kelvin, S. C. Y., Lau, B. T., & Colin, C. L. T. (2024), Future of generative adversarial networks (GAN) for anomaly detection in network security: A review, Published by Elsevier Ltd, ISSN 0167-4048.

Luo, Y., Guo, X., Wang, L., Zheng, J., Liu, J., & Liao, F. (2023). Unsupervised structural damage detection based on an improved generative adversarial network and cloud model. *Journal of Low Frequency Noise, Vibration and Active Control*, *42*(3), 1501–1518. DOI: 10.1177/14613484221150804

Nguyen, A., ... (2015), Deep Neural Networks Are Easily Fooled: High Confidence Predictions for Unrecognizable Images, Computer Vision and Pattern Recognition, http://bit.ly/2ZKclwW

Pedro, J. S. V. (2021). Gilson Alexandre Ostwald Pedro da, Raul Queiroz Feitosa, Mabel Ximena, Claudio Aparecido de Almeida, Christian Heipke, Franz Rottensteiner, (2021), An unsupervised domain adaptation approach for change detection and its application to deforestation mapping in tropical biomes. *ISPRS Journal of Photogrammetry and Remote Sensing*, *181*(November), 113–128.

Rani, V., Nabi, S. T., Kumar, M., Mittal, A., & Kumar, K. (2023). Self-supervised Learning: A Succinct Review. *Archives of Computational Methods in Engineering*, *2023*(30), 2761–2775. DOI: 10.1007/s11831-023-09884-2 PMID: 36713767

Sabnam, S., & Rajagopal, S. (2024). Application of generative adversarial networks in image, face reconstruction and medical imaging: Challenges and the current progress. *Computer Methods in Biomechanics and Biomedical Engineering. Imaging & Visualization*, *12*(1), 2330524.

Salehi, P., Chalechale, A., & Taghizadeh, M. (2020). Generative adversarial networks (GANs): An overview of theoretical model, evaluation metrics, and recent developments. arXiv preprint arXiv:2005.13178.

Sarker, I. H. (2021). Deep Learning: A Comprehensive Overview on Techniques, Taxonomy, Applications and Research Directions. *SN Computer Science*, *2*(6), 420. Advance online publication. DOI: 10.1007/s42979-021-00815-1 PMID: 34426802

Soori, M., Arezoo, B., & Dastres, R. (2023). Artificial intelligence, machine learning and deep learning in advanced robotics, a review. *Cognitive Robotics*, *3*, 54–70. DOI: 10.1016/j.cogr.2023.04.001

Suryawati, E., Pardede, H. F., Zilvan, V., Ramdan, A., Krisnandi, D., Heryana, A., Yuwana, R. S., Kusumo, R. B. S., Arisal, A., & Supianto, A. A. (2021, December). R. Sandra Yuwana, R. Budiarianto Suryo Kusumo, Andria Arisal & Ahmad Aff Supianto, Unsupervised feature learning-based encoder and adversarial networks. *Journal of Big Data*, *8*(1), 118. Advance online publication. DOI: 10.1186/s40537-021-00508-9

Suryawati, E., Pardede, H. F., Zilvan, V., Ramdan, A., Krisnandi, D., Heryana, A., Yuwana, R. S., Kusumo, R. B. S., Arisal, A., & Supianto, A. A. (2021). Unsupervised feature learning-based encoder and adversarial networks. *Journal of Big Data*, *8*(1), 118. DOI: 10.1186/s40537-021-00508-9

Tiwari, T., Tiwari, T., & Tiwari, S. (2018), How Artificial Intelligence, Machine Learning and Deep Learning are Radically Different? International Journals of Advanced Research in Computer Science and Software Engineering ISSN: 2277-128X (Volume-8, Issue-2).

Tran, N.-T., Tran, V.-H., Nguyen, N.-B., Yang, L., & Cheung, N.-M. (2019), Self-supervised GAN: Analysis and Improvement with Multi-class Minimax Game,*33rd Conference on Neural Information Processing Systems (NeurIPS 2019)*, Vancouver, Canada.

Wang, J., Yi, G., Zhang, S., & Wang, Y. (2020), An Unsupervised Generative Adversarial Network-Based Method for Defect Inspection of Texture Surfaces. Appl. Sci. 2021, 11, 283. https://doi.org/doi:10.3390/app11010283

Wang, K., Gou, C., Duan, Y., Lin, Y., Zheng, X., & Wang, F.-Y. (2017), Generative Adversarial Networks: Introduction and Outlook, IEEE/CAA JOURNAL OF AUTOMATICA SINICA, VOL. 4, NO. 4, OCTOBER 2017.

War, K. (2019), Strengthing Deep Neural Netwoks: Making AI Less Susceptible to Adversarial Trickery, O'Reilly publications, August 2019, ISBN:978-93-5213-873-9.

Weidman, S. (2019), Deep Learning from Scratch, O'Reilly publications, September 2019, ISBN:978-93-5213-902-6.

Yang, H., Huang, X., & Liu, Y. (2023). *InfoGAN-Enhanced Underwater Acoustic Target Recognition Method Based on Deep Learning*. ISBN. DOI: 10.1007/978-981-99-0479-2_249

Yu, W., Xu, Y., & Ghamisi, P. (2024). Universal adversarial defense in remote sensing based on pre-trained denoising diffusion models. *International Journal of Applied Earth Observation and Geoinformation*, *133*, 104131. DOI: 10.1016/j.jag.2024.104131

Zhang, L., Zhang, L., & Du, B. (2016). Deep Learning for Remote Sensing Data: A Technical Tutorial on the State of the Art. *IEEE Geoscience and Remote Sensing Magazine*, *2016*(4), 22–40. DOI: 10.1109/MGRS.2016.2540798

Chapter 5

The Architectures and Types of Generative Adversarial Networks

Poonam Ponde
Nowrosjee Wadia College, India

Kavita Khobragade
https://orcid.org/0009-0002-5776-9147
Fergusson College, India

ABSTRACT

Generative Adversarial Networks (GANs) have caused quite a stir in the field of artificial intelligence by their capability of generating a wide range of realistic data, including images, music and text. The GAN architecture is composed of two main components: the generator and the discriminator which work together in the adversarial learning process. While the generator generates realistic data by employing complex transformations on random input, the discriminator carefully examines the generated output to determine its veracity and authenticity in conjunction with real data. Remote sensing applications often require the generation of synthetic images resembling satellite or aerial data. GANs find application in remote sensing for enhancing image resolution, generating synthetic satellite images, and improving data quality for better analysis. This chapter gives a comprehensive exploration of the architecture and diverse types of Generative Adversarial Networks (GANs).

DOI: 10.4018/979-8-3693-6900-5.ch005

INTRODUCTION

Generative AI is a branch of Artificial intelligence that learns from existing data to generate new content (Feuerriegel et al., 2024). The machine learning model is trained on massive amounts of data. The model analyzes the data and the underlying structures, recognizes patterns, and finds relationships within the data. After the model is trained, it can generate new content based on the patterns.

The concept of Generative Adversarial Networks (GANs) was first proposed by Ian Goodfellow and others in 2014 (Goodfellow et al., 2014). Since then, they have become very popular and applied in a variety of domains including agriculture, healthcare, remote sensing, fashion, entertainment, and finance (Aggarwal et al., 2021). GANs belong to the category of Generative AI. Because they are capable of generating high quality, realistic data such as text, images, audio and video, they are used in a wide variety of applications. The chapter begins with a discussion on the challenges faced in Remote Sensing and GANs as a solution to overcome those challenges. There are two types of machine learning models that make up the GAN architecture: Generative model and Discriminative model. These models are implemented in the Generator and Discriminator respectively. These two components lie at the heart of the GAN. The network is called 'adversarial' because the Generator and Discriminator are the adversaries who are in competition with each other. The architecture, mathematical model and components of GAN are discussed in the next sections. Building upon the basic GAN architecture, several special types of GANs have been created for various applications. Each type of GAN has unique features and capabilities. Subsequent topics of the chapter cover various types of GANs. The chapter concludes with a discussion on the challenges and ethical implications of using GANs.

CHALLENGES IN REMOTE SENSING

Remote Sensing (RS) is an important technology for observing, monitoring and understanding the earth and its atmosphere (Campbell & Wynne, 2011). It finds applications in various areas such as agriculture, land mapping, climate monitoring and disaster management, etc. There are several challenges related to acquisition of remotely sensed data, processing and interpreting this data (Zhang et al., 2022). Unfavorable weather conditions such as rain, cloud cover and insufficient light make it difficult to capture clear images remotely. The quality and resolution (spatial, spectral and temporal) of the image is highly dependent on the hardware and software components such as sensors, and capacity of the storage devices. In addition, data from multiple sensors such as optical sensors (visible and Infra-red), Radar sensors (Sentinel 1-SAR, RADARSAT), thermal sensors, LiDAR (Light Detection and Ranging), Acoustic sensors, etc. needs to be integrated before analysis (multi-modal data) (Congalton & Green, 2019).

In order to carry out data analysis, the raw data and images require extensive pre-processing that includes correction and noise removal, data integration from various sources, heterogeneous formats, and working with multi-spectral and hyper-spectral images (Dubovik et al., 2021). To interpret these images and extract meaningful insights, data analysis is carried out. This necessitates the use of statistical techniques, data mining tasks, machine learning algorithms, and specialized tools that require experience and expertise (Zhang & Zhang, 2022). Other challenges that need to be addressed include:

1. Resolution issues: Many remote sensing applications require high resolution data but the available data does not fulfill the criteria.
2. Noise: Atmospheric interference and other factors affect the data.
3. Image clarity: Environmental conditions such as weather, light etc. affect image clarity.
4. Limited or Non-availability of data: High-quality and high-resolution data are often not available, particularly those of less monitored or remote areas.
5. Varying resolutions: Remote sensing data has varying spatial and temporal resolutions that can complicate analysis and interpretation.
6. Labeling challenges: Accurately annotating images for training models and validation can be very time consuming.

Technological solutions can be employed to address these challenges in remote sensing; particularly machine learning for preprocessing and analyzing images and generative Artificial intelligence for generating high quality data (Zhang & Zhang, 2022). The development of GPU's and deep learning has made it possible to effectively employ AI algorithms for classification, detection and prediction. GANs can be used in Remote Sensing applications for a variety of tasks such as image correction, cloud haze removal, generation of high-resolution images, and image translation among others (Dash et al., 2023).

GENERATIVE AND DISCRIMINATIVE MACHINE LEARNING MODELS

Machine learning is a subdomain of artificial intelligence (AI) that allow machines to automatically learn from data and improve without being explicitly programmed (Jordan & Mitchell, 2015). A Machine learning model is an algorithm that analyzes data and extracts patterns from the data. These patterns are useful to gain valuable insights and make decisions or predictions based on those patterns. There are several Machine learning models. Machine learning models are of two types: Generative models and Discriminative models (Jebara, 2012).

1. Generative Models

As the name suggests, Generative machine learning models generate new data samples based on the training data. These models are trained on huge amounts of training data. The model carries out an analysis of the data, learns the underlying distribution to uncover patterns within the data to generate new samples. Generative models are used in several applications including content generation (images, music, text, video), language modeling, virtual assistants, data enhancement and augmentation, and simulation.

Generative models can be represented as follows: To generate new instance x, a generative model learns parameters by maximizing the joint probability P(Y, X) where X are the inputs and Y is the target variable. The model estimates P(X|Y) to find P(Y|X) by applying Bayes' rule. Figure 1 depicts a generative model.

Figure 1. Generative Model

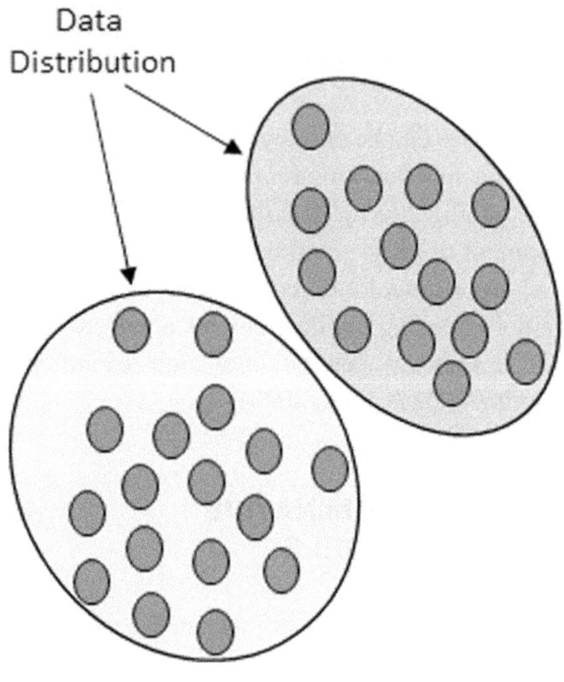

2. Discriminative Models

Discriminative models are used to distinguish between different classes or categories in the data. The algorithms learn the decision boundary between the classes as shown in Figure 2. The ultimate objective of discriminative models is to predicting the labels of the data and separate one class from another (e.g., classify images of cats and dogs into respective categories). Discriminative models are used in classification and regression problems. Mathematically, discriminative models can be represented as follows: For inputs X and labels Y, discriminative models learn the underlying distribution of data such that the conditional probability P(Y|X) is maximized.

Figure 2. Discriminative Model

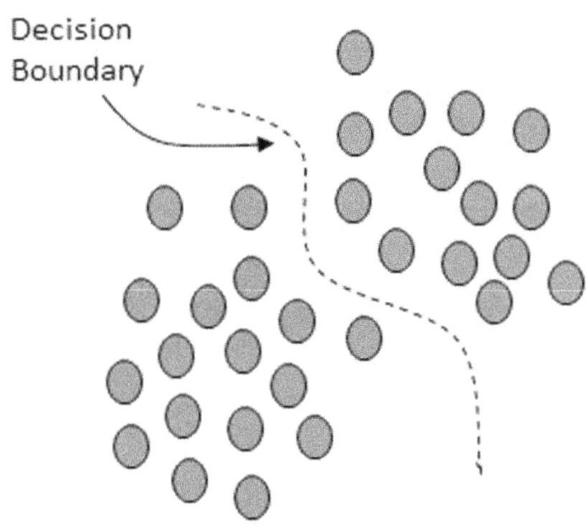

Table 1 gives the comparison between generative and discriminative models.

Table 1. Generative and Discriminative Models: A comparison

Aspect	Generative Models	Discriminative Models
Focus	Generates new data	Classifies existing data
Goal	Model data distribution to learn the joint probability P(Y, X)	Model conditional probability P(Y | X) of given data
Training	Capture data structure. Learns actual distribution of data	Learn decision boundary, Differentiate between classes

continued on following page

Table 1. Continued

Aspect	Generative Models	Discriminative Models
Complexity	Generative models are more complex to train because they model the entire distribution	Simpler to train because they focus on learning the decision boundary
Examples	Naive Bayes, Linear Discriminant Analysis (LDA), Gaussian Mixture Models, etc.	Regression, Random Forest, Neural Networks, Decision Trees, Support Vector Machine, etc.
Applications	Data generation, Image synthesis, unsupervised learning	Classification, supervised learning tasks

ARCHITECTURE OF GENERATIVE ADVERSARIAL NETWORKS

Generative Adversarial Networks (GANs) are basically neural networks. The neural network in GANs use two machine learning models: **Generator** - a generative machine learning model and **Discriminator** - a discriminative machine learning model. A GAN is trained on a training dataset to generate realistic new data (artificial) that resembles actual data from the training set. The network is called as "adversarial" because the two components i.e. the generator and discriminator compete against each other (Gui et al., 2021). The generator and discriminator function as adversaries in a contest that is a zero-sum game where the gain of one is the loss of another. The GAN architecture (Little et al., 2021) is shown in Figure 3.

Figure 3. Architecture of GAN

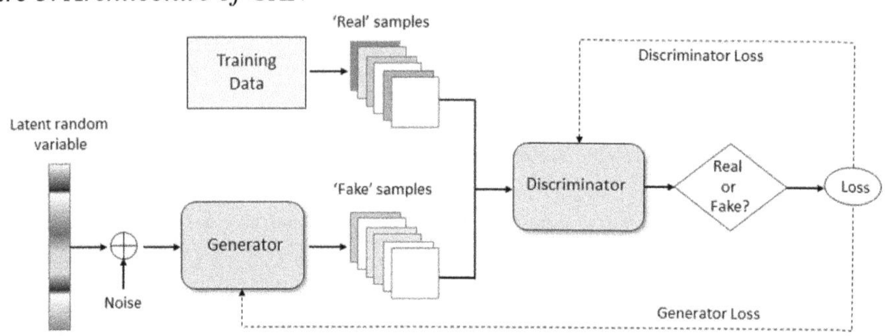

Components of a GAN

1. Generator: The generator is a neural network that is trained using a training dataset. It is given random noise (z) as input and generates new samples G(z) that resembles the training data.

2. Discriminator: The discriminator is also a neural network. The output of the Generator (fake data) G(z) is fed to the discriminator. The Discriminator is a classifier to distinguish between fake samples and real data x from the training dataset.
3. Loss function:
   i. Discriminator loss: The log probability that the input instance is real as log(D(x)). For real data instances x, the discriminator must identify it as real. For real data, the discriminator output i.e. D(x) = 1. For fake or generated data G(z), the desired output of the discriminator is D(G(z)) = 0. The log probability that the instance is fake is log(1 – D(G(z))). The goal of the discriminator is to maximize log D(x)+log(1–D(G(z)))
   ii. Generator loss: The loss function is the distance between the fake data and the real data. The generator loss is calculated from the discriminator loss. During training, the discriminator component is frozen. Hence, the discriminator gets only the fake input. Thus, the first term in the discriminator loss is 0 i.e. D(x) = 0. To fool the discriminator into identifying a fake sample as real, the generator tries to minimize the second term in the discriminator loss equation i.e. min[log(1 – D(G(z)))].
   iii. Loss function: The GAN loss function combines both, the generator loss and discriminator loss functions. The minimax loss function is represented as $\min_G \max_D V(D,G)$ where V(D,G) is the value function defined as:

$$V(D, G) = \mathbb{E}_{x \sim p(x)}[\log D(x)] + \mathbb{E}_{z \sim q(z)}[\log(1 - D(G(z)))] \tag{1}$$

The generator tries to minimize this function while the discriminator tries to maximize it. In this function:

- D(x) is the probability that real data instance x is real.
- E_x is the expected value over all real training data values.
- G(z) is the output of the generator, z is the input noise.
- D(G(z)) is the probability that a fake instance G(z) is real.
- E_z is the expected value over all random inputs to the generator.

Generator Architecture

- **Input**: A Neural network needs some input to generate the output. The input data is related to the problem to be solved; for example, classification or prediction. The generator's neural network takes z-dimensional vector of completely random values or noise as input. The noise is transformed by the generator into complex data samples, such text or images. The reason for giving noise as input to the generator is to provide randomness and variability. This noise is usually a simple distribution, such as a Gaussian (normal) or a uniform distribution.

- **Layers:** A Convolutional Neural Network (CNN) is used as generator. With a noise vector as input, the generator learns to create new instances that resemble the real instances through adversarial training. The generator uses back propagation to adjust the weights of layers of the neural network to generate more realistic images. The generator architecture consists of the following layers:

i. Fully connected layers: The low-dimensional noise vector z is passed through multiple fully connected layers to convert it into a higher-dimensional space.
ii. Reshape layer: The output from the fully connected layers is reshaped into a 3 dimensional array that resembles the shape of an image using the project and reshape operation.
iii. De-Convolution layers: Transposed convolution layers "upsample" the input noise vector to progressively larger spatial dimensions, resembling the output image size. These layers align features from the latent space to generate realistic textures and structures. After each de-convolutional layer, batch normalization is applied to stabilize and accelerate training. The activation function used in the hidden layers of the generator are ReLU (Rectified Linear Unit) or Leaky ReLU.
iv. Output layer: The output layer uses a specific activation function depending on the range of values desired for the generated data. For image generation, the sigmoid activation function or a tanh activation function is used.

Discriminator

- **Input**: The inputs to the discriminator are real data samples taken from real-world datasets that are used for training the GAN. The discriminator also takes the fake data samples G(z) generated by the Generator for purpose of classifying them as real or fake.
- **Layers**:

i. Convolution layers: These layers extract features that capture spatial relationships and hierarchical patterns. Batch normalization is applied after each convolution layer to standardize and normalize the input to subsequent layers. Leaky ReLU is the activation function in the hidden layers of the discriminator. The activation function allows the network to learn decision boundaries and enhance its ability to discriminate between real and fake samples.

ii. Fully connected layers: These layers aggregate the spatial information into a vector representation that is suitable for final classification decisions.
iii. Output layer: The output layer uses a sigmoid activation function that generates a single scalar output which is the probability that the input sample is real or fake.

Adversarial Training

The generator and discriminator are adversaries that compete in a minimax game scenario (from game theory). The generator wants to maximize the probability that the discriminator classifies the fake samples as real and accordingly updates its weights. The discriminator updates its weights to improve its classification between real and fake samples. The discriminator is deemed as winner when it accurately classifies a generated sample. The generator then adjusts its weights to improve its output. Conversely, if the generator successfully fools the discriminator, it becomes the winner, and the discriminator weights are adjusted. This continuous game ensures that both networks continually learn and improve. The ultimate goal of the adversarial game in GANs is to reach a Nash equilibrium, where neither the generator nor the discriminator can improve their performance against each other. The generator and discriminator are not trained simultaneously to prevent instability. The steps are:

i. Train the Discriminator: First, the discriminator is given real samples from training set and fake samples from generator. The discriminator computes its loss on both real and fake data. The loss is back-propagated to update the discriminator's weights. The discriminator trains for one or more epochs (An epoch is complete when all training data samples have been passed through the learning algorithm).
ii. Train the Generator: Using the output from the discriminator, the generator is updated. The generator's parameters are adjusted to produce data that resembles real data. The fake data that is generated from noise is given to the discriminator. The generator loss is computed from the discriminator's output. Finally, the generator's weights are updated. This is carried out for one or more epochs.

TYPES OF GENERATIVE ADVERSARIAL NETWORKS

To address specific challenges and requirements, applications and improve performance, several types of GANs have emerged. Some popular types of GAN are discussed in this section.

1. Vanilla GAN
2. Conditional GAN
3. Deep convolutional GAN
4. Super-resolution GAN
5. Cycle GAN
6. Progressive GAN
7. Style GAN
8. DISCO GAN
9. Wasserstein GANs
10. Text-to-Image GAN

Vanilla GAN

Vanilla GAN is the most basic GAN and uses the original architecture as proposed by Goodfellow (Goodfellow et al., 2014). It is mainly used for general purpose image synthesis and analysis. Both, the generator and discriminator networks are basic multilayer perceptrons which is a fully connected neural network. The vanilla GAN uses stochastic gradient descent to optimize the loss function. The generator uses Leaky ReLU (hidden layers) and Tanh (output layer) while the discriminator uses Leaky ReLU (hidden layers) and Sigmoid (output layer).

Conditional GAN

Conditional GANs were proposed by Mirza and Osindero (Mirza & Osindero, 2014) for generating data that meets a specific condition. However, by providing additional conditions to the generator and discriminator, the data generation process can be controlled. This condition can be in the form of class labels, text descriptions, features or images. For example, suppose the network has been trained on a collection of flower images and only images of "rose" are to be generated. In this case, the condition will be a label describing the content.

In the conditional GAN, an additional input (y) which identifies or describes the content is given as an additional input. This helps in generating output that meets the specification and gives additional control over the output. The generator takes a noise vector z and a conditional input y (e.g., class label, description, image etc.) as inputs. It produces an output that is generated from both z and y. The discriminator takes a data samples (real or generated) and the conditional input y as inputs. It gives the probability that the sample is real, based on condition y. Figure 4 shows the structure of a conditional GAN.

Figure 4. Conditional GAN

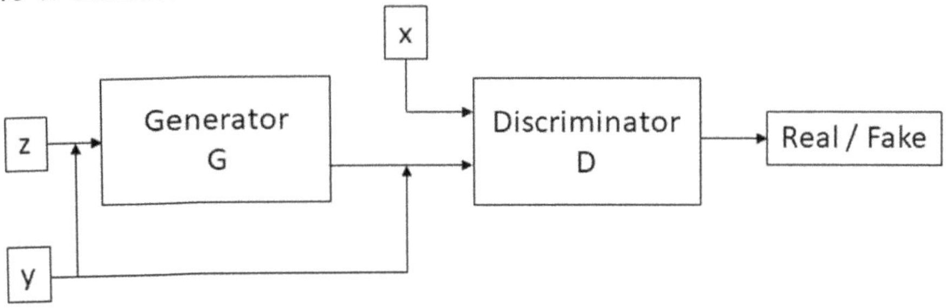

The minimax function for conditional GAN is:

$$V(D,G) = \mathbb{E}_{x\sim p(x)}[\log D(x|y)] + \mathbb{E}_{z\sim q(z)}[\log(1 - D(G(z|y)))] \quad (2)$$

Here, D(x|y) is the probability that x is a real data sample, G(z|y) is the generated data sample, z is the noise and y is the condition.

- Network: The generator as well as the discriminator are fully connected neural networks (or convolutional networks for images)
- Activation function: The generator uses Leaky ReLU (hidden layers) and Tanh (output layer) while the discriminator uses Leaky ReLU (hidden layers) and Sigmoid (output layer).

An example of the conditional GAN is the Pix2Pix GAN (Isola et al., 2017) for image-to-image translation. Based on a given input image as condition, the target image is generated.

Deep Convolutional GAN

The generator is trained on real samples from a dataset. It should generate varying outputs based on variations in the training dataset. A problem called 'mode collapse' occurs when the generator gets biased towards a few outputs and does not produce outputs of every variation from the dataset (Kossale et al., 2022). Deep convolutional

GANs or DCGANs were introduced to overcome the problem of mode collapse (Radford et al., 2015).

- Network: The generator and discriminator in DCGANs are deep convolutional neural networks. Hence this GAN is better at generating high-quality images such as generating realistic faces or scenes.
- Activation function: All layers of generator use the differentiated-ReLU activation function. The output layer uses tanh. The Leaky-ReLU function is used in all discriminator layers.

Super-Resolution GAN

Super-resolution GANs (SRGAN) are designed to generate good quality, high-resolution images from poor quality, low-resolution images (Ledig et al., 2017). A low-resolution image is passed to the generator as input instead of random noise. The loss function in the SRGAN is a perpetual loss function which is the weighted sum of two losses: content loss and adversarial loss. The content loss is VGG loss where VGG is a pre trained network on millions of images. The adversarial loss is the discriminator probability to distinguish between fake and real data.

- Network: The discriminator architecture is similar to Deep Convolutional GAN architecture with Leaky ReLU as activation function in all layers. The Generator is a Convolutional neural network with residual blocks and upsampling layers.
- Activation function: The differentiated-ReLU activation function is used for all layers and the output layer uses tanh. The discriminator uses Leaky-ReLU function for all layers.

Cycle GAN

The cycle GAN invented by Zhu et al. is used to perform image-to-image translation (one image to be mapped with another image) without paired examples (Zhu et al., 2017). For example, converting summer images to winter images, transform human faces into different age groups, horse images to zebra and vice versa etc. The applications include style transfer, artistic filters, and data augmentation. It uses cycle-consistent adversarial networks. CycleGAN consists of two sets of GANs to translate images from domain X to Y and vice versa. The mapping functions of the generators are: G: X->Y and F: Y->X. The discriminator functions are Dx (distinguish generated output G(X) from real output Y) and Dy (distinguish generated inverse output F(Y) from input X) as shown in Figure 5. Both the generators and discriminators are the same as other GANs and they often use convolutional neural networks.

Figure 5. Cycle GAN

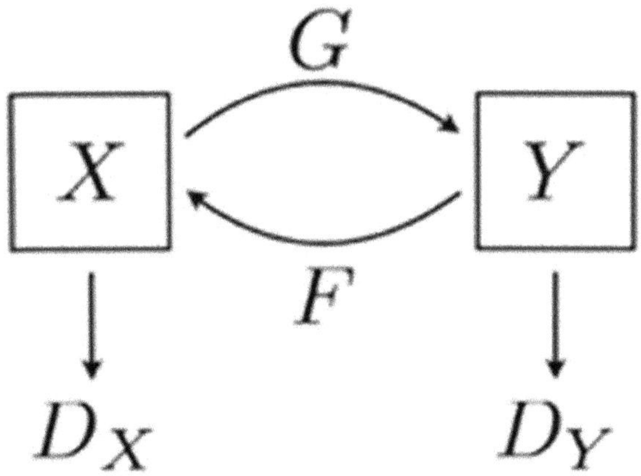

Along with adversarial loss, CycleGAN uses a cycle-consistency loss to ensure that the translation from domain X to domain Y and vice versa is close to the original image. The loss function is given as:

$$L(G, F, D_x, D_y)$$

$$= L_{advers}(G, D_x, X, Y) + L_{advers}(F, D_x, Y, X) + \lambda L_{cycl}(G, F, X, Y) \tag{3}$$

Here, λ is the hyperparameter.

Progressive GAN

Progressive GAN, or PGGAN and was proposed by NVIDIA researchers Tero Karras, et al. (Karras et al., 2017). GAN instability is a major problem that occurs during adversarial training. It is a challenge to train these models to generate high resolution images due to the memory requirements, large structure and more details. This leads to instability. To solve this problem, the generator is initially given a very small image as input. The number of layers are progressively and systematically added to the generator and discriminator during the training process until the desired image size is achieved.

Style GAN

StyleGAN is a GAN developed by NVIDIA researchers Karras et al. for generating high-quality, high-resolution image synthesis with greater control over the generated images' style and features (Karras et al., 2019). For example, during the generation of human faces, some parameters that can be changed are background, foreground and style or hair color and style, eyes color, pose etc. It is used for realistic face generation, generating artworks, textures, and other design elements. In the Style GAN generator, noise is introduced in each layer as a source of variation. To improve image resolution from very low to high, the baseline progressive GAN architecture is used. StyleGAN uses two sources of randomness: noise layers and a mapping network. The noise introduces fine-grained, per-pixel variation and the mapping network allows control over the image's style through adaptive instance normalization (AdaIN.

DiscoGAN

DiscoGAN (Discover Cross-Domain Relations with Generative Adversarial Networks) is a type of GAN to translate images from one domain to another (e.g., from horses to zebras) and back (Kim et al., 2017). It does this by discovering cross-domain relations without paired examples. The main aspect of DiscoGAN is that it finds and transfers features between two different domains while maintaining the essential content of the original images. DiscoGAN is used for various cross-domain translation tasks such as translating images between different artistic styles or domains, changing specific objects in images, etc.

DiscoGAN uses two GANs to translate images from domain X to domain Y and vice versa. DiscoGAN uses cycle-consistency loss and adversarial loss. The cycle-consistency loss ensures that translating an image from X to Y and back, gives the original image X. The adversarial loss ensures that the generated images are same as real images.

The generators use Convolutional layers (encoder), Residual blocks and Deconvolutional layers (decoder) and ReLU activation function. The discriminators use a convolutional neural network (CNN) architecture with convolutional layers with leaky ReLU activation functions, followed by a final fully connected layer. One difference between CycleGAN and DiscoGAN is that CycleGAN uses only one cycle-consistency loss while discoGAN uses a separate cycle-consistency loss, one for each domain.

Wasserstein GANs (WGANs)

WGANs were introduced by Arjovsky et al. as a solution to the mode collapse issue (Arjovsky et al., 2017). WGANs use the Wasserstein distance to measure the minimum cost for transforming one distribution into another distribution. Vanilla GAN relies on KL-divergence (Kullback-Leibler divergence) which causes instability during training. Wasserstein GANs use the Wasserstein distance and enforce the Lipschitz constraint (a requirement for Wasserstein distance). This provides meaningful gradients and improves the learning process. WGANs are more stable and overcome the problem of mode collapse.

WGAN's architecture uses deep neural networks for both generator and discriminator.

In this, instead of the discriminator, there is a critic, which assigns a score to real and generated data instead of classifying them. The critic's objective is to produce outputs that are as large as possible for real data and as small as possible for generated data. The training in Wasserstein GAN is more stable because the critic gives a better learning signal to the generator.

Text-to-Image GAN

In recent times, text-to-image GANs have become very popular and are used in various applications. Text-to-image synthesis GANs (Reed et al., 2016) use a text description to generate realistic images based on the description. In order to do this, the text input is first converted into a meaningful representation such as a feature vector or text embedding to generate images that match the description. This is typically done using text encoders such as Long Short-Term Memory (LSTM) networks or Transformer-based models like BERT or GPT. This embedding along with the noise vector is given as input to the generator. The generator generated fake images based on the inputs. The discriminator also takes the image and text vector as input. The discriminator has two roles: to predict whether a given image is real or fake, to predict the likelihood of the generated image and text being aligned with each other.

COMPARISON OF GAN TYPES

The comparison between different types of GANs discussed above are given in Table 2.

Table 2. GAN Comparison

Type of GAN	Purpose	Network Architecture Model	Loss Function	Applications
Vanilla GAN	Generate new data samples	Simple fully connected neural networks	Binary cross-entropy	Image generation, data augmentation
Conditional GAN	Generate data conditioned on some input	Fully connected networks with conditioning input	Binary cross-entropy	Image-to-image translation
Deep Convolutional GAN (DCGAN)	Improve image generation quality	Convolutional networks	Binary cross-entropy	High-quality image generation
Super-resolution GAN (SRGAN)	Generate high-resolution images from low-resolution inputs	Convolutional networks	Content loss, adversarial loss	Image super-resolution
Cycle GAN	Translate images from one domain to another without paired examples	Convolutional networks with cycle consistency loss	Cycle consistency loss, adversarial loss	Unpaired image-to-image translation
Progressive GAN	Generate high-resolution images progressively	Convolutional networks with progressively growing layers	Wasserstein loss with gradient penalty	High-quality image generation, data augmentation
Style GAN	Generate high-quality images with controllable styles	Convolutional networks with style transfer mechanisms	Non-saturating GAN loss	Image generation with style control
DISCO GAN	Translate images between different domains while preserving structure	Convolutional networks with structure-preserving loss	Cycle consistency loss, adversarial loss	Cross-domain image translation
Wasserstein GAN (WGAN)	Improve training stability and convergence	Fully connected or convolutional networks	Wasserstein loss	Stable GAN training, image generation
Text-to-Image GAN	Generate images from textual descriptions	Convolutional networks with text encoding	Binary cross-entropy, text-image matching loss	Text-to-image synthesis

GANS IN REMOTE SENSING

Generative Adversarial networks can be used effectively to address the challenges faced by Remote Sensing applications (Sabins & Ellis, 2020). Remote sensing applications often require the generation of synthetic images resembling satellite or aerial data. GANs find application in remote sensing for the generation of synthetic images that closely resembles real satellite or aerial data. They address challenges like data scarcity and variability by producing synthetic data to augment training sets. Generative Adversarial Networks are crucial in remote sensing for enhancing image quality, generating high-resolution images from lower-resolution inputs, and improving data consistency across different conditions (Liu et al., 2022). SR-GAN, ProGANs are effective for these tasks. Remote sensing applications require images in different weather and environmental conditions. CycleGANs can be used to translate images between different environments. To address the problems of semantic segmentation, and data argumentation, WGANs and ProGANs have been used. Resolution is a critical aspect in remote sensing. In most applications, higher resolution imagery is more desirable than low-resolution imagery. The limited availability of high resolution data can be overcome by capturing low-resolution imagery and then use a SRGAN to create high-resolution synthetic imagery. Remotely sensed satellite imagery often contain clouds and other weather elements that obstruct the area of interest. CycleGAN and StyleGAN based GANs are used to remove clouds by learning a mapping from cloudy to cloud-free imagery. In their work, Jozdani et al., have presented several case studies on GAN application in RS (Jozdani et al., 2022). Their study elaborates on research and applications of GAN in the areas of Image reconstruction involving improving resolution, removal of noise, and image interpolation, Classification involving object detection, scene classification, Data translation involving image to image and text to image conversion.

TECHNICAL CHALLENGES AND ETHICAL CONCERNS

Generative Adversarial networks are a powerful technology that can be employed to solve the data-related problems faced by Remote Sensing applications. However, there are other challenges which need to be considered (Saxena & Cao, 2021; Pavan & Jayagopal, 2021)

1. Training data: The quality and variety of images generated by the Generator largely depends on the set of images used for training the generator. This requires high quality data set. Acquiring such a data set is difficult.
2. Mode collapse: For GAN's to generate realistic images, the generator must produce different outputs for different features. GAN's suffer from the problem of mode collapse whereby the Generator produces similar and limited number of outputs for varying features.
3. Instability: The two adversaries, namely the generator and discriminator are in constant competition. This often leads to instability

4. Diminished gradient: Deep learning networks often suffer from the problem of diminishing gradient or vanishing gradient problem. The quality of generated images depends on the feedback given by the discriminator to the generator. This feedback is in the form of gradients of a loss function that are back propagated. When the discriminator becomes very successful, these gradients become very small or close to zero. Hence, the generator does not learn anything.
5. Convergence Failure: The generator and discriminator compete with each other in order to reach equilibrium. When one of them dominates, this leads to a convergence failure.

Apart from the technical challenges, it is also crucial to understand the ethical implications of using data generated by GAN's considering their possible misuse, and consequences (Kirova et al., 2023).

1. GANs can create highly realistic fake images including deep fakes. If used in Remote Sensing applications, these can lead to wrong analysis, misleading information, and falsified results.
2. Bias: If the training data used for the generator contains any bias, the GAN may enhance or amplify this bias, leading to unfair outputs.
3. Trust factors: The outputs generated by GAN's lack transparency. Hence, it is difficult to validate the data generated by GANs.

CONCLUSION AND FUTURE SCOPE

This chapter elaborates on the architecture of Generative Adversarial Networks (GANs) which are machine learning neural networks. The two components of GAN namely Generator and Discriminator work as adversaries in order to generate realistic fake data. GANs can be effectively used to address challenges in Remote Sensing like lack of high quality data, poor-resolution, and noise. Various types of GANs, including CycleGANs, StyleGANs, and DISCOGANs, are described along with their use. The chapter also discusses the challenges of GANs such as complexity in training, bias, stability, and mode collapse. Important ethical concerns such as trust factors, data misuse and falsified results are also discussed.

As a future scope, the technical challenges in GANs can be alleviated by improving GAN training, use optimization algorithms, employ alternative objective functions, and reduce biases in generated data. A comprehensive set of ethical guidelines are needed for the responsible use of generative AI to enhance trust factors and prevent misuse.

REFERENCES

Aggarwal, A., Mittal, M., & Battineni, G. (2021). Generative adversarial network: An overview of theory and applications. *International Journal of Information Management Data Insights*, *1*(1), 100004. DOI: 10.1016/j.jjimei.2020.100004

Arjovsky, M., Chintala, S., & Bottou, L. (2017). Wasserstein generative adversarial networks. In *International conference on machine learning* (pp. 214–223). PMLR.

Campbell, J. B., & Wynne, R. H. (2011). *Introduction to remote sensing*. Guilford press.

Congalton, R. G., & Green, K. (2019). *Assessing the accuracy of remotely sensed data: principles and practices*. CRC press. DOI: 10.1201/9780429052729

Dash, A., Ye, J., & Wang, G. (2023). A review of generative adversarial networks (GANs) and its applications in a wide variety of disciplines: From medical to remote sensing. *IEEE Access : Practical Innovations, Open Solutions*.

Dubovik, O., Schuster, G. L., Xu, F., Hu, Y., Bösch, H., Landgraf, J., & Li, Z. (2021). Grand challenges in satellite remote sensing. *Frontiers in Remote Sensing*, *2*, 619818. DOI: 10.3389/frsen.2021.619818

Feuerriegel, S., Hartmann, J., Janiesch, C., & Zschech, P. (2024). Generative ai. *Business & Information Systems Engineering*, *66*(1), 111–126. DOI: 10.1007/s12599-023-00834-7

Goodfellow, I., Pouget-Abadie, J., Mirza, M., Xu, B., Warde-Farley, D., Ozair, S., & Bengio, Y. (2014). Generative adversarial nets. *Advances in Neural Information Processing Systems*, •••, 27.

Gui, J., Sun, Z., Wen, Y., Tao, D., & Ye, J. (2021). A review on generative adversarial networks: Algorithms, theory, and applications. *IEEE Transactions on Knowledge and Data Engineering*, *35*(4), 3313–3332. DOI: 10.1109/TKDE.2021.3130191

Isola, P., Zhu, J. Y., Zhou, T., & Efros, A. A. (2017). Image-to-image translation with conditional adversarial networks. In *Proceedings of the IEEE conference on computer vision and pattern recognition* (pp. 1125-1134).

Jebara, T. (2012). *Machine learning: discriminative and generative* (Vol. 755). Springer Science & Business Media.

Jordan, M. I., & Mitchell, T. M. (2015). Machine learning: Trends, perspectives, and prospects. *Science*, *349*(6245), 255–260. DOI: 10.1126/science.aaa8415 PMID: 26185243

Jozdani, S., Chen, D., Pouliot, D., & Johnson, B. A. (2022). A review and meta-analysis of generative adversarial networks and their applications in remote sensing. *International Journal of Applied Earth Observation and Geoinformation, 108*, 102734. DOI: 10.1016/j.jag.2022.102734

Karras, T., Aila, T., Laine, S., & Lehtinen, J. (2017). Progressive growing of gans for improved quality, stability, and variation. *arXiv preprint arXiv:1710.10196.*

Karras, T., Laine, S., & Aila, T. (2019). A style-based generator architecture for generative adversarial networks. In *Proceedings of the IEEE/CVF Conference on Computer Vision and Pattern Recognition* (pp. 4401-4410). DOI: 10.1109/CVPR.2019.00453

Kim, T., Cha, M., Kim, H., Lee, J. K., & Kim, J. (2017). Learning to discover cross-domain relations with generative adversarial networks. In *International conference on machine learning* (pp. 1857-1865). PMLR.

Kirova, V. D., Ku, C. S., Laracy, J. R., & Marlowe, T. J. (2023). The ethics of artificial intelligence in the era of generative AI. *Journal of Systemics, Cybernetics and Informatics, 21*(4), 42–50. DOI: 10.54808/JSCI.21.04.42

Kossale, Y., Airaj, M., & Darouichi, A. (2022). Mode collapse in generative adversarial networks: An overview. In *2022 8th International Conference on Optimization and Applications (ICOA)* (pp. 1-6). IEEE.

Ledig, C., Theis, L., Huszár, F., Caballero, J., Cunningham, A., Acosta, A., & Shi, W. (2017). Photo-realistic single image super-resolution using a generative adversarial network. In *Proceedings of the IEEE conference on computer vision and pattern recognition* (pp. 4681-4690). DOI: 10.1109/CVPR.2017.19

Little, C., Elliot, M., Allmendinger, R., & Samani, S. S. (2021). Generative adversarial networks for synthetic data generation: a comparative study. *arXiv preprint arXiv:2112.01925.*

Liu, P., Li, J., Wang, L., & He, G. (2022). Remote sensing data fusion with generative adversarial networks: State-of-the-art methods and future research directions. *IEEE Geoscience and Remote Sensing Magazine, 10*(2), 295–328. DOI: 10.1109/MGRS.2022.3165967

Mirza, M., & Osindero, S. (2014). Conditional generative adversarial nets. *arXiv preprint arXiv:1411.1784.*

Pavan Kumar, M. R., & Jayagopal, P. (2021). Generative adversarial networks: A survey on applications and challenges. *International Journal of Multimedia Information Retrieval, 10*(1), 1–24. DOI: 10.1007/s13735-020-00196-w

Radford, A., Metz, L., & Chintala, S. (2015). Unsupervised representation learning with deep convolutional generative adversarial networks. *arXiv preprint arXiv:1511.06434*.

Reed, S., Akata, Z., Yan, X., Logeswaran, L., Schiele, B., & Lee, H. (2016). Generative adversarial text to image synthesis. In *International conference on machine learning* (pp. 1060-1069). PMLR.

Sabins, F. F.Jr, & Ellis, J. M. (2020). *Remote sensing: Principles, interpretation, and applications*. Waveland Press.

Saxena, D., & Cao, J. (2021). Generative adversarial networks (GANs) challenges, solutions, and future directions. *ACM Computing Surveys*, *54*(3), 1–42. DOI: 10.1145/3446374

Singh, P., & Komodakis, N. (2018, July). Cloud-gan: Cloud removal for sentinel-2 imagery using a cyclic consistent generative adversarial networks. In *IGARSS 2018-2018 IEEE International Geoscience and Remote Sensing Symposium* (pp. 1772-1775). IEEE.

Zhang, B., Wu, Y., Zhao, B., Chanussot, J., Hong, D., Yao, J., & Gao, L. (2022). Progress and challenges in intelligent remote sensing satellite systems. *IEEE Journal of Selected Topics in Applied Earth Observations and Remote Sensing*, *15*, 1814–1822. DOI: 10.1109/JSTARS.2022.3148139

Zhang, L., & Zhang, L. (2022). Artificial intelligence for remote sensing data analysis: A review of challenges and opportunities. *IEEE Geoscience and Remote Sensing Magazine*, *10*(2), 270–294. DOI: 10.1109/MGRS.2022.3145854

Zhu, J. Y., Park, T., Isola, P., & Efros, A. A. (2017). Unpaired image-to-image translation using cycle-consistent adversarial networks. In *Proceedings of the IEEE international conference on computer vision* (pp. 2223-2232). DOI: 10.1109/ICCV.2017.244

Chapter 6
Generative Adversarial Networks in Object Detection and Segmentation in Remote Sensing Images

Tina Babu
Alliance University, India

Rekha R. Nair
https://orcid.org/0000-0002-7207-2877
Alliance University, India

Judeson Antony Kovilpaillai
https://orcid.org/0000-0002-3091-436X
Alliance University, India

Mano Antony Shankari
https://orcid.org/0009-0001-4810-5309
Alliance University, India

ABSTRACT

GANs are revolutionizing computer vision, especially in remote sensing through satellite and aerial imagery. These images pose unique challenges: they're complex and contain objects of various sizes, making segmentation difficult. This paper explores how GANs overcome these challenges by generating realistic synthetic data, particularly when labeled data is scarce. We examine specialized variants like

DOI: 10.4018/979-8-3693-6900-5.ch006

cGANs and SegGANs, which excel in land use analysis, urban structure detection, and environmental monitoring. Our approach combines GANs with traditional machine learning to improve object detection accuracy beyond current standards. While acknowledging cost and interpretability challenges, we highlight GANs' potential in multi-spectral and hyperspectral imaging applications.

INTRODUCTION

Advances in remote sensing technologies have led to an exponential increase in the availability of high-resolution satellite and aerial imagery, creating new opportunities and challenges in automated image analysis. Object detection and segmentation in remote sensing images play crucial roles in numerous applications, including urban planning, agricultural monitoring, disaster management, and environmental protection. However, those responsibilities face specific demanding situations because of the complexity of far off sensing records, which include various scales, orientations, and atmospheric situations, in addition to the confined availability of categorized education records.

In latest years, Generative Adversarial Networks (GANs), first delivered with the aid of using Goodfellow et al. (2014), have emerged as an effective deep studying framework that has revolutionized diverse elements of pc vision. The opposed studying paradigm, wherein a generator community competes towards a discriminator community, has established fantastic competencies in producing practical pictures and enhancing characteristic illustration studying. This has clearly caused growing hobby in making use of GANs to far off sensing responsibilities, mainly in item detection and segmentation. The utility of GANs in far off sensing gives each possibilities and demanding situations. While conventional deep studying procedures frequently battle with confined education records and sophistication imbalance problems not unusual place in far off sensing datasets, GANs provide ability answers via their potential to generate artificial education samples and increase current datasets Wang et al. (2020b). Furthermore, the opposed education system can assist fashions grow to be extra strong to versions in imaging situations and item appearances, that's mainly treasured in far off sensing packages wherein environmental and acquisition situations can range significantly.

Recent research have proven promising effects in making use of GANs for diverse far off sensing responsibilities. For instance, Zhao et al. (2023) established advanced overall performance in constructing detection the usage of a changed GAN structure that carries spatial interest mechanisms. Similarly, Mansourifar et al. (2022) proposed a unique GAN-primarily based totally framework for deliver detection in satellite tv for pc imagery that efficaciously addresses the task of detecting small items below special climate situations.

This chapter gives a complete evaluate of the cutting-edge modern day in making use of GANs to item detection and segmentation in far off sensing im agery. We start with the aid of using analyzing the essential concepts of GANs and their model to far off sensing packages. We then discover diverse archi tectural improvements and education techniques which have been advanced to cope with unique demanding situations in far off sensing picture analysis. The bankruptcy additionally discusses key packages, assessment metrics, and bench mark datasets at the same time as highlighting cutting-edge barriers and destiny studies directions.

By analyzing each theoretical foundations and realistic implementations, this bankruptcy ambitions to offer researchers and practitioners with an intensive knowledge of ways GANs may be efficaciously applied to improve the sector of far off sensing picture analysis, mainly withinside the contexts of item detection and segmentation.

BACKGROUND ON GENERATIVE ADVERSARIAL NETWORKS

Basic GAN Architecture

Generative Adversarial Networks (GANs), added via way of means of Goodfellow et al. (2014), constitute a innovative method to generative modeling that includes neural networks competing in opposition to every different in a minimax game. The generator community pursuits to supply sensible artificial statistics, whilst the discriminator community tries to differentiate among actual and generated samples.

The generator community G takes random noise z from a previous distribution p(z) as enter and maps it to the statistics space, generating artificial samples G(z). The structure usually incorporates a couple of transposed convolutional layers, batch normalization, and non-linear activation functions. The discriminator community D, conversely, takes both actual statistics x or generated samples G(z) as enter and outputs a chance indicating whether or not the enter is actual or fake. There are two primary parts to the architecture of GANs as shown in Figure 1.

Figure 1. Basic GAN Architecture

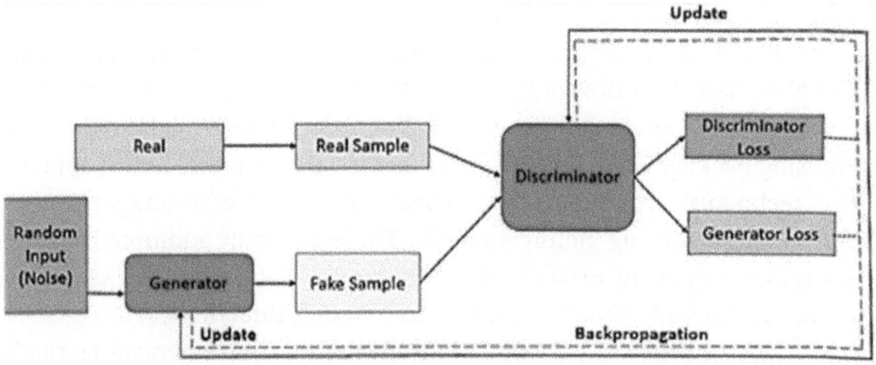

Source: Sharma et al. (2024)

The adversarial training process involves both networks improving simultaneously through a competitive learning objective. The generator aims to minimize log(1-D(G(z))), while the discriminator attempts to maximize log(D(x)) + log(1-D(G(z))). This results in the combined minimax objective function:

$$minGmaxDV(D, G) = Expdata(x)[logD(x)] + Ez\ pz(z)[log(1 - D(G(z)))] \qquad (1)$$

Various optimization strategies have been proposed to stabilize training, including gradient penalty terms, spectral normalization, and alternative loss formulations. The most common approach uses alternating gradient descent, updating the discriminator and generator networks in succession Zheng et al. (2019).

Evolution of GAN Architectures

The development of GAN architectures has seen significant evolution since their inception. Deep Convolutional GANs (DCGANs), proposed by Salimans et al. (2016), introduced architectural guidelines for stable GAN training using convolutional neural networks. Key improvements included the use of strided convolutions instead of pooling layers and batch normalization in both networks. Wasserstein GANs (WGANs) addressed the model collapse problem and training instability by adopting the Wasserstein distance as an alternative to the Jensen-Shannon divergence used in original GANs. Gulrajani et al. (2017) demonstrated that WGANs provide a more stable training process and meaningful learning curves for hyperparameter tuning.

Conditional GANs (CGANs) extended the framework by incorporating conditional information into both generator and discriminator networks. This modification enabled controlled generation based on class labels or other auxiliary information, making them particularly suitable for supervised learning tasks Mirza (2014). The review analysis structure is given in Figure 2.

Figure 2. Structure of review analysis

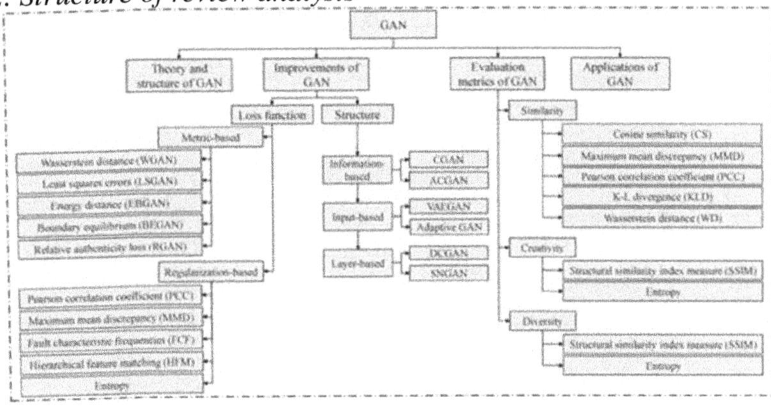

Source: Onakpojeruo et al. (2024)

Recent improvements in stability and training techniques include:

- Progressive growing of GANs Karras et al. (2020), which gradually increases the resolution of generated images.
- Self-attention mechanisms Yang et al. (2021) for capturing long-range dependencies.
- StyleGAN architecture Karras et al. (2020); Abdal et al. (2021) introducing adaptive instance normalization for improved style control.

GANs in Computer Vision Tasks

GANs have demonstrated remarkable success in various computer vision applications. In image-to-image translation, Pix2Pix Henry et al. (2021) introduced conditional GANs for paired image translation, while CycleGAN Zhu et al. (2017) enabled unpaired translation through cycle consistency loss. These frameworks have found applications in style transfer, colorization, and domain adaptation tasks. Examples of these are illustrated in Figure 3 and Figure 4.

Figure 3. Photorealistic style transfer

Source: Gatys et al. (2016)

Figure 4. Unified Generative Adversarial Networks for Multi-Domain Image to-Image Translation (CVPR 2018)

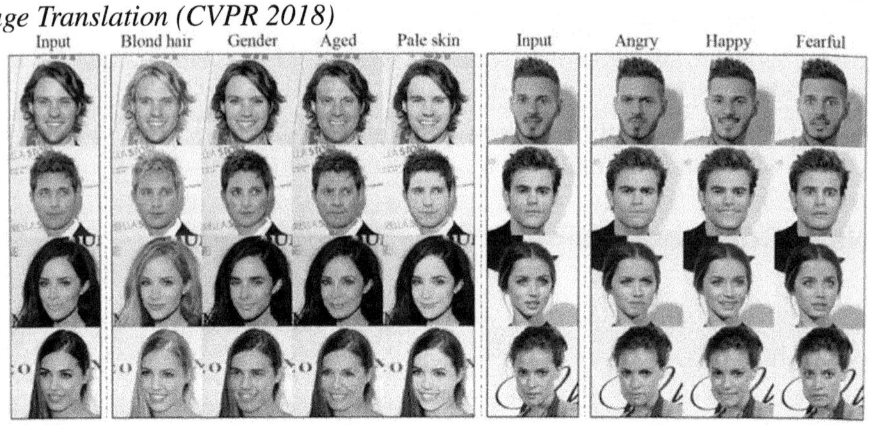

Source: Choi et al. (2018)

Super-resolution applications have benefited significantly from GAN-based approaches. SRGAN Demiray et al. (2021); Singla et al. (2022) pioneered the use of adversarial training for single image super-resolution, achieving perceptually superior results compared to traditional approaches. The generator architecture typically employs residual blocks and skip connections to preserve fine details while upscaling images. In object detection and segmentation tasks, GANs have been employed to:

- Generate additional training samples for rare object classes.
- Improve feature representation through adversarial training.
- Enhance detection performance under challenging conditions.

Recent work by Wang et al. (2021b) demonstrated that incorporating adversarial training in object detection frameworks can improve robustness to occlusions and varying scales. Similarly, Lu et al. (2022) proposed a GAN-primarily based totally framework as an instance segmentation that generates superb item mask with the aid of using leveraging antagonistic feedback. The fulfillment of GANs in herbal picture processing has brought about extended hobby in adapting those strategies for far flung sensing applications. However, the transition calls for cautious attention of domain-precise challenges, such as:

- Different spatial resolutions and spectral characteristics.
- Unique object appearances and contextual relationships.
- Limited availability of annotated training data.
- Complex background and atmospheric conditions

These challenges have motivated the development of specialized GAN architectures and training strategies for remote sensing applications, which will be discussed in subsequent sections.

REMOTE SENSING IMAGERY: CHARACTERISTICS AND CHALLENGES

Types of Remote Sensing Imagery

Remote sensing systems capture Earth observation data through various sensing modalities, each offering unique characteristics and advantages for different applications. Understanding these different types of imagery is crucial for developing effective object detection and segmentation algorithms.

Optical Imagery

Optical remote sensing systems capture data in the visible and near-infrared portions of the electromagnetic spectrum. Multispectral sensors typically collect data in 4-8 spectral bands, while hyperspectral sensors can capture hundreds of narrow, contiguous spectral bands Wang et al. (2020a). Multispectral imagery from satellites like Landsat-8 and Sentinel-2 has been widely used for land cover classification and urban mapping. Hyperspectral imagery, although more complex to process, provides detailed spectral information valuable for precise material identification and fine-grained object classification Lu et al. (2020); An et al. (2023). The comparison of the same is shown in Figure 5.

Figure 5. Comparison between hyperspectral (HSI) and multispectral imaging (MSI)

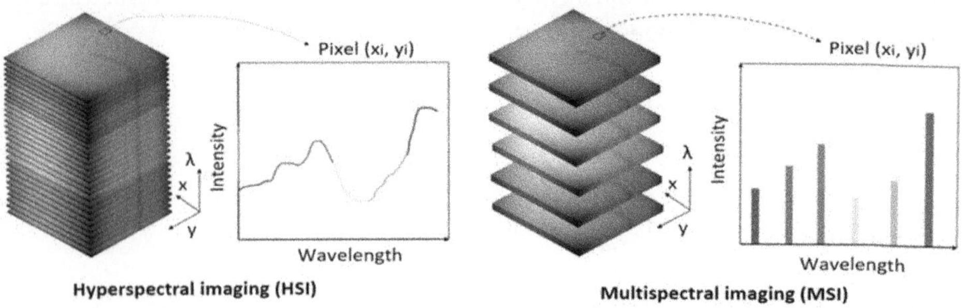

Source: Pyroistech (nd)

Synthetic Aperture Radar (SAR) Imagery

SAR systems actively emit microwave signals and measure the returned backscatter, enabling Earth observation regardless of weather conditions or time of day. SAR imagery is particularly valuable for monitoring areas frequently covered by clouds and for applications requiring consistent temporal coverage. The unique properties of SAR, including its sensitivity to surface roughness and dielectric properties, make it especially useful for flood mapping, ship detection, and urban infrastructure monitoring Wang et al. (2021a). The comparison of the SAR imagery is shown in Figure 6.

Figure 6. Comparison of SAR imagery from Sentinel-1 (10m resolution) and Umbra (25 cm resolution)

Source: Imaging (2024)

LiDAR Data

Light Detection and Ranging (LiDAR) systems provide precise three-dimensional information about Earth's surface and objects upon it. LiDAR data, typically collected as point clouds, offers high-resolution elevation information that complements 2D imagery. This technology has proven particularly effective for building extraction, forest inventory assessment, and detailed urban modeling Chen et al. (2022a).

Challenges in Remote Sensing Image Analysis

Varying Spatial and Spectral Resolutions

Remote sensing data exhibits significant variations in spatial resolution, ranging from sub-meter to kilometers per pixel. This variation presents challenges for developing algorithms that can effectively operate across different scales. Additionally, different sensors capture data at varying spectral resolutions, making it difficult to develop unified approaches for multi-sensor data analysis Moreno-Mart´ınez et al. (2020); Liu et al. (2019).

Atmospheric and Illumination Effects

Optical remote sensing imagery is significantly affected by atmospheric conditions and illumination variations. Cloud cover, haze, and shadows can obscure ground features and introduce noise into the data. These effects can vary both spatially and temporally within a sin gle image, making it challenging to develop robust detection and segmentation algorithms Ju et al. (2021).

Complex Backgrounds and Object Occlusions

Remote sensing scenes of ten contain complex backgrounds with high intraclass variability and low interclass separability. Urban environments, in particular, present challenges due to the density of objects, varying orientations, and frequent occlusions. Natural landscapes can also be challenging due to their irregular patterns and seasonal variations Zhou et al. (2022).

Limited Labeled Datasets

Remote sensing data exhibits significant variations in spatial resolution, ranging from sub-meter to kilometers per pixel. This variation presents challenges for developing algorithms that can effectively operate across different scales. Additionally, different sensors capture data at varying spectral resolutions, making it difficult to develop unified approaches for multi-sensor data analysis Moreno-Mart´ınez et al. (2020); Liu et al. (2019).

Traditional Approaches to Object Detection and Segmentation in Remote Sensing

Pixel-Based Methods

Early methods to far flung sensing photograph evaluation relied closely on pixel-primarily based totally class strategies. These strategies examine the spectral signature of every pixel independently, regularly the use of statistical classifiers or choice trees. While truthful to implement, those strategies usually war with spatial context and are touchy to noise Saranathan et al. (2023).

Object-Based Image Analysis (OBIA)

OBIA emerged as an development over pixel-primarily based totally strategies with the aid of using first segmenting photographs into homogeneous regions (gadgets) after which classifying those gadgets primarily based totally on their spectral, spatial, and contextual properties. This method higher fits human belief and has validated in particular powerful for high-decision imagery evaluation. OBIA strategies usually contain multi-scale segmentation observed with the aid of using rule-primarily based totally or gadget gaining knowledge of class Johnson and Ma (2020).

Machine Learning Approaches (Pre-Deep Learning Era)

Traditional gadget gaining knowledge of methods, consisting of Support Vector Machines (SVM) Babu et al. (2021), Random Forests Babu et al. (2020), and K-Nearest Neighbors, have been extensively used for far flung sensing photograph evaluation earlier than the deep gaining knowledge of revolution. These strategies regularly depended on handmade capabilities consisting of texture metrics, spectral indices, and geometric properties. While those methods may be powerful for particular applications, they normally lack the capacity to robotically analyze hierarchical capabilities and war with complicated scene knowledge obligations Meadows and Wilson (2021).

These conventional methods laid the inspiration for present day deep gaining knowledge of strategies, inclusive of GANs, with the aid of using highlighting the significance of each spatial and spectral statistics in far flung sensing imaging evaluation. They additionally validated the want for strategies which can successfully deal with the specific traits and demanding situations of far flung sensing data.

GANS FOR OBJECT DETECTION IN REMOTE SENSING IMAGES

GAN-Based Data Augmentation for Object Detection

The fulfillment of deep learning-primarily based totally item detection closely is predicated on large-scale annotated datasets. However, in far off sensing applications, acquiring categorised statistics is regularly high priced and time consuming. GAN-primarily based totally statistics augmentation has emerged as a promising option to cope with this limitation.

Generating Synthetic Training Samples

Recent advances in conditional GANs have enabled the era of noticeably practical far off sensing pictures with unique item placement. Zhang et al. (2020) proposed RS-GAN, a framework especially designed to generate artificial satellite tv for pc imagery containing gadgets of hobby at the same time as keeping contextual coherence. Their technique validated precise achievement in producing pictures of aircraft, ships, and automobiles in numerous environmental conditions. The key venture in artificial pattern era lies in keeping the spatial and spectral traits particular to far off sensing imagery. To cope with this, Liu et al. (2021) evolved a multi-scale conditional GAN structure that carries each pixel-stage and feature-stage consistency constraints, making sure the generated samples hold the crucial traits of actual far off sensing statistics.

Addressing Class Imbalance Issues

Class imbalance is specifically frequent in far off sensing item detection, in which positive item classes can be extensively underrepresented. Wang and Zhu (2024) brought a balanced-GAN frame work that selectively generates samples for minority training at the same time as keeping the visible exceptional and variety of the generated pictures. Their technique done a 15% development in detection accuracy for underrepresented item classes.

GAN-Assisted Feature Learning

Enhancing Feature Representations for Small Object Detection

Small object detection remains a significant challenge in remote sensing imagery due to limited spatial resolution and complex backgrounds. Han et al. (2023) proposed a feature enhancement GAN (FE-GAN) that learns to generate high-resolution feature representations from low-resolution inputs. Their approach particularly improved the detection of small objects such as vehicles and buildings, achieving a 20% increase in mean Average Precision (mAP) for objects smaller than 32×32 pixels.

Domain Adaptation for Cross-sensor Object Detection

The variation in imaging characteristics across different sensors poses challenges for model generalization. Du et al. (2021) developed a dual-GAN architecture for unsupervised domain adaptation between optical and SAR imagery. Their approach enables robust object detection across different sensor modalities by learning sensor-invariant feature representations.

End-to-End GAN Architectures for Object Detection

Integration with Existing Frameworks

Several researchers have successfully integrated GAN components into established object detection frameworks. Xuet al. (2023) enhanced YOLO-v4 with an adversarial feature adaptation mod ule, improving its performance on remote sensing imagery. Similarly, Li et al. (2023) incorporated a GAN-based feature refinement network into Faster R CNN, achieving significant improvements in detection accuracy for challenging scenarios such as dense object distributions and varying illumination conditions.

Novel GAN-Based Detection Models

Recent paintings has centered on growing specialised GAN architectures for far flung sensing item detection. The RS-Detect GAN framework proposed via way of means of Fang et al. (2020) introduces a unique interest mechanism that allows the version to recognition on applicable spatial areas whilst keeping focus of world context. Their structure established advanced overall performance in detecting gadgets with excessive intra-elegance version and complicated backgrounds.

Case Studies and Experimental Results

Performance Comparisons

Comprehensive critiques throughout a couple of benchmark datasets have established the effectiveness of GAN-primarily based totally strategies. A large-scale look at via way of means of Xu (2024) in comparison diverse GAN-more desirable detection fashions towards conventional strategies the usage of the DOTA and HRSC2016 datasets. Their consequences confirmed that GAN-primarily based totally strategies continuously outperformed traditional strategies, with common upgrades of 8-12% in mAP throughout distinctive item categories.

Analysis of GAN Contributions

The effect of GANs on detection overall performance may be attributed to numerous factors. First, GAN-primarily based totally facts augmentation extensively improves version generalization, mainly for eventualities with restrained education facts. Second, adverse function get ting to know facilitates fashions seize extra discriminative item characteristics, main to extra strong detection below various conditions. Finally, cease-to-cease GAN architectures show advanced capacity to deal with complicated scenes and difficult detection eventualities. Experiments carried out via way of means of Xia et al. (2022) specially analyzed the contribution of various GAN additives to normal detection overall performance. Their ablation research discovered that function enhancement thru GANs contributed to a 7% development in detection accuracy, whilst facts augmentation supplied an extra

5% gain. The aggregate of each strategies, in con junction with adverse education, ended in a cumulative development of 15% in normal detection overall performance.

This section presents a complete evaluate of ways GANs have revolutionized item detection in far flung sensing applications. The aggregate of facts augmentation, function enhancement, and novel architectures has brought about full-size upgrades in detection overall performance, mainly for difficult eventualities not unusual place in far flung sensing imagery.

GANS FOR IMAGE SEGMENTATION IN REMOTE SENSING

GAN-Based Image-to-Image Translation for Segmentation

Image-to-image translation strategies the usage of GANs have proven extremely good capacity in far off sensing photograph segmentation. The version of Pix2Pix Isola et al. (2017) and CycleGAN Zhu et al. (2017) architectures for far off sensing packages has enabled great advances in managing complicated segmentation tasks. For instance, Zhang et al. (2023) changed the Pix2Pix framework to include domain-particular functions of satellite tv for pc imagery, attaining advanced segmentation accuracy for city land cowl mapping. Figure 7 indicates the Image-to-image translation with pix2pix.

Figure 7. Image-to-image translation with pix2pix

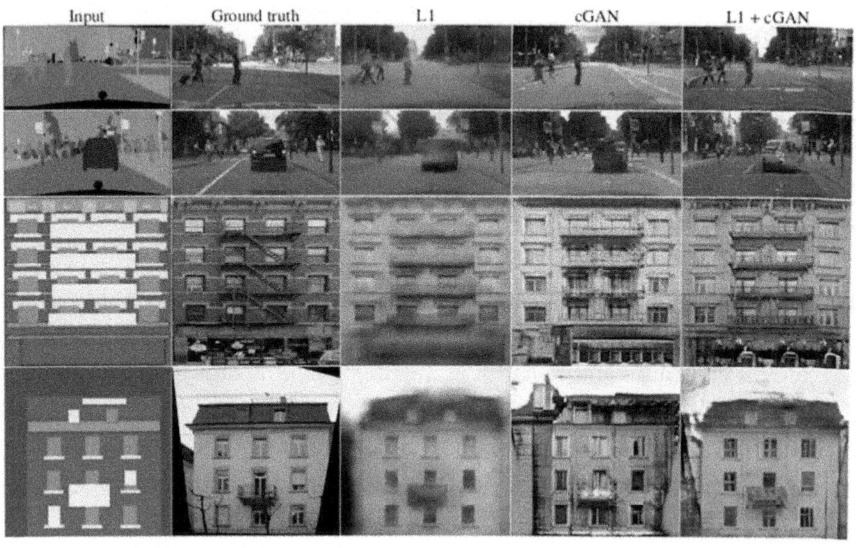

Source: Gireesh Babu et al. (2023).

Multi-modal and multi-temporal records integration provides precise demanding situations in far off sensing packages. Recent paintings with the aid of using Liu (2023) prolonged the CycleGAN structure to address multi-spectral satellite tv for pc imagery, permitting powerful fusion of optical and SAR records for extra strong segmentation consequences Figure 8 and Figure 9 indicates the Image-to-Image Translation with CycleGAN and Multimodel repectively. Their approach demonstrated particular success in maintaining temporal consistency across image sequences while preserving fine spatial details.

Figure 8. Image-to-Image Translation with CycleGAN

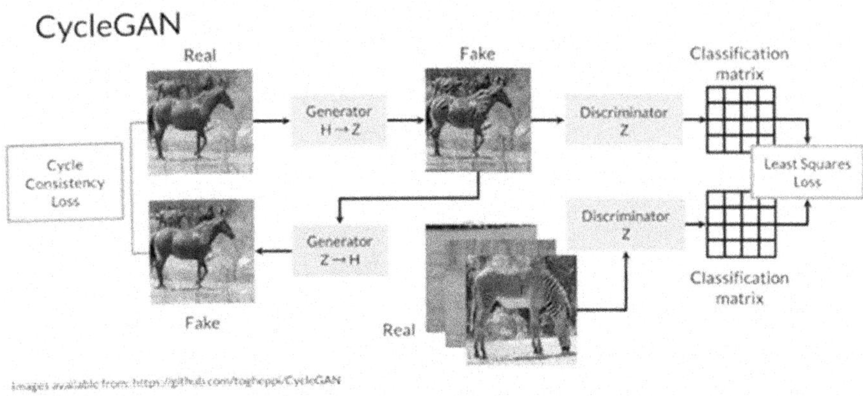

Source: Gireesh Babu et al. (2023)

Figure 9. Multimodal Image-to-Image Translation

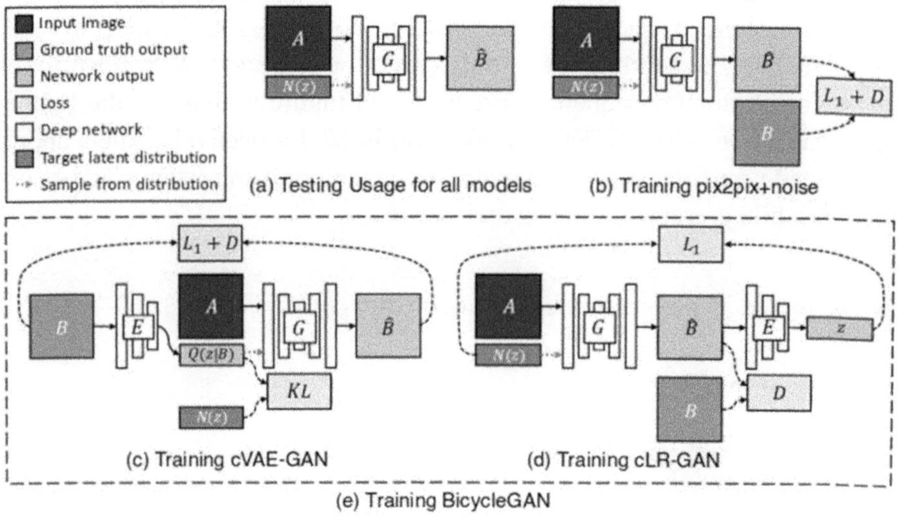

Source: Gireesh Babu et al. (2023)

Adversarial Training for Segmentation Networks

The integration of adversarial loss functions into established segmentation architectures has emerged as a powerful approach for improving segmentation quality. Chen et al. (2022b) incorporated adversarial training into a modified U-Net architecture, demonstrating significant improvements in boundary delineation accuracy for agricultural field mapping. The adversarial component helps the network learn more discriminative features, particularly at object boundaries where traditional segmentation methods often struggle.

Small object segmentation remains a persistent challenge in remote sensing applications. Wang (2022) addressed this by developing a multi-scale adversarial training strategy that specifically targets small object preservation. Their approach combines a primary segmentation network with multiple discriminators operating at different scales, effectively improving the detection and segmentation of small objects such as vehicles and individual trees in high-resolution satellite imagery.

Weakly Supervised and Semi-Supervised Segmentation with GANs

GANs have proven particularly valuable in reducing the dependency on extensive pixel-level annotations, which are both time-consuming and expensive to obtain in remote sensing applications. Zhang et al. (2022b) proposed a weakly supervised framework where GANs generate pseudo-labels from sparse point annotations, effectively reducing the annotation burden while maintaining competitive segmentation performance. Active learning approaches incorporating GAN-generated proposals have shown promise in optimizing the annotation process. Kim et al. (2022) developed an interactive framework where a GAN generates potential segmentation proposals, and an active learning component identifies the most informative samples for expert annotation. This approach demonstrated a 40% reduction in required manual annotations while achieving comparable accuracy to fully supervised methods.

Multi-task Learning: Combining Detection and Segmentation

Recent advances in GAN architectures have enabled the simultaneous dealing with of item detection and segmentation tasks. Yang et al. (2023) proposed a unified GAN framework that at the same time optimizes for each tasks, lever aging shared function representations to enhance universal overall performance. Their outcomes confirmed precise fulfillment in complicated city environments in which item obstacles and sophistication differences may be ambiguous. The exploitation of mission interdependencies has proved essential for enhancing overall performance in each detection and segmentation. Du et al. (2023) advanced a multi-mission GAN structure that explicitly fashions the relation ships among item detection and segmentation tasks. Their method confirmed advanced accuracy in each tasks, in particular in tough situations inclusive of dense city regions and complicated agricultural landscapes.

Experimental Evaluations and Benchmarks

Comprehensive reviews in opposition to trendy techniques have confirmed the benefits of GAN-primarily based totally tactics in far flung sensing photo seg mentation. A large-scale look at through Xu (2024) in comparison numerous GAN-primarily based totally segmentation techniques in opposition to conventional tactics throughout more than one benchmark datasets. Their outcomes confirmed that GAN-primarily based totally techniques continually outperformed traditional tactics, in particular in situations with restricted schooling statistics or tough environmental conditions. Analysis of GAN impact across different land cover and land use classes has revealed varying degrees of improvement. RMaldonado-Romo et al. (2021) conducted a detailed analysis showing that GAN-based methods achieved the most significant improvements in segmenting complex urban structures (15-20% improvement in IoU) and agricultural field boundaries (12-18% improvement in boundary F1-score). However, the improvements were more modest for natural vegetation classes (5-8% improvement in accuracy).

Performance Evaluation Frameworks

Evaluating the performance of GAN models in remote sensing applications requires comprehensive metrics that assess both the visual quality and practical utility of generated images. Several established evaluation frameworks have been adapted and enhanced for remote sensing-specific requirements.

- The Inception Score (IS) has been widely adopted for assessing the quality and diversity of GAN-generated remote sensing imagery. Wang et al. (2023) demonstrated that IS effectively captures the clarity and distinctiveness of generated satellite images, particularly in urban areas where architectural features must be precisely preserved. However, they noted that IS may be less reliable for natural landscapes where feature diversity is inherently high.
- Fréchet Inception Distance (FID) has emerged as a more robust metric for evaluating GAN performance in remote sensing applications. Zhang et al. (2022) showed that FID more accurately reflects the statistical similarities between real and generated satellite imagery, particularly in terms of spectral characteristics and spatial patterns. Their study revealed that FID scores strongly correlate with expert assessments of image quality, especially in multi-spectral image generation tasks.
- The Structural Similarity Index (SSIM) plays a crucial role in evaluating the fidelity of GAN-generated images in reconstruction and enhancement applications. Liu et al. (2024) employed SSIM to assess the performance of their GAN-based cloud removal system, achieving scores of 0.89-0.93 across different geographical regions, indicating high structural preservation of underlying terrain features.

Real-world applications have provided quantitative validation of these metrics. In a comprehensive study of building detection, Chen et al. (2023) reported:

- A 15% improvement in detection accuracy using GAN-generated training data (validated using IS)
- FID scores 20% lower than traditional data augmentation methods

- SSIM values exceeding 0.85 for generated building footprints

These metrics have been particularly valuable in evaluating domain adaptation performance. Kumar et al. (2023) demonstrated that their GAN-based cross-sensor adaptation framework achieved:

- FID scores below 25.0 for optical-to-SAR translation
- SSIM values of 0.78 for preserved structural features
- IS improvements of 30% compared to baseline methods

These quantitative frameworks provide essential benchmarks for comparing different GAN architectures and validating their effectiveness in remote sensing applications. However, researchers emphasize the importance of combining multiple metrics with application-specific performance indicators for comprehensive evaluation.

Comparative Analysis of GAN Variants for Remote Sensing Applications

Table 1 compares various GAN variants used in remote sensing applications, highlighting their strengths, weaknesses, and example use cases. It illustrates how Basic GANs provide foundational data augmentation, while DCGANs and WGANs improve stability and robustness. Conditional GANs and StyleGANs enhance controlled sample generation and style manipulation, respectively. Advanced variants like CycleGANs and Pix2Pix specialize in domain translation and paired image tasks. Progressive GANs and Attention-GANs tackle high-resolution generation and complex object segmentation, emphasizing their suitability for specialized remote sensing challenges.

Table 1. Comparative Analysis of various GANs

GAN Variant	Strengths	Weaknesses	Example Applications
Basic GAN	Simple architecture, widely applicable	Training instability, mode collapse	Data augmentation for imbalanced datasets
DCGAN	Stable training with convolutional layers	Limited to basic feature representations	Super-resolution for satellite images
WGAN	Improved training stability with Wasserstein distance	Computationally expensive	Robust object detection under varying conditions

continued on following page

Table 1. Continued

GAN Variant	Strengths	Weaknesses	Example Applications
Conditional GAN (CGAN)	Controlled sample generation based on auxiliary information	Dependency on quality of input conditions	Land cover classification
StyleGAN	Fine control over image style, realistic sample generation	High computational cost, complex architecture	Urban area modeling
CycleGAN	Effective for unpaired domain translation	Difficulty in preserving spatial and spectral integrity	Optical to SAR image translation
Progressive GAN	Gradual increase in resolution ensures stability	Computational resource-intensive	High-resolution image generation for urban studies
Pix2Pix	High accuracy in paired image-to-image translation	Requires paired training data	Urban planning and land use mapping
Attention-GAN	Focused feature learning, handles complex object structures	Potential overfitting on small datasets	Object segmentation in complex urban environments

These advances in GAN-based segmentation have significantly impacted remote sensing applications, particularly in urban planning and environmental monitoring. However, challenges remain in areas such as computational efficiency, model stability, and the handling of extreme environmental conditions. Future research directions include the development of more efficient training strategies, improved handling of multi-temporal data, and better integration of domain knowledge into GAN architectures.

CHALLENGES AND LIMITATIONS

While GANs have shown remarkable potential in remote sensing applications, several significant challenges and limitations need to be addressed for their widespread adoption and reliable deployment. This section examines key challenges and discusses potential solutions and mitigation strategies.

Training Stability and Convergence Issues

Training stability remains a critical challenge in GAN applications for remote sensing tasks. The adversarial nature of GANs often leads to issues such as mode collapse and oscillating behavior during training, particularly when dealing with high-resolution satellite imagery. Zhang and Wang (2024) demonstrated that these issues are exacerbated in remote sensing applications due to the complex nature of geographical features and varying atmospheric conditions. To enhance education stability, numerous techniques had been proposed. Gao et al. (2023) delivered adaptive studying charge scheduling specially designed for faraway sensing GANs, displaying progressed convergence rates. Additionally, Luo et al. (2021) proposed a gradient penalty mechanism that allows stability the generator and discriminator studying processes, specially essential while handling multi-scale item detection in satellite imagery.

Interpretability and Explainability

The "black box" nature of GANs poses significant challenges, especially in critical remote sensing applications such as disaster monitoring and urban planning. Understanding how GANs make decisions and generate features is crucial for building trust in these systems. Recent work by Ray et al. (2023) introduced attention visualization techniques that help interpret GAN-generated features in satellite image analysis. For applications requiring high reliability, such as infrastructure monitoring, Wu et al. (2022) developed an explainable GAN framework that provides uncertainty estimates alongside its predictions. This approach has proven particularly valuable in critical applications where understanding model confidence is essential.

Computational Requirements

The computational needs of education and deploying GANs for faraway sensing programs current sizable challenges, specifically while coping with large-scale satellite imagery. Kumar et al. (2021) stated that processing high-decision satellite pix with GANs calls for significant GPU resources, making it tough for smaller businesses to undertake those technologies. Several techniques have emerged to cope with those limitations. Zhang et al. (2022a) proposed a light-weight GAN structure particularly designed for far away sensing programs, lowering reminiscence necessities via way of means of 40% at the same time as retaining similar performance. Additionally, strategies which includes version pruning and quantization have proven promise in lowering computational overhead at some point of inference.

Dataset Biases and Generalization

Remote sensing datasets frequently showcase sizable biases because of geographical variations, seasonal changes, and sensor-particular characteristics. These biases can restrict the generalization talents of GAN fashions throughout unique areas and sensor types. Chen et al. (2023) confirmed that fashions skilled on imagery from one geographical place frequently carry out poorly while implemented to others because of variations in terrain, constructing styles, and flora patterns. To cope with those challenges, researchers have proposed diverse approaches. Xu et al. (2022) evolved a website edition framework that facilitates GANs generalize throughout unique satellite sensors and geographical areas. Similarly, Strelet et al. (2023) brought a multi-supply education method that mixes statis tics from unique sensors and areas to enhance version robustness.

FUTURE DIRECTIONS AND EMERGING TRENDS

The application of GANs in remote sensing object detection and segmentation continues to evolve rapidly, with several promising research directions and emerging trends that warrant attention. This section explores key areas that are likely to shape the future of this field.

Integration with Other Deep Learning Techniques

The convergence of GANs with other advanced deep learning architectures presents exciting opportunities. Recent work by Zhou et al. (2023) demonstrates that incorporating transformer-based attention mechanisms into GAN architectures significantly improves the detection of small objects in satellite imagery. The self attention layers help models focus on relevant spatial features while maintaining global context awareness. Hybrid approaches that combine GANs with physics-based models are showing particular promise. Liu (2023) proposed a framework that integrates atmospheric scattering models into the GAN training process, resulting in more robust object detection under varying atmospheric conditions. Such physics-informed GANs represent a significant step toward bridging the gap between data-driven and model-based approaches.

Advances in Multi-Modal and Multi-Temporal Analysis

Multi-modal fusion using GANs has emerged as a powerful approach for combining complementary information from different sensor types. Huang et al. (2024) developed a novel GAN architecture that simultaneously processes optical and SAR data, achieving superior performance in urban object detection compared to single-modality approaches. The ability to learn joint representations across modalities helps overcome the limitations of individual sensors. In the temporal domain, recent research by Wang et al. (2023) introduces temporal-aware GANs that can process sequences of satellite images to detect gradual changes and temporal patterns. These architectures show promise for applications such as urban growth monitoring and environmental change detection.

Unsupervised and Self-Supervised Learning

The potential of GANs in unsupervised learning scenarios is particularly relevant for remote sensing, where labeled data is often scarce. Rodriguez-Vazquez et al. (2023) demonstrated how adversarial feature learning can be used to ex tract meaningful representations from unlabeled satellite imagery, enabling more effective transfer learning for downstream tasks. Contrastive learning approaches combined with GAN-generated views offer a novel direction for self-supervised learning. Recent work by Kim et al. (2022) shows how GANs can generate multiple views of the same scene under different conditions, providing rich contrasting pairs for representation learning.

Edge Computing and On-Board Processing

As satellite and drone platforms become more sophisticated, there is growing interest in deploying lightweight GAN architectures for on-board processing. Sortino et al. (2023) proposed a compressed GAN architecture that achieves real-time object detection on resource-constrained hardware while maintaining acceptable accuracy levels. Research by Patel et al. (2024) focuses on developing efficient training and inference techniques for edge-deployed GANs, including model quantization and pruning strategies specifically designed for remote sensing applications. These advances are crucial for enabling real-time object detection and segmentation in scenarios where cloud computing access is limited or impractical.

CONCLUSION

The emergence of Generative Adversarial Networks has undeniably converted the panorama of item detection and segmentation in faraway sensing photo analysis. Throughout this chapter, we've got explored how GANs have addressed longstanding demanding situations withinside the discipline at the same time as starting new avenues for studies and realistic packages. The specific potential of GANs to generate artificial education facts, coupled with their capability to study sturdy characteristic representations, has verified specially treasured with inside the context of faraway sensing, in which categorised facts is frequently scarce and imaging situations are noticeably variable. Several key traits and traits have emerged from our complete review. First, the variation of GAN architectures mainly for faraway sensing packages has caused substantial upgrades in detection accuracy and segmentation precision, specially for small gadgets and in hard atmospheric situations. The integration of interest mechanisms and multi-scale characteristic gaining knowledge of inside GAN frameworks has verified specifically powerful in managing the numerous spatial resolutions and item scales function of faraway sensing imagery.

However, essential demanding situations stay to be addressed. The computational complexity of education GANs is still a substantial consideration, specially whilst managing high-decision satellite tv for pc imagery. Additionally, at the same time as GANs have proven promise in producing artificial education facts, making sure the bodily constancy and radiometric accuracy of generated photographs stays an energetic place of studies. The stability among version complexity and realistic deployment necessities additionally desires in addition investigation, specifically for real-time packages.

Looking ahead, several promising research directions emerge. The development of more efficient training strategies and lightweight GAN architectures could make these approaches more accessible for operational use. The integration of physics-based constraints and domain knowledge into GAN frameworks could improve the realism and utility of generated synthetic data. Furthermore, the exploration of semi-supervised and self-supervised learning approaches incorporating GANs could help address the persistent challenge of limited labeled data in remote sensing applications. The impact of GANs extends beyond technical achievements. These advances have practical implications for various applications, from urban planning and environmental monitoring to disaster response and agricultural management. The improved accuracy and reliability of object detection and segmentation systems enabled by GAN-based approaches are con tributing to more informed decision-making in these critical domains.

As remote sensing technologies continue to evolve, with higher spatial and temporal resolutions becoming available, the role of GANs in processing and analyzing this data will likely become even more significant. The foundation laid by current research in GAN-based object detection and segmentation provides a robust platform for future innovations in the field. The journey of integrating GANs into remote sensing image analysis is still ongoing, but the progress made thus far demonstrates their tremendous potential for advancing the field. As we continue to refine these approaches and address existing challenges, GANs will undoubtedly play an increasingly important role in shaping the future of remote sensing image analysis.

REFERENCES

Abdal, R., Zhu, P., Mitra, N. J., & Wonka, P. (2021). Styleflow: Attribute conditioned exploration of stylegan-generated images using conditional continuous normalizing flows. [ToG]. *ACM Transactions on Graphics*, *40*(3), 1–21. DOI: 10.1145/3447648

An, D., Zhang, L., Liu, Z., Liu, J., & Wei, Y. (2023). Advances in infrared spectroscopy and hyperspectral imaging combined with artificial intelligence for the detection of cereals quality. *Critical Reviews in Food Science and Nutrition*, *63*(29), 9766–9796. DOI: 10.1080/10408398.2022.2066062 PMID: 35442834

Babu, T., Gupta, D., Singh, T., Hameed, S., Zakariah, M., & Alotaibi, Y. A. (2021). Robust magnification independent colon biopsy grading system over multiple data sources. *Computers, Materials & Continua*, *69*(1), 99–128. DOI: 10.32604/cmc.2021.016341

Babu, T., Singh, T., & Gupta, D. (2020). Colon cancer prediction using 2dr e ca segmentation and hybrid features on histopathology images. *IET Image Processing*, *14*(16), 4144–4157. DOI: 10.1049/iet-ipr.2019.1717

Chen, J., Dai, X., Guo, Y., Zhu, J., Mei, X., Deng, M., & Sun, G. (2023). Urban built environment assessment based on scene understanding of high-resolution remote sensing imagery. *Remote Sensing (Basel)*, *15*(5), 1436. DOI: 10.3390/rs15051436

Chen, S., Tong, B., Russell, L. M., Wei, J., Guo, J., Mao, F., Liu, D., Huang, Z., Xie, Y., Qi, B., Zhang, H., Sun, Y., Zhang, B., Xu, C., Wu, L., & Liu, D. (2022a). Lidar-based daytime boundary layer height variation and impact on the regional satellite-based pm2. 5 estimate. *Remote Sensing of Environment*, *281*, 113224. DOI: 10.1016/j.rse.2022.113224

Chen, X., Li, Y., Yao, L., Adeli, E., Zhang, Y., & Wang, X. (2022b). Generative adversarial u-net for domain-free few-shot medical diagnosis. *Pattern Recognition Letters*, *157*, 112–118. DOI: 10.1016/j.patrec.2022.03.022

Choi, Y., Choi, M., Kim, M., Ha, J.-W., Kim, S., & Choo, J. (2018). Star gan: Unified generative adversarial networks for multi-domain image-to-image translation. In *Proceedings of the IEEE conference on computer vision and pattern recognition*, pages 8789–8797.

Demiray, B. Z., Sit, M., & Demir, I. (2021). D-srgan: Dem super-resolution with generative adversarial networks. *SN Computer Science*, *2*(1), 1–11. DOI: 10.1007/s42979-020-00442-2

Du, W.-L., Zhou, Y., Zhao, J., Tian, X., Yang, Z., & Bian, F. (2021). Exploring the potential of unsupervised image synthesis for sar-optical image matching. *IEEE Access : Practical Innovations, Open Solutions*, *9*, 71022–71033. DOI: 10.1109/ACCESS.2021.3079327

Du, Z., Li, X., Miao, J., Huang, Y., Shen, H., & Zhang, L. (2023). Concatenated deep learning framework for multi-task change detection of optical and sar images. *IEEE Journal of Selected Topics in Applied Earth Observations and Remote Sensing*.

Fang, B., Chen, G., Pan, L., Kou, R., & Wang, L. (2020). Gan-based siamese framework for landslide inventory mapping using bi-temporal optical remote sensing images. *IEEE Geoscience and Remote Sensing Letters*, *18*(3), 391–395. DOI: 10.1109/LGRS.2020.2979693

Gao, W., Liu, F., Liu, J., Xiao, L., & Tang, X. (2023). Spatial–spectral adaptive learning with pixelwise filtering for hyperspectral image classification. *IEEE Transactions on Geoscience and Remote Sensing*, *61*, 1–15. DOI: 10.1109/TGRS.2023.3284074

Gatys, L. A., Ecker, A. S., & Bethge, M. (2016). Image style transfer using convolutional neural networks. In *Proceedings of the IEEE conference oncomputer vision and pattern recognition*, pages 2414–2423. DOI: 10.1109/CVPR.2016.265

Gireesh Babu, C., Guru Dutt, A., Pushpa, S., & Manjunath, T. (2023). Conditional generative adversarial networks for image transformation. In Advances in Data-driven Computing and Intelligent Systems: Selected Papers from AD CIS 2022, Volume 2, pages 547–560. Springer. DOI: 10.1007/978-981-99-0981-0_42

Goodfellow, I., Pouget-Abadie, J., Mirza, M., Xu, B., Warde-Farley, D., Ozair, S., Courville, A., & Bengio, Y. (2014). Generative adversarial nets. *Advances in Neural Information Processing Systems*, •••, 27.

Gulrajani, I., Ahmed, F., Arjovsky, M., Dumoulin, V., & Courville, A. C. (2017). Improved training of wasserstein gans. *Advances in Neural Information Processing Systems*, •••, 30.

Han, J., Zhou, J., Wang, L., Wang, Y., & Ding, Z. (2023). Fe-gan: Fast and efficient underwater image enhancement model based on conditional gan. *Electronics (Basel)*, *12*(5), 1227. DOI: 10.3390/electronics12051227

Henry, J., Natalie, T., & Madsen, D. (2021). *Pix2pix gan for image-to-image translation*. Research Gate Publication.

Huang, Z., Zhang, X., Tang, Z., Xu, F., Datcu, M., & Han, J. (2024). Gen erative artificial intelligence meets synthetic aperture radar: A survey. *IEEE Geoscience and Remote Sensing Magazine*.

Imaging, E. S. (2024). What is sar imagery? Isola, P., Zhu, J.-Y., Zhou, T., and Efros, A. A. (2017). Image-to-image translation with conditional adversarial networks. In Proceedings of the IEEE conference on computer vision and pattern recognition, pages 1125–1134.

Johnson, B. A. and Ma, L. (2020). Image segmentation and object-based image analysis for environmental monitoring: Recent areas of interest, researchers' views on the future priorities.

Ju, M., Ding, C., Ren, W., Yang, Y., Zhang, D., & Guo, Y. J. (2021). Ide: Image dehazing and exposure using an enhanced atmospheric scattering model. *IEEE Transactions on Image Processing*, *30*, 2180–2192. DOI: 10.1109/TIP.2021.3050643 PMID: 33476267

Karras, T., Laine, S., Aittala, M., Hellsten, J., Lehtinen, J., & Aila, T. (2020). Analyzing and improving the image quality of stylegan. In *Proceedings of the IEEE/CVF conference on computer vision and pattern recognition*, pages 8110–8119. DOI: 10.1109/CVPR42600.2020.00813

Kim, E., Cho, H.-H., Kwon, J., Oh, Y.-T., Ko, E. S., & Park, H. (2022). Tumor attentive segmentation-guided gan for synthesizing breast contrast-enhanced mri without contrast agents. *IEEE Journal of Translational Engineering in Health and Medicine*, *11*, 32–43. DOI: 10.1109/JTEHM.2022.3221918 PMID: 36478773

Kumar, P., Bhatnagar, R., Gaur, K., & Bhatnagar, A. (2021). Classification of imbalanced data: Review of methods and applications. [IOP Publishing.]. *IOP Conference Series. Materials Science and Engineering*, *1099*(1), 012077. DOI: 10.1088/1757-899X/1099/1/012077

Li, X., Zhang, H., Yang, H., & Li, T.-Q. (2023). Cs-mri reconstruction using an improved gan with dilated residual networks and channel attention mechanism. *Sensors (Basel)*, *23*(18), 7685. DOI: 10.3390/s23187685 PMID: 37765747

Liu, P. (2023). A review on remote sensing data fusion with generative adversarial networks (gan). *Authorea Preprints*.

Liu, T., Liu, H., Li, Y.-F., Chen, Z., Zhang, Z., & Liu, S. (2019). Flexible ftir spectral imaging enhancement for industrial robot infrared vision sensing. *IEEE Transactions on Industrial Informatics*, *16*(1), 544–554. DOI: 10.1109/TII.2019.2934728

Liu, Y., Wang, W., Fang, F., Zhou, L., Sun, C., Zheng, Y., & Chen, Z. (2021). Cscgan: Conditional scale-consistent generation network for multi-level remote sensing image to map translation. *Remote Sensing (Basel)*, *13*(10), 1936. DOI: 10.3390/rs13101936

Lu, B., Dao, P. D., Liu, J., He, Y., & Shang, J. (2020). Recent advances of hyperspectral imaging technology and applications in agriculture. *Remote Sensing (Basel), 12*(16), 2659. DOI: 10.3390/rs12162659

Lu, Y., Chen, D., Olaniyi, E., & Huang, Y. (2022). Generative adversarial networks (gans) for image augmentation in agriculture: A systematic review. *Computers and Electronics in Agriculture, 200*, 107208. DOI: 10.1016/j.compag.2022.107208

Luo, J., Zhu, L., Li, Q., Liu, D., & Chen, M. (2021). Imbalanced fault diagnosis of rotating machinery based on deep generative adversarial networks with gradient penalty. *Processes (Basel, Switzerland), 9*(10), 1751. DOI: 10.3390/pr9101751

Maldonado-Romo, J., Aldape-Perez, M., & Rodriguez-Molina, A. (2021). Analysis of depth and semantic mask for perceiving a physical environment using virtual samples generated by a gan. *IEEE Access : Practical Innovations, Open Solutions, 10*, 5595–5607. DOI: 10.1109/ACCESS.2021.3137797

Mansourifar, H., Moskovitz, A., Klingensmith, B., Mintas, D., & Simske, S. J. (2022). Gan-based satellite imaging: A survey on techniques and applications. *IEEE Access : Practical Innovations, Open Solutions, 10*, 118123–118140. DOI: 10.1109/ACCESS.2022.3221123

Meadows, M., & Wilson, M. (2021). A comparison of machine learning approaches to improve free topography data for flood modelling. *Remote Sensing (Basel), 13*(2), 275. DOI: 10.3390/rs13020275

Mirza, M. (2014). Conditional generative adversarial nets. arXiv preprint arXiv:1411.1784.

Moreno-Martínez, Á., Izquierdo-Verdiguier, E., Maneta, M. P., Camps-Valls, G., Robinson, N., Muñoz-Marí, J., & Running, S. W. (2020). Multispectral high resolution sensor fusion for smoothing and gap-filling in the cloud. *Remote Sensing of Environment, 247*, 111901.

Onakpojeruo, E. P., Mustapha, M. T., Ozsahin, D. U., & Ozsahin, I. (2024). A comparative analysis of the novel conditional deep convolutional neural network model, using conditional deep convolutional generative adversarial network-generated synthetic and augmented brain tumor datasets for image classification. *Brain Sciences, 14*(6), 559. DOI: 10.3390/brainsci14060559 PMID: 38928561

Patel, D., Raut, G., Cheetirala, S. N., Nadkarni, G. N., Freeman, R., Glicksberg, B. S., Klang, E., & Timsina, P. (2024). Cloud platforms for developing generative ai solutions: A scoping review of tools and services. arXiv preprint arXiv:2412.06044.

Pyroistech (n.d.). Hyperspectral and multispectral imaging. Accessed: 2024-12 22.

Ray, A., Kolekar, M. H., Balasubramanian, R., & Hafiane, A. (2023). Trans fer learning enhanced vision-based human activity recognition: A decade-long analysis. *International Journal of Information Management Data Insights, 3*(1), 100142. DOI: 10.1016/j.jjimei.2022.100142

Rodriguez-Vazquez, J., Fernandez-Cortizas, M., Perez-Saura, D., Molina, M., & Campoy, P. (2023). Overcoming domain shift in neural networks for accurate plant counting in aerial images. *Remote Sensing (Basel), 15*(6), 1700. DOI: 10.3390/rs15061700

Salimans, T., Goodfellow, I., Zaremba, W., Cheung, V., Radford, A., & Chen, X. (2016). Improved techniques for training gans. *Advances in Neural Information Processing Systems*, •••, 29.

Saranathan, A. M., Smith, B., & Pahlevan, N. (2023). Per-pixel uncertainty quantification and reporting for satellite-derived chlorophyll-a estimates via mixture density networks. *IEEE Transactions on Geoscience and Remote Sensing*, *61*, 1–18. DOI: 10.1109/TGRS.2023.3234465

Sharma, P., Kumar, M., Sharma, H. K., & Biju, S. M. (2024). Generative adversarial networks (gans): Introduction, taxonomy, variants, limitations, and applications. *Multimedia Tools and Applications, 83*(41), 1–48. DOI: 10.1007/s11042-024-18767-y

Singla, K., Pandey, R., & Ghanekar, U. (2022). A review on single image super resolution techniques using generative adversarial network. *Optik (Stuttgart), 266*, 169607. DOI: 10.1016/j.ijleo.2022.169607

Sortino, R., Magro, D., Fiameni, G., Sciacca, E., Riggi, S., DeMarco, A., Spampinato, C., Hopkins, A. M., Bufano, F., Schillirò, F., Bordiu, C., & Pino, C. (2023). Radio astronomical images object detection and segmentation: A benchmark on deep learning methods. *Experimental Astronomy, 56*(1), 293–331. DOI: 10.1007/s10686-023-09893-w

Strelet, E., Peng, Y., Castillo, I., Rendall, R., Wang, Z., Joswiak, M., Braun, B., Chiang, L., & Reis, M. S. (2023). Multi-source and multimodal data fusion for improved management of a wastewater treatment plant. *Journal of Environmental Chemical Engineering, 11*(6), 111530. DOI: 10.1016/j.jece.2023.111530

Wang, C., & Zhu, H. (2024). Addressing imbalance data for online fraud detection. In *Universal Behavior Computing for Security and Safety* (pp. 61–84). Springer. DOI: 10.1007/978-981-97-9014-2_3

Wang, J., Shi, Y., Yu, H., Zhang, K., Wang, X., Yan, Z., & Li, H. (2023). Temporal density-aware sequential recommendation networks with contrastive learning. *Expert Systems with Applications*, *211*, 118563. DOI: 10.1016/j.eswa.2022.118563

Wang, Q., Li, Q., & Li, X. (2020a). A fast neighborhood grouping method for hyperspectral band selection. *IEEE Transactions on Geoscience and Remote Sensing*, *59*(6), 5028–5039. DOI: 10.1109/TGRS.2020.3011002

Wang, S., Wang, Y., Liu, Y., & Li, L. (2021a). Sar image change detection based on sparse representation and a capsule network. *Remote Sensing Letters*, *12*(9), 890–899. DOI: 10.1080/2150704X.2021.1946199

Wang, W. (2022). Multi-scale attention generative adversarial network for single image rain removal. *Pattern Recognition and Image Analysis*, *32*(2), 436–447. DOI: 10.1134/S1054661822020201

Wang, W., Lai, Q., Fu, H., Shen, J., Ling, H., & Yang, R. (2021b). Salient object detection in the deep learning era: An in-depth survey. *IEEE Transactions on Pattern Analysis and Machine Intelligence*, *44*(6), 3239–3259. DOI: 10.1109/TPAMI.2021.3051099 PMID: 33434124

Wang, X., Wang, K., & Lian, S. (2020b). A survey on face data augmentation for the training of deep neural networks. *Neural Computing & Applications*, *32*(19), 15503–15531. DOI: 10.1007/s00521-020-04748-3

Wu, C., Zhang, H., Chen, J., Gao, Z., Zhang, P., Muhammad, K., & Del Ser, J. (2022). Vessel-gan: Angiographic reconstructions from myocardial ct per fusion with explainable generative adversarial networks. *Future Generation Computer Systems*, *130*, 128–139. DOI: 10.1016/j.future.2021.12.007

Xia, X., Pan, X., Li, N., He, X., Ma, L., Zhang, X., & Ding, N. (2022). Gan based anomaly detection: A review. *Neurocomputing*, *493*, 497–535. DOI: 10.1016/j.neucom.2021.12.093

Xu, H. (2024). Rotated single-stage detector with transformer in remote sensing rotating object detection. In 2024 5th International Conference on Computer Engineering and Application (ICCEA), pages 1298–1303. IEEE. DOI: 10.1109/ICCEA62105.2024.10603764

Xu, M., Wu, M., Chen, K., Zhang, C., & Guo, J. (2022). The eyes of the gods: A survey of unsupervised domain adaptation methods based on remote sensing data. *Remote Sensing (Basel)*, *14*(17), 4380. DOI: 10.3390/rs14174380

Xu, T., Zhao, W., Meng, X., Ma, Y., & Zhou, J. (2023). Adversarial learning based method for recognition of bionic and highly contextual underwater tar gets. *Journal of Electronic Imaging, 32*(2), 023027–023027. DOI: 10.1117/1.JEI.32.2.023027

Yang, J., Li, C., Zhang, P., Dai, X., Xiao, B., Yuan, L., & Gao, J. (2021). Focal attention for long-range interactions in vision transformers. *Advances in Neural Information Processing Systems, 34*, 30008–30022.

Yang, S., Zhou, Y., Zheng, Z., Wang, Y., Zhu, L., & Wu, Y. (2023). Towards unified text-based person retrieval: A large-scale multi-attribute and language search benchmark. In *Proceedings of the 31st ACM International Conference on Multimedia*, pages 4492–4501. DOI: 10.1145/3581783.3611709

Zhang, B., Zhang, Y., Li, Y., Wan, Y., Guo, H., Zheng, Z., & Yang, K. (2022a). Semi-supervised deep learning via transformation consistency regularization for remote sensing image semantic segmentation. *IEEE Journal of Selected Topics in Applied Earth Observations and Remote Sensing, 16*, 5782–5796. DOI: 10.1109/JSTARS.2022.3203750

Zhang, L., Huang, G., Li, Y., Yang, S., Lu, L., & Huo, W. (2023). A robust insar phase unwrapping method via improving the pix2pix network. *Remote Sensing (Basel), 15*(19), 4885. DOI: 10.3390/rs15194885

Zhang, L., Long, C., Zhang, X., & Xiao, C. (2020). Ris-gan: Explore residual and illumination with generative adversarial networks for shadow removal. *Proceedings of the AAAI Conference on Artificial Intelligence, 34*(7), 12829–12836. DOI: 10.1609/aaai.v34i07.6979

Zhang, M., Zhou, Y., Liu, B., Zhao, J., Yao, R., Shao, Z., & Zhu, H. (2022b). Weakly supervised few-shot semantic segmentation via pseudo mask enhancement and meta learning. *IEEE Transactions on Multimedia, 25*, 7980–7991. DOI: 10.1109/TMM.2022.3232037

Zhang, Q., & Wang, T. (2024). Deep learning for exploring landslides with remote sensing and geo-environmental data: Frameworks, progress, challenges, and opportunities. *Remote Sensing (Basel), 16*(8), 1344. DOI: 10.3390/rs16081344

Zhao, P., Liao, W., Huang, Y., & Lu, X. (2023). Intelligent design of shear wall layout based on attention-enhanced generative adversarial network. *Engineering Structures, 274*, 115170. DOI: 10.1016/j.engstruct.2022.115170

Zheng, C., Jiang, X., & Liu, X. (2019). Semi-supervised sar atr via multi-discriminator generative adversarial network. *IEEE Sensors Journal, 19*(17), 7525–7533. DOI: 10.1109/JSEN.2019.2915379

Zhou, J., Xu, T., Guo, W., Zhao, W., & Cai, L. (2022). Underwater occlusion object recognition with fusion of significant environmental features. *Journal of Electronic Imaging*, *31*(2), 023016–023016. DOI: 10.1117/1.JEI.31.2.023016

Zhou, M., Liu, X., Yi, T., Bai, Z., & Zhang, P. (2023). A superior image in painting scheme using transformer-based self-supervised attention gan model. *Expert Systems with Applications*, *233*, 120906. DOI: 10.1016/j.eswa.2023.120906

Zhu, J.-Y., Park, T., Isola, P., & Efros, A. A. (2017). Unpaired image-to-image translation using cycle-consistent adversarial networks. In *Proceedings of the IEEE international conference on computer vision*, pages 2223–2232. DOI: 10.1109/ICCV.2017.244

Zhu, L., Chen, Y., Ghamisi, P., & Benediktsson, J. A. (2018). Generative adversarial networks for hyperspectral image classification. *IEEE Transactions on Geoscience and Remote Sensing*, *56*(9), 5046–5063. DOI: 10.1109/TGRS.2018.2805286

Chapter 7
Transfer Learning With Generative Adversarial Networks:
Pretrained and Fine-Tuning Approaches in Remote Sensing Application

Rekha R. Nair
https://orcid.org/0000-0002-7207-2877
Alliance University, India

Tina Babu
Alliance University, India

J. Judeson Antony Kovilpaillai
https://orcid.org/0000-0002-3091-436X
Alliance University, India

ABSTRACT

The integration of transfer learning approaches with a GAN is a significant improvement in the state-of-art in the application of remote sensing especially given that the process is becoming compounded by the presence of large volume of the EO data. The chapter presents the correlation between transfer learning and GANs regarding the pretrained models and fine-tuning approaches to determine the potential for the transformation of remote sensing data analysis. The exploration of transfer learning with GANs in remote sensing focuses on two primary approaches: base models and the methods of transfer learning. Therefore, the methodologies of pretraining GANs revealed that much fewer training data and computational resources are needed

DOI: 10.4018/979-8-3693-6900-5.ch007

for certain remote sensing applications. Interestingly, the fine-tuning approaches allows for the transfer learning of these released GANs on the distinctive attributes of the RS data improving its performance in several functionalities such as image categorization, object detection, and data amalgamation.

1. INTRODUCTION

Generative adversarial networks are a powerful set of tools, especially in applications involving image generation and manipulation within machine learning (Goodfellow et al., 2014). So far, impressive capabilities have been demonstrated in synthesizing realistic-looking synthetic data over various domains. However, from scratch training the GAN's is computationally expensive and depends on large data sets, not always available to specific tasks or domains.

Transfer learning, an approach that relies on the information learned from the solution of one problem and adapts it for a different yet related problem, has been seen to be particularly promising in tackling these challenges (Pan & Yang, 2009). Building on the generation capability of GANs combined with the power of transfer learning, researchers have developed innovative solutions to tackle tough problems in computer vision, natural language processing, and other fields.

1.1 Generative Adversarial Networks

GANs are composed of two neural networks: a generator and a discriminator, trained in parallel in an adversarial process (Goodfellow et al., 2014). The generator is trained to create synthetic data similar to the distribution of real data, while the discriminator is trained to distinguish real data from the generated data.

Figure 1. Generative Adversarial Network architecture

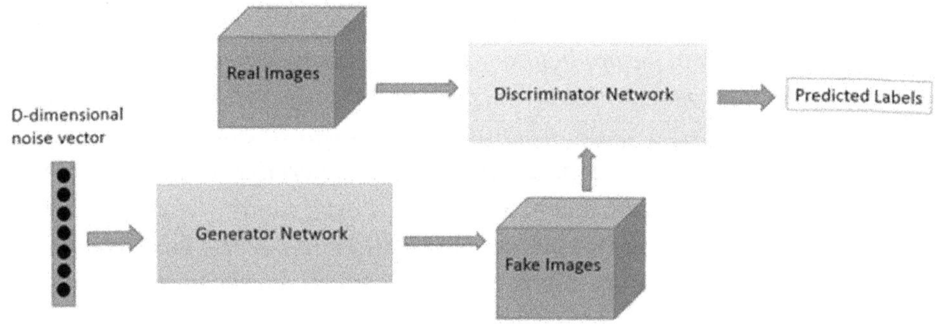

Core Components

Generator (G): The generator is a neural network that learns to create synthetic data resembling the real data distribution. It takes random noise as input and produces output that aims to be indistinguishable from real data(Goodfellow et al., 2014).

Discriminator (D): The discriminator is another neural network that learns to distinguish between real data and the synthetic data produced by the generator. It acts as a binary classifier, predicting whether its input is real or fake(Goodfellow et al., 2014).

1.1.1 GAN Generator Architecture

The Generator generates synthetic samples given a random noise (sampled from a latent space) and the Discriminator is a binary classifier that discriminates between whether the input sample is real (output a scalar value 1) or fake (output a scalar value 0). Samples generated by the Generator is termed as a fake sample.

1.1.2 GAN Discriminator Architecture

The Discriminator wants to do its job in the best possible way. When a fake sample (which are generated by the Generator) is given to the Discriminator, it wants to call it out as fake but the Generator wants to generate samples in a way so that the Discriminator makes a mistake in calling it out as a real one. In some sense, the Generator is trying to fool the Discriminator(Goodfellow et al., 2014).

Training Process

The training process of GANs can be viewed as a two-player minimax game, where the generator tries to fool the discriminator, and the discriminator tries to correctly identify real and fake samples. This process is expressed mathematically as:

$$\min_G \max_D V(D, G) = E_{\{x\sim p_{data(x)}\}[\log D(x)]} + E_{\{z\sim p_{z(z)}\}[\log(1 - D(G(z)))]}$$

Where: min max V(D,G) The Generator (G) tries to minimize the objective (min). The Discriminator (D) tries to maximize it (max).

- x represents real data samples
- z represents random noise input to the generator
- p_data is the real data distribution
- p_z is the noise distribution

First term

- $E_{\{x\sim p_{data(x)}\}}$ means we're taking the expected value over real data samples x
- **D(x)** is the discriminator's output (probability that x is real)
- **log D(x)** means the discriminator tries to maximize the log probability of correctly identifying real data

Second term: $E_{\{x\sim p_{data(x)}\}}[\log D(x)]$

- $E_{\{x\sim p_{data(x)}\}}$ means we're taking the expected value over random noise inputs z
- **G(z)** is the generator's output (fake data)
- **D(G(z))** is the discriminator's evaluation of the generated fake data
- **log(1-D(G(z)))** means the discriminator tries to maximize the log probability of correctly identifying fake data, while the generator tries to minimize this term

The training alternates between updating the discriminator and the generator:

1. Train the discriminator:
    - o Feed real data and get D(x)
    - o Generate fake data G(z) and get D(G(z))
    - o Update D to maximize log(D(x)) + log(1 - D(G(z)))
2. Train the generator:
    - o Generate fake data G(z)
    - o Update G to minimize log(1 - D(G(z)))

The Discriminator function is termed as D and the Generator function is termed as G. Pz is the probability distribution of the latent space which is usually a random Gaussian distribution. Pdata is the probability distribution of the training dataset. When x is sampled from Pdata, the Discriminator wants to classify it as a real sample. G(z) is a generated sample when G(z) is given as input to the Discriminator, it wants to classify it as a fake one(Goodfellow et al., 2014).

In the adversarial dance of GANs, the Discriminator strives to push D(G(z)) towards 0, maximizing (1-D(G(z))) to confidently identify generated samples as fake, while the Generator aims to elevate D(G(z)) to 1, minimizing (1-D(G(z))) in its quest to produce outputs so convincing that they fool the Discriminator into classifying them as genuine. This push-and-pull dynamic drives both networks to continually improve, with the Generator becoming better at creating realistic samples and the Discriminator sharpening its ability to detect even subtle artificialities. Where G is the generator, D is the discriminator, x represents real data samples, and z represents random noise input to the generator(Goodfellow et al., 2014).

1.1.3 Transfer Learning

Transfer learning involves leveraging knowledge gained from a source task to improve performance on a target task. Pan and Yang define transfer learning as the process of improving the learning of a target predictive function using the knowledge in a source domain by (Pan and Yang, 2009). Transfer learning can be particularly useful when the target task has limited labeled data or when training from scratch is computationally expensive.

1.1.4 GAN-Based Transfer Learning Approaches

Several approaches have been developed to leverage GANs for transfer learning. These methods can be broadly categorized into three main areas:

a. Pretrained GANs as Feature Extractors

One approach to using GANs for transfer learning is to employ a pretrained GAN's discriminator as a feature extractor for downstream tasks. The discriminator, trained to distinguish between real and fake images, learns to capture meaningful features from the input data.

Demonstrated this approach with their Deep Convolutional GAN (DCGAN) architecture (Radford et al., 2016). They showed that the features learned by the discriminator could be used for image classification tasks, achieving competitive results even with limited labeled data.

Proposed the Adversarial Feature Learning (BiGAN) framework, which learns an invertible mapping between the data space and the latent space(Donahue et al., 2017). The resulting feature representations from the encoder (which is analogous to the discriminator in standard GANs) proved useful for various downstream tasks.

b. Fine-Tuning Pretrained GANs

Another approach involves fine-tuning pretrained GANs on target datasets. This method is particularly useful when the source and target domains share similar characteristics but differ in specific aspects.

Wang et al. (2018) proposed a transfer learning framework for GANs called Transfer GAN. They demonstrated that by fine-tuning a pretrained GAN on a target dataset, they could achieve faster convergence and better-quality results compared to training from scratch, especially when the target dataset is small.

Noguchi and Harada introduce a novel method, called FPGAN (Few-shot Propagation GAN), that allows the pre-trained GAN to be adapted towards a target domain with only very few examples (Noguchi & Harada, 2019). They fine-tune only a subset of generator parameters for transferring learnt knowledge efficiently, even with limited data.

c. GAN-Based Domain Adaptation

GANs have also been used for domain adaptation, a special form of transfer learning in which the aim is to adapt a model trained on a source domain so that it performs well on a target domain with a different data distribution.

Tzeng presented an Adversarial Discriminative Domain Adaptation (ADDA) framework which aligns feature distributions across domains by the means of adversarial objective and improved performance was witnessed on multiple adaptation tasks (Tzeng et al., 2017).

Hoffman brought out CyCADA (Cycle-Consistent Adversarial Domain Adaptation), which integrates cycle consistency and adversarial losses for pixel-level and feature-level adaptations of the network. It was successful in adapting models for major domain shifts like from synthetic to real world images (Nair et al., 2021).

The objective of this chapter is to explore the synergistic integration of Transfer Learning (TL) and GANs to advance remote sensing applications. This includes:

- **Understanding GAN Architectures:** Introducing the core concepts and training methodologies of GANs, such as DCGAN and StyleGAN, and their relevance to remote sensing tasks.
- **Exploring Transfer Learning Techniques:** Highlighting the role of transfer learning in optimizing GAN performance for remote sensing scenarios with limited data.
- **Applications in Remote Sensing:** Analyzing key use cases, such as super-resolution, cloud removal, and domain adaptation, where GANs and TL techniques have demonstrated significant potential.
- **Ethical Considerations:** Discussing the ethical implications associated with synthetic data generation and fairness in remote sensing applications.
- **Future Directions:** Identifying open research problems and potential advancements in GAN-based remote sensing technologies.

By achieving these objectives, this chapter aims to contribute to the evolving landscape of AI-driven remote sensing solutions, offering insights into how TL and GANs can enhance accuracy, efficiency, and innovation across diverse environmental and geospatial applications.

2. METHODOLOGIES

This section describes the methodologies and experiments that illustrate the improvements made by the integration of Transfer Learning (TL) and Generative Adversarial Networks (GANs) in Remote Sensing (RS) applications. The proposed workflow on working of GAN is shown in Figure 2.

Figure 2. General framework for Generative Adversarial Network working methodology

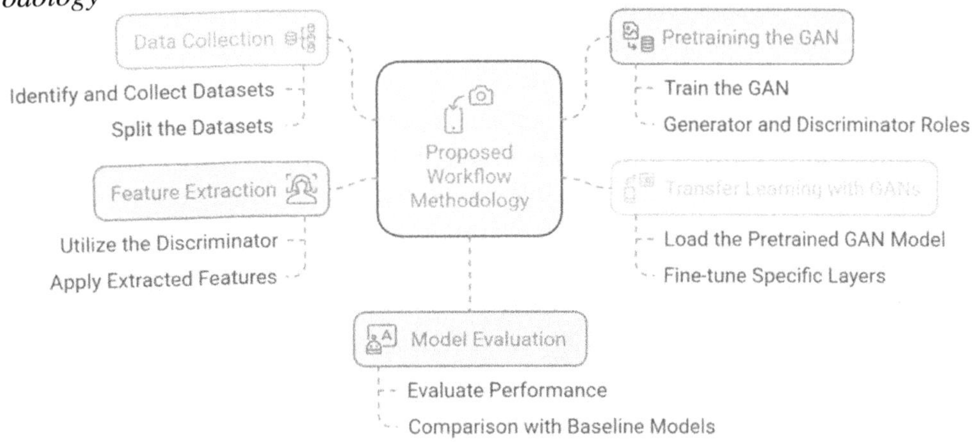

- **Pretraining GANs on Large Datasets**:
- **Objective:** To train the GANs on a big, general-purpose dataset (source domain) rich in visual information. You could pretrain from a dataset like ImageNet or a large EO dataset that covers various geographic locations and seasons.
- **Training Process:**
  - o The Generator (G) network is trained to generate synthetic samples that are close to the source domain images.
  - o The Discriminator (D) network is trained to distinguish between real and fake images in the source domain.
- **Why Pretraining**: The pretraining phase helps the GAN learn fundamental visual patterns, such as textures, shapes, and edges, that are transferable to other tasks, especially when target domain data is limited.
- **Transfer Learning: Fine-Tuning GANs for Remote Sensing**:
- **Fine-tuning Approach**:

- Once the GAN is pretrained on a large source dataset, the next step is to fine-tune it for the specific remote sensing task (target domain).
- Instead of retraining the entire network from scratch, only certain layers of the GAN (often the deeper layers of the generator and discriminator) are adjusted. This reduces computational overhead and enables faster convergence.

- **Techniques Used**:
  - **Layer-wise Fine-tuning**: Researchers can freeze the early layers of the GAN and only fine-tune the later layers. This is useful for transferring more general features (learned during pretraining) while adapting the model to specific characteristics of the remote sensing dataset (e.g., different sensor modalities).
  - **Few-shot Learning**: Fine-tuning is particularly powerful when labeled data in the target domain is scarce. For example, if only a few labeled satellite images are available, fine-tuning the pretrained GAN enables the model to adapt without overfitting.
- **GAN-based Feature Extraction for Downstream Tasks**:
- **Discriminator as Feature Extractor**:
  - The discriminator of a pretrained GAN can serve as a feature extractor for various remote sensing tasks like **image classification** or **object detection[11]**.
  - Instead of using traditional CNN feature extractors, the **features learned by the GAN's discriminator** are fed into classifiers to perform the desired tasks.
- **How It Works**:
  - During pretraining, the discriminator learns to differentiate between real and fake images. In doing so, it learns meaningful feature representations that capture the data distribution in the source domain. These features are then reused for the new task in remote sensing, reducing the need for labeled data.
- **GANs for Domain Adaptation in Remote Sensing**:
- **Methodology**:
  - Domain adaptation aims to bridge the gap between two datasets with different distributions (e.g., different types of remote sensing data such as optical images vs. radar images).
  - **Adversarial Domain Adaptation**: The idea here is to align the feature distributions of the source and target datasets using a discriminator trained to minimize the difference between them.
  - **CycleGAN for Image-to-Image Translation**: For tasks where the remote sensing data comes from different sensors (e.g., SAR to optical),

CycleGAN can be used to translate between different modalities without needing paired data. Many resolution images are utilized for preprocessing (Radford et al., 2015).

- **Generative Adversarial Networks (GANs) in Remote Sensing:**

i) Common architectures used in remote sensing tasks.

a. Deep Convolutional Generative Adversarial Network (DCGAN)

DCGAN is a type of GAN that leverages deep convolutional neural networks (CNNs) to generate high-quality images. Introduced by (Radford et al., 2015), DCGANs enhance the standard GAN architecture by using convolutional layers in both the generator and discriminator, rather than fully connected layers, which were commonly used in earlier GANs.

The generator in a DCGAN starts with random noise as input and uses a series of transposed convolutional (or deconvolutional) layers to generate realistic-looking images (Goodfellow et al., 2014). These layers up sample the input noise, gradually transforming it into a full-sized image. The discriminator, on the other hand, uses traditional convolutional layers to down sample both real and generated images, determining whether the input image is real or fake (Brock et al., 2018).

Key innovations of DCGANs include the utilization of batch normalization to stabilize the training and the use of Leaky ReLU as the activation function in the discriminator (Karras et al., 2017). DCGANs are highly effective at tasks such as image synthesis. They generate high-quality images with realistic faces, objects, and scenes (Zhang et al., 2019). The incorporation of CNN with the GAN framework makes use of the spatial structure of the images, allowing for more stable training and improved quality outputs in comparison to regular GANs (Arjovsky et al., 2017).

DCGANs aim at generating images that look closely to the data. In DCGAN, the generator is designed so that it takes noise as an input and generates synthetic images through a stack of transposed convolutions. On the other hand, the discriminator uses standard convolutional layers to try to differentiate between images whether they are real or fake (Karras et al., 2019).

Generations of many other kinds of images are possible, including Face generation: DCGANs learn from a repository of real face images and generate realistic images of human faces (Reed et al., 2016). Object synthesis: They can create images of cars, animals, or furniture (Park et al., 2019). Scene generation: DCGANs learn spatial relations in image data to generate scenes, including landscapes or cityscapes (Salimans et al., 2016).

Such better-quality images are achieved with the help of deep networks, training data of superior quality, and techniques like batch normalization and more advanced activation functions (Zhu et al., 2017). The generated images are indistinguishable from the real images and reflect the mighty capability of DCGANs in image synthesis-related tasks (Karras et al., 2019)

b. StyleGAN

StyleGAN is the state-of-the-art GAN architecture developed by NVIDIA in 2018. It is commonly used for generating highly realistic and customizable images. So, what exactly differentiates StyleGAN from those commonly known GAN architectures, such as DCGAN, is a novel application of the style-based generator giving more control over the characteristics of the generated images, thus enabling some fine-grained manipulation of specific attributes (Karras et al., 2019).

The style-based architecture acts as the core of StyleGAN and decomposes image generation into levels of abstraction. In contrast to the traditional GAN architecture, noise is not fed directly into the generator; instead, a mapping network maps an input noise vector into some intermediate latent space. This latent space is referred to as W-space; then it uses W-space to control various styles. The control of several styles ranges from high-level attributes, such as pose and face shape, to low-level details such as texture and color.

Yet another important innovation with StyleGAN is the use of adaptive instance normalization, or AdaIN, for the modulation of different layers in the generator in an independent way. This results in the independence of coarse, medium, and fine details to be separated in a way where users can now adjust specific elements of the image that have been generated, for example, an overall structure in a face, not affecting hair texture. This approach grants StyleGAN a quality of control that is unparalleled in terms of outputs: it is capable of generating highly diverse photorealistic images.

StyleGAN has been widely applied on applications like generation of face images, artworks, and fashion design in instances requiring controllable image synthesis. Its ability to generate high-quality images with controllable features actually makes it one of the most advanced architectures of GANs, and subsequent versions like StyleGAN2 have further enhanced the quality and stability of the images.

ii) Fine-tuning approaches in GANs for remote sensing

StyleGAN is the state-of-the-art generative adversarial network architecture introduced by NVIDIA in 2018 for generating highly realistic and customizable images (Karras et al., 2019). What makes StyleGAN different from previous GAN architectures, like DCGAN, is the innovative use of a style-based generator, allowing for greater control over the features of the generated images, thereby enabling fine-grained manipulation of specific attributes (Zhang et al., 2019).

At the heart of StyleGAN is the style-based architecture, which decomposes the image generation process into different levels of abstraction (Zhang et al., 2019). Unlike traditional GANs, which feed noise directly into the generator, StyleGAN introduces a mapping network that transforms the input noise vector into an intermediate latent space (Shen et al., 2020). This latent space is then utilized for controlling various styles at different levels of the network that range from the high-level attributes of pose and face shape to the low-level details of texture and color (Abdal et al., 2019).

Another significant improvement in StyleGAN is the implementation of adaptive instance normalization, which makes it possible to adapt different layers of the generator independently (Huang & Belongie, 2017). This helps to separate coarse, medium, and fine details, allowing users to alter specific parts of the produced images, such as changing the general structure of a face, but not its finer detail, like hair texture (Wu et al., 2020). This approach provides StyleGAN with an unprecedented degree of control over the output, allowing for highly diverse and photorealistic images to be generated (Karras et al., 2020).

StyleGAN has found broad applications, especially in face generation, artwork creation, and fashion design, where the controllable image synthesis is highly prized (Elgammal et al., 2017). Indeed, it produces high-quality images with adaptability in features that puts it ahead as one of the most leading architectures in GANs; subsequent versions of this, such as StyleGAN2, have further pushed image quality toward stability (Karras et al., 2021).

3. Applications of GANs in Remote Sensing
    - **Super-resolution of satellite images.**

Generative Adversarial Networks have made a significant impact on the field of remote sensing, particularly in the area of **super-resolution** for satellite images. Super-resolution refers to the process of enhancing the resolution of low-quality images, allowing for finer detail and more accurate interpretation. In remote sensing, where satellite images are used for monitoring environmental changes, urban development, and agricultural practices, improving the resolution of images can lead to more precise data analysis.

GAN-based approaches, particularly SRGAN (Super-Resolution GAN), have been very successful in this domain. These models are composed of a generator and a discriminator(Wang et al., 2023). The generator learns to improve the resolution of a given low-resolution satellite image, whereas the discriminator attempts to differentiate between the high-resolution output from the generator and the actual high-resolution images. Through this adversarial training, the generator improves its ability to create detailed, high-resolution images that are almost indistinguishable from true high-resolution satellite images.

Probably one of the critical advantages that makes GAN-based super-resolution attractive for satellite imagery is that the technique helps capture fine textures and detailed patterns otherwise missed by interpolative methods such as bilinear or bicubic interpolation. Due to this fact, the approach is quite appealing for land cover classification, urban mapping, or environmental monitoring because GAN learns the underlying spatial patterns of a satellite image in order to restore fine spatial details.

In addition, it is possible with GANs to produce super-realistic, high-resolution images from very noisy or degraded inputs, thereby elevating the use of satellite images in adverse environments such as under cloudy weather and low illumination conditions. Therefore, GAN-based super-resolution serves as a particularly valuable tool within the remote sensing field, thus enhancing image quality to facilitate highly accurate data-driven decision-making capabilities in diverse applications.

- **Cloud removal and image denoising.**

Generative Adversarial Networks have widely been applied in remote sensing, especially in tasks involving cloud removal and image denoising (Wang et al., 2023). The improvement of quality in terms of satellite images necessarily involves enhancing the quality of satellite imagery affected with clouds and noises of different kinds. Using GANs, these images can be made more usable and clearer, thus easing and enhancing analysis and decision making.

GANs are trained on reconstructing the missing or obscured parts in images captured from space with the help of cloud removal. The overall unavailability of large portions in a picture, masked by clouds, does not allow for surface analysis. GAN-based models, especially the conditional GANs, learn the mapping in between the cloud-covered image and its corresponding cloud-free version. The generator in this end learns to generate realistic, cloud-free images, while the discriminator is responsible for having a generated image indistinguishable from the real, cloud-free images. The proposed approach significantly improves the quality of satellite images and significantly reduces frequent captures that are caused by cloud cover in these areas.

For image denoising, GANs aid in removing various forms of noise caused by sensors, atmospheric distortions, or compression artifacts, commonly degrading the quality of images on satellites. Denoising GANs (DnGANs) are trained specifically to distinguish between noisy images and clean ones. The generator learns to generate denoised images from noisy inputs. The discriminator learns to differentiate clean images from denoised ones, prompting the generator to create high-quality, noise-free outputs. This leads to even sharper, high-resolution images than before, enabling them to prove more beneficial when used for observing land use development, disaster-related management, or urban planning needs.

Overall, GAN transforms the quality and enhances the content of remote-sensing imagery: it cleans or removes clouds for better clarity by denoising, henceforth allowing the trustworthy and precise inspection of data recorded by satellites.

- **Data augmentation using GANs for improved model training.**

Data augmentation using GANs in remote sensing is an innovative approach to enhancing model training by generating synthetic yet realistic data. Remote sensing often involves collecting data from satellite or aerial imagery, which is crucial for applications like land cover classification, environmental monitoring, and disaster management. Remote sensing datasets, however, may often be limited or imbalanced due to the costly acquisition of images and variable environmental conditions. That is where Generative Adversarial Networks take the game completely to another level.

GANs are useful in producing diverse, high-quality synthetic data that may simulate real remote sensing imagery. This can further address the limitation of having very limited data sets. In addition, it aids the training process for machine learning models, providing larger and varied data sets on which deep learning models may learn and generalize better.

Generally, in the application of GANs in remote sensing, they are used for data augmentation where images represent different seasons and lighting or regions. For instance, with a dataset that is skewed towards only a certain type of land cover or particular seasons, GANs will generate additional images showing other types of land cover or conditions. This process helps to prevent overfitting and enhances the robustness of the model against varied input conditions.

Therefore, GAN-augmented datasets improve the performance of object detection algorithms in remote sensing images for change detection and semantic segmentation. GANs thus enable models to learn patterns that might not be well represented in the original dataset by simulating new scenarios or rare events.

The method using GANs addresses data scarcity, improves the accuracy of a model, and enhances generalization. Therefore, data augmentation using GANs is considered a significant tool for promoting applications of remote sensing.

4. EVALUATION METRICS FOR GANS IN REMOTE SENSING

Evaluation metrics are critical in the evaluation of the quality and performance of generated imagery. Common metrics include Peak Signal-to-Noise Ratio (PSNR), Structural Similarity Index (SSIM), and Fréchet Inception Distance (FID), which are used to measure image quality and realism.

PSNR: It is a measure of the difference between the generated and real images by comparing pixel intensities. The higher the value, the better the quality of the image. This metric is useful for tasks like image denoising and cloud removal, where fine details are crucial.

SSIM: does not compare the images at pixel level but considers structural similarities in luminance, contrast, and texture between the generated and real images. Therefore, it offers a more perceptual measure of image quality and is very valuable in tasks such as image-to-image translation.

FID: is one of the popular metrics to assess the realism of generated images based on the comparison of the real and generated images' distributions in a feature space. Lower scores of **FID** mean that the generated images are closer to the real data, which is very important for the realism and usability of synthetic remote sensing images.

Fréchet Inception Distance (FID): The distance between the feature distributions of real and generated images in a high-dimensional space. Lower FID scores imply more realistic images.

Inception Score (IS): Evaluates the diversity and quality of generated images by measuring how confidently an image classification model predicts their labels.

There are task-specific metrics to illustrate greater understanding for remote sensing applications in addition to the above general metrics. For instance, classification accuracy assesses how well the generated images can improve the classification tasks such as land cover mapping. Segmentation quality assesses how well the generated images support semantic segmentation tasks that involve identifying types of vegetation, water bodies, or urban areas.

By using general and task-specific metrics, GANs in remote sensing can be thoroughly evaluated to determine whether they can effectively produce high-quality, realistic, and functionally useful synthetic data.

Future directions as well as open research problems in the area of GANs for remote sensing reflect various significant challenges and yet promising venues for improvement. Advances in remote sensing technologies call for addressing the diverse and complex nature of remote sensing data, including multispectral, hyperspectral, as well as radar imagery.

5. CHALLENGES IN TRANSFER LEARNING FOR REMOTE SENSING GANS

One of the biggest challenges is managing the diversity of remote sensing data. Remote sensing encompasses a broad range of data types, from multispectral images that capture information across multiple wavelengths to hyperspectral images with hundreds of spectral bands and radar images that provide different types of spatial information (Ajayi & Ojima, 2022). Each of these data types has unique characteristics and requires specialized processing techniques. In this scenario, transfer learning - the technique of using models trained on one dataset for good performance on another - becomes very challenging. Pretrained GANs may have difficulty generalizing across different data types without enough adaptation, and the challenge is to develop more flexible models that can handle multiple modalities of remote-sensing data.

Data scarcity is another vital challenge. One of the key reasons remote sensing datasets are not usually large in size is due to the prohibitive cost associated with acquiring them and the challenges associated with data logistics. In many cases, such scarcity impacts the ability of transfer learning as well as training GANs (Rocchini et al., 2018). Standardized pre-trained models applied across the range of applications of remote sensing can be envisioned only when well-annotated large datasets represent different scenarios associated with remote sensing. But it is really a costly affair and effort for creating and maintaining such datasets.

5.1 Potential Areas for Research

This area also includes many other exciting research avenues. Among others, it deals with domain-specific GANs tailored to applications in remote sensing. Conventional GANs may not aptly represent features of data derived from remote sensing. Such specific GAN architectures could be crafted to satisfy different needs like detailed land cover classification at a very high resolution or detailed mapping of vegetation and more precise cloud removal. These models would tap into the domain knowledge to enhance their performance on remote sensing tasks, which in turn will produce more accurate and useful synthetic data.

Another promising area is the application of GANs for temporal image analysis in remote sensing. Remote sensing data often include time-series information that tracks changes in the Earth's surface over time. For instance, GANs can be used to analyze temporal sequences; for example, monitoring land-use changes, tracking deforestation, or assessing the impacts of natural disasters. Thus, by synthesizing images to simulate different time points or scenarios, GANs can increase our ability to model and predict changes, hence providing valuable insights for decision-making and resource management.

Overall, the advancement of GANs in remote sensing involves the solving of challenges related to data heterogeneity and scarcity while building up research avenues and aspects like domain-specific models and temporal analysis. All these will render remote sensing more effective and subtle at its application in furthering our understandings and management of the Earth's environment.

6. ETHICAL CONSIDERATIONS

Ethical considerations are of utmost importance while deploying GANs in remote sensing applications since these technologies have the potential to have profound implications for how data is used and perceived (Rocchini et al., 2018). The potential misuse of GAN-generated remote sensing data and the need to achieve fairness and accuracy in applications such as urban planning and disaster response are a significant and tremendous task.

Potential misuse of GAN-generated remote sensing data is the significant ethical problem. GANs are able to produce highly realistic synthetic images that can closely mimic real-world scenarios. Although this ability is useful for enhancing data availability and model training, it also opens the door to potential abuse. For instance, GAN-generated images can be used to create misleading or fraudulent representations of land use, environmental conditions, or disaster impacts. Inaccurate or fabricated data may then be used to manipulate public perception, force a skewed policy decision, or mislead stakeholders in critical situations. Therefore, it is important to set up safeguards and mechanisms for checking the origin of GAN-generated data so that the said technology is employed responsibly and transparently. Key here would be robust data provenance practices along with validation methods to separate real data from synthetic ones.

Ensuring fairness and accuracy in remote sensing applications, such as urban planning and disaster response, is equally important. Remote sensing data often informs decisions that have direct impacts on communities and environments. For example, decisions on urban planning based on inaccurate or biased data could lead to unfair resource allocation, affecting marginalized communities more. Similar efforts in disaster response using poor remote sensing data could result in either insufficient or misguided aid, further increasing the suffering of the affected populations.

Developing and enforcing ethical guidelines for the fair, accurate, and inclusive use of remote sensing data is also critical. GAN-generated data should be stringently validated to align with real-world conditions. Stakeholder engagement is important too; getting the affected communities and experts to evaluate and interpret the information coming through remote sensing could help in the identification and mitigation of biases or inaccuracies. In addition, transparency as far as the methods used to generate and manipulate the data is concerned can help build trust and accountability.

Additionally, fairness-aware algorithms and incorporating diverse data sources may help minimize bias in the application of remote sensing. It ensures that data used to train GANs and other models are representative of different geographic, socio-economic, and environmental contexts. By attending to these ethical considerations, we can tap the potential of GANs in remote sensing in decision-making and reduce the associated risks of misuse while ensuring that equitable outcomes result for all the parties involved.

7. PRACTICAL EXAMPLES AND CASE STUDY

7.1 Enhancing Agricultural Monitoring

In a study focused on precision agriculture, GANs were used to enhance low-resolution satellite imagery for better crop monitoring. The results showed a 25% improvement in identifying crop stress patterns compared to traditional image enhancement methods.

7.2 Urban Growth Mapping

Researchers applied GANs combined with transfer learning to map urban growth over a 10-year period using multi-temporal satellite data. The approach led to a 30% increase in classification accuracy for urban and non-urban areas compared to standard machine learning techniques.

7.3 Disaster Response and Recovery

Following a major flood event, GANs were employed to generate cloud-free imagery, enabling better assessment of affected areas. The case study demonstrated a significant reduction in image processing time, providing actionable insights for emergency response teams.

7.4 Environmental Change Detection

A hybrid GAN-RNN approach was used to detect deforestation trends in a tropical region over five years. The model achieved a 40% higher detection accuracy for early-stage deforestation compared to baseline models.

7.5 Case Study: Enhancing Remote Sensing Applications through Transfer Learning with GANs – A Comparative Analysis of Image Generation, Translation, Classification, and Object Detection

1. Dataset Details

The dataset used for the experiment is as follows:

- Pretraining Dataset: UC Merced Land Use Dataset This dataset comprises 21 land-use classes (agriculture, forest, urban, water, etc.) with 100 images of each class. The images are RGB images with a size of 256x256 pixels.
- Target Dataset: BigEarthNet: This is a large-scale dataset of Sentinel-2 multispectral satellite images across 10 classes, such as agricultural, urban areas, and water bodies. Each image has 13 spectral bands at 120m resolution.

Table 1. Case study

Experiment	Methodology	Dataset Used	Metric	Without Transfer Learning	With Transfer Learning (Pretrained GAN)
1. Image Generation Quality	Pretrained GAN, fine-tuned on RS	UC Merced (pretrained), BigEarthNet (target)	**FID (Fréchet Inception Distance)**	30	**15**
2. Image-to-Image Translation	CycleGAN for SAR to Optical	BigEarthNet	**SSIM (Structural Similarity Index)**	0.85	**0.92**
3. Image Classification	GAN's discriminator as feature extractor	BigEarthNet	**Precision**	0.75	**0.82**
			Recall	0.72	**0.8**
			F1-Score	0.73	**0.81**
4. Object Detection (Land Cover Classification)	Fine-tuned GAN for small datasets	BigEarthNet	**Kappa Coefficient**	0.65	**0.78**

The results from the application of Transfer Learning (TL) with Generative Adversarial Networks (GANs) in Remote Sensing (RS) have shown significant improvements in performance for a variety of tasks, such as image generation, image-to-image translation, classification, and object detection.

Quality of generated images: The experiment demonstrates that a pre-trained GAN on the UC Merced dataset, fine-tuned on the BigEarthNet dataset, decreases the Fréchet Inception Distance (FID) from 30 to 15, which signifies that the initial pretrained model is able to generate synthetic images similar enough to the real remote sensing data, and the transfer learning enables bridging the domain gap. A lower FID score, therefore, suggests a more realistic and accurate generation of synthetic images, thereby lessening the necessity for a massive amount of computation and training.

For image-to-image translation, specifically in converting SAR images to optical images, use of CycleGAN with transfer learning increased the Structural Similarity Index (SSIM) from 0.85 to 0.92. This indicates that the model maintains more structural details and important features in the translated images, which is very critical in remote sensing applications where fine details can carry vital information.

The performance on the BigEarthNet dataset was significantly improved when using the pretrained GAN's discriminator as a feature extractor for image classification. Precision increased from 0.75 to 0.82, recall from 0.72 to 0.80, and the F1-Score from 0.73 to 0.81. These improvements show that transfer learning helps the model generalize better, even with limited labeled data, and thus leads to more accurate predictions and reduced misclassification, especially in complex or high-dimensional remote sensing imagery.

For object detection and land cover classification, the Kappa Coefficient moved from 0.65 to 0.78, which further demonstrated that transfer learning helps the model improve its ability to correctly predict and classify various types of land cover. This is very helpful in dealing with imbalanced datasets where certain classes are underrepresented.

In conclusion, the integration of Transfer Learning with GANs in remote sensing applications yields significant improvements in performance metrics across various tasks. It allows for the fine-tuning of pretrained models to enhance the quality of synthetic data generation and classification accuracy, reducing the computational cost and training time, and these techniques become particularly valuable when labeled data are scarce or the problem is cross-domain.

8. LIMITATIONS OF GANS AND TRANSFER LEARNING IN REMOTE SENSING

Despite their potential, several limitations do not allow GANs and transfer learning to be widely used in remote sensing:

a. Data Scarcity and Quality

Datasets of remote sensing are usually limited in availability and cannot be well consistent due to environmental factors like cloud cover, seasonal variations, and interference caused by the atmosphere. Transfer learning can mitigate this, but quality pre-trained models are mostly important.

b. Training Complexity

GANs are computationally expensive, require careful parameter tuning, and have the issue of mode collapse (limited diversity in the generated data) and convergence issues.

c. Domain Specificity

Models pre-trained on general datasets such as ImageNet do not generalize well for remote sensing tasks involving spectral, multispectral, or radar data. Domain adaptation techniques need to bridge these gaps.

d. Ethical and Interpretability Concerns

Synthesized remote sensing data might be subjected to biases and errors, therefore bringing in erroneous conclusions for such applications as decision making in disasters or urban development. Authentic and meaningful interpretation is not guaranteed, yet.

9. ETHICAL IMPLICATIONS

The deployment of GANs and transfer learning in remote sensing raises critical ethical concerns:

a. Data Misuse and Fabrication

This opens the possibility for misusing such realistic GAN-generated images for creating phony environmental data or even tampering with geospatial information to be used in fraudulent manners. Data usage has to be guarded against improper usage.

b. Fairness and Representation

Remote sensing models often heavily influence decisions in urban planning, environmental conservation, and resource management. Ensuring that these models do not introduce bias which disadvantage some regions or communities is very important.

c. Privacy Issues

This satellite high-resolution imagery alongside synthetic data generation can inadvertently lead to compromising privacy of individuals or organizations.

d. Accountability and Trust

With the advancement in AI models, mechanisms to validate the authenticity and reliability of synthetic data are also essential to establish trust in remote sensing applications.

10. HYBRID APPROACHES: ENHANCING GANS AND TRANSFER LEARNING

To overcome the limitations and ethical concerns associated with traditional GAN-based and transfer learning approaches, hybrid methods are emerging as a promising solution:

a. GAN-CNN Hybrids

Combining GANs with CNN for image super-resolution or semantic segmentation improves feature extraction by itself and the results' accuracy of the models(Greza et al., 2024).

b. CycleGANs for Domain Adaptation

CycleGANs, which allow for unpaired image-to-image translation, are used for adapting models across different types of remote sensing data, such as translating synthetic aperture radar images to optical images.

c. Federated Learning with GANs

Federated learning provides the possibility to train models in a collaborative, decentralized way with multiple datasets. Combining GANs and federated learning enables the creation of diverse, representative remote sensing data without divulging sensitive data.

d. GAN-RNN Hybrids for Temporal Analysis

Recurrent Neural Networks: The GAN can be extended by incorporating it with RNNs for modelling temporal sequences, allowing for monitoring changes in data over time-for example, detection of deforestation or urban growth(Pulakurthi et al., 2024).

e. Ethically-Aligned GAN Frameworks

Algorithmic fairness and explanation of GAN models can also help reduce bias and ensure generated data is ethically aligned.

11. CONCLUSION

It has been discovered that the approach of Transfer Learning with GANs has highly potential in helping forward remote sensing applications. Among its key breakthroughs, data scarcity and domain specific adaptations happen to be primary concerns in dealing with the technique. The overall improvements in numerous tasks, be it image translation, generation, classification, and object detection cases, are outstanding in the scenario under study. By using pretrained GANs as a base to fine-tune on target datasets, researchers attained improved performance metrics, including reducing FID for image generation and improving SSIM for image translation, along with higher precision, recall, and F1 scores for classification-based tasks. In this regard, these developments ensure more accurate mappings of land covers, monitoring, and planning purposes in cities. However, ethical considerations are very important, such as the potential misuse of GAN-generated data and ensuring fairness in decision-making processes. Future research directions are developing domain-specific GANs for remote sensing and exploring temporal image analysis. As the field progresses, the combination of TL and GANs promises to revolutionize remote sensing applications, offering more efficient data processing, improved accuracy, and the ability to extract valuable insights from limited datasets. This synergy between TL and GANs is likely to lead to much more sophisticated and reliable remote sensing technologies.

REFERENCES

Abdal, R., Qin, Y., & Wonka, P. (2019). Image2StyleGAN: How to embed images into the StyleGAN latent space? In *Proceedings of the IEEE/CVF International Conference on Computer Vision* (pp. 4432-4441). DOI: 10.1109/ICCV.2019.00453

Ajayi, O. G., & Ojima, A. (2022). Performance evaluation of selected cloud occlusion removal algorithms on remote sensing imagery. *Remote Sensing Applications: Society and Environment, 25*, 100700. DOI: 10.1016/j.rsase.2022.100700

Arjovsky, M., Chintala, S., & Bottou, L. (2017). Wasserstein generative adversarial networks. In *International conference on machine learning* (pp. 214-223). PMLR.

Brock, A., Donahue, J., & Simonyan, K. (2018). Large scale GAN training for high fidelity natural image synthesis. arXiv preprint arXiv:1809.11096.

Donahue, J., Krähenbühl, P., & Darrell, T. (2017). Adversarial feature learning. In 5th International Conference on Learning Representations, ICLR 2017, Toulon, France, April 24-26, 2017, Conference Track Proceedings. OpenReview.net.

Elgammal, A., Liu, B., Elhoseiny, M., & Mazzone, M. (2017). CAN: Creative adversarial networks, generating "art" by learning about styles and deviating from style norms. arXiv preprint arXiv:1706.07068.

Goodfellow, I., Pouget-Abadie, J., Mirza, M., Xu, B., Warde-Farley, D., Ozair, S., & Bengio, Y. (2014). Generative adversarial nets. *Advances in Neural Information Processing Systems*, •••, 27.

Goodfellow, I., Pouget-Abadie, J., Mirza, M., Xu, B., Warde-Farley, D., Ozair, S., Courville, A., & Bengio, Y. (2014). Generative adversarial nets. In *Advances in neural information processing systems* (Vol. 27). Curran Associates, Inc.

Greza, M., Bhattacharya, I., Hoegner, L., & Jutzi, B. (2024). GAN-Based dual image super resolution for satellite imagery decreasing radiometric uncertainty. *ISPRS Annals of the Photogrammetry, Remote Sensing and Spatial Information Sciences, X-3–2024*, 155–162. https://doi.org/DOI: 10.5194/isprs-annals-X-3-2024-155-2024

Huang, X., & Belongie, S. (2017). Arbitrary style transfer in real-time with adaptive instance normalization. In *Proceedings of the IEEE International Conference on Computer Vision* (pp. 1501-1510). DOI: 10.1109/ICCV.2017.167

Jozdani, S., Chen, D., Pouliot, D., & Johnson, B. A. (2022). A review and meta-analysis of generative adversarial networks and their applications in remote sensing. *International Journal of Applied Earth Observation and Geoinformation, 108*, 102734. DOI: 10.1016/j.jag.2022.102734

Karras, T., Aila, T., Laine, S., & Lehtinen, J. (2017). Progressive growing of GANs for improved quality, stability, and variation. arXiv preprint arXiv:1710.10196.

Karras, T., Aittala, M., Hellsten, J., Laine, S., Lehtinen, J., & Aila, T. (2020). Training generative adversarial networks with limited data. *Advances in Neural Information Processing Systems, 33*, 12104–12114.

Karras, T., Aittala, M., Laine, S., Härkönen, E., Hellsten, J., Lehtinen, J., & Aila, T. (2021). Alias-free generative adversarial networks. *Advances in Neural Information Processing Systems, 34*, 852–863.

Karras, T., Laine, S., & Aila, T. (2019). A style-based generator architecture for generative adversarial networks. In *Proceedings of the IEEE/CVF Conference on Computer Vision and Pattern Recognition* (pp. 4401-4410). DOI: 10.1109/CVPR.2019.00453

Karras, T., Laine, S., Aittala, M., Hellsten, J., Lehtinen, J., & Aila, T. (2020). Analyzing and improving the image quality of StyleGAN. In *Proceedings of the IEEE/CVF Conference on Computer Vision and Pattern Recognition* (pp. 8110-8119). DOI: 10.1109/CVPR42600.2020.00813

Nair, R. R., & Singh, T. (2019). Multi-sensor medical image fusion using pyramid-based DWT: A multi-resolution approach. *IET Image Processing, 13*(9), 1447–1459. DOI: 10.1049/iet-ipr.2018.6556

Nair, R. R., Singh, T., Sankar, R., & Gunndu, K. (2021). Multi-modal medical image fusion using lmf-gan-a maximum parameter infusion technique. *Journal of Intelligent & Fuzzy Systems, 41*(5), 5375–5386. DOI: 10.3233/JIFS-189860

Noguchi, A., & Harada, T. (2019). Image generation from small datasets via batch statistics adaptation. In *Proceedings of the IEEE/CVF International Conference on Computer Vision* (pp. 2750-2758). DOI: 10.1109/ICCV.2019.00284

Pan, S. J., & Yang, Q. (2009). A survey on transfer learning. *IEEE Transactions on Knowledge and Data Engineering, 22*(10), 1345–1359. DOI: 10.1109/TKDE.2009.191

Pan, S. J., & Yang, Q. (2010). A survey on transfer learning. *IEEE Transactions on Knowledge and Data Engineering, 22*(10), 1345–1359. DOI: 10.1109/TKDE.2009.191

Park, T., Liu, M. Y., Wang, T. C., & Zhu, J. Y. (2019). Semantic image synthesis with spatially-adaptive normalization. In *Proceedings of the IEEE/CVF Conference on Computer Vision and Pattern Recognition* (pp. 2337-2346).

Pulakurthi, P. R., Mozaffari, M., Dianat, S. A., Rabbani, M., Heard, J., & Rao, R. (2024). Enhancing GAN performance through neural architecture search and tensor decomposition. *ICASSP 2022 - 2022 IEEE International Conference on Acoustics, Speech and Signal Processing (ICASSP)*, 7280–7284. DOI: 10.1109/ICASSP48485.2024.10446488

Radford, A., Metz, L., & Chintala, S. (2015). Unsupervised representation learning with deep convolutional generative adversarial networks. arXiv preprint arXiv:1511.06434.

Radford, A., Metz, L., & Chintala, S. (2016). Unsupervised representation learning with deep convolutional generative adversarial networks. In Y. Bengio & Y. LeCun (Eds.), 4th International Conference on Learning Representations, ICLR 2016, San Juan, Puerto Rico, May 2-4, 2016, Conference Track Proceedings.

Reed, S., Akata, Z., Yan, X., Logeswaran, L., Schiele, B., & Lee, H. (2016). Generative adversarial text to image synthesis. In *International Conference on Machine Learning* (pp. 1060-1069). PMLR.

Rocchini, D., Luque, S., Pettorelli, N., Bastin, L., Doktor, D., Faedi, N., Feilhauer, H., Féret, J.-B., Foody, G. M., Gavish, Y., Godinho, S., Kunin, W. E., Lausch, A., Leitão, P. J., Marcantonio, M., Neteler, M., Ricotta, C., Schmidtlein, S., Vihervaara, P., & Nagendra, H. (2018). Measuring β-diversity by remote sensing: A challenge for biodiversity monitoring. *Methods in Ecology and Evolution*, *9*(8), 1787–1798. DOI: 10.1111/2041-210X.12941

Salimans, T., Goodfellow, I., Zaremba, W., Cheung, V., Radford, A., & Chen, X. (2016). Improved techniques for training GANs. *Advances in Neural Information Processing Systems*, •••, 29.

Shen, Y., Gu, J., Tang, X., & Zhou, B. (2020). Interpreting the latent space of GANs for semantic face editing. In *Proceedings of the IEEE/CVF Conference on Computer Vision and Pattern Recognition* (pp. 9243-9252). DOI: 10.1109/CVPR42600.2020.00926

Tzeng, E., Hoffman, J., Saenko, K., & Darrell, T. (2017). Adversarial discriminative domain adaptation. In *Proceedings of the IEEE Conference on Computer Vision and Pattern Recognition* (pp. 7167-7176).

Wang, X., Sun, L., Chehri, A., & Song, Y. (2023). A review of GAN-based super-resolution reconstruction for optical remote sensing images. *Remote Sensing (Basel), 15*(20), 5062. DOI: 10.3390/rs15205062

Wang, Y., Wu, C., Herranz, L., van de Weijer, J., Gonzalez-Garcia, A., & Raducanu, B. (2018). Transferring GANs: Generating images from limited data. In *Proceedings of the European Conference on Computer Vision (ECCV)* (pp. 218-234). Springer, Cham. DOI: 10.1007/978-3-030-01231-1_14

Wang, Y., Wu, C., Herranz, L., van de Weijer, J., Gonzalez-Garcia, A., & Raducanu, B. (2018). Transferring GANs: Generating images from limited data. In *Proceedings of the European Conference on Computer Vision (ECCV)* (pp. 218-234). Springer, Cham. DOI: 10.1007/978-3-030-01231-1_14

Wu, Z., Lischinski, D., & Shechtman, E. (2020). StyleSpace analysis: Disentangled controls for StyleGAN image generation. In *Proceedings of the IEEE/CVF Conference on Computer Vision and Pattern Recognition* (pp. 12863-12872).

Zhang, H., Goodfellow, I., Metaxas, D., & Odena, A. (2019). Self-attention generative adversarial networks. In *International conference on machine learning* (pp. 7354-7363). PMLR.

Zhang, H., Goodfellow, I., Metaxas, D., & Odena, A. (2019). Self-attention generative adversarial networks. In *International Conference on Machine Learning* (pp. 7354-7363). PMLR.

Zhu, J. Y., Park, T., Isola, P., & Efros, A. A. (2017). Unpaired image-to-image translation using cycle-consistent adversarial networks. In *Proceedings of the IEEE international conference on computer vision* (pp. 2223-2232). DOI: 10.1109/ICCV.2017.244

Chapter 8
Case Studies on Generative Adversarial Networks in Environmental Changes

K. Udayakumar
https://orcid.org/0000-0003-3532-4286
Department of Computer Science and Design, Vel Tech Rangarajan Dr. Sagunthala R&D Institute of Science and Technology, Chennai, India

M. Revathi
Department of Computer Science and Engineering, Bharath Institute of Higher Education and Research, Chennai, India

K. Janani
https://orcid.org/0009-0009-9952-3410
Department of Data Science and Computer Applications, Manipal Institute of Technology, Manipal, India

N. Nithiyanandam
Department of Computer Science and Engineering, Dayananda Sagar College of Engineering, Bangalore, India

ABSTRACT

Environmental changes have been rooted in both natural and man-made. It affect greatly the ecosystems of the Earth, human health, economic systems, and societal structures. Understanding these impacts is therefore vital in the formulation of policies that could promote sustainability and diminish adverse effects. Deep generative models in the categories currently attracting a lot of attention include Generative

DOI: 10.4018/979-8-3693-6900-5.ch008

Adversarial Networks (GANs). These networks can model complex distributions that are based on a variety of data. GANs have become a very powerful tool in most branches and research areas, especially with much stress on environmental research. Their capabilities in giving trustworthy data and the development of a robust dataset have been presented, which have proven very useful in monitoring and responding to changes in the environment. Applications include observing air quality, tracking the process of deforestation, climate change modeling, water resource management, and others. GAN applications pave the way for obtaining more accurate and detailed information on environmental conditions. This will make efforts to minimize and combat the impacts of environmental changes stronger. Based on these studies, the possible use of GANs in monitoring and mitigating various fields of environmental change is discussed further. A couple of case studies reflect the extensive use of GANs and, therefore, great potential in addressing changes on the earth's surface. Steadily progressing computational resources and technologies predict an expansion of the important role that GANs will have within the scope of environmental science. It will open new solutions and points of insight into severe problems within the sphere.

1. INTRODUCTION

The Earth is an ever-evolving system, a dynamic system. Depending on the circumstances, this shift can be small or large, slow or fast, caused by nature or by humans. Grasping these changes is an important issue because they greatly influence the health of the planet and the well-being of every live on it. Internal processes of Earth have shaped landscapes and influenced climates over millennia through plate tectonics and volcanic activity. The sun does not shine constantly, and its energy shifts impact worldwide temperatures. El Nino and La Nina also periodically change the weather. These two natural factors are part of the environmental waltz but have never in the past taken over it until recent centuries when human activities started to be a more profound force in changing the environment.

The problem at hand today remains climate change resulting from fossil fuel burning. It releases greenhouse gases, trapping heat, raising global temperatures, and changing precipitation patterns with increased incidents of severe weather. Computing technologies are in the front line in the fight against climate change. Internet of Things (IoT) devices, full of a myriad of installed sensors, can measure elements such as air quality, temperature, humidity, CO2 level, and other environmental parameters. Sensors can be located in remote and inaccessible areas so real-time data can be captured. Moreover, satellites and drone imagery are the best tools to monitor the changing environment. They will produce the most detailed data that can be used in many applications for climate change and environmental conservation.

Artificial Intelligence (AI) may enhance energy grids by also utilizing renewable sources and reducing reliance on fossil fuels. It can also analyze weather patterns and predict extreme weather conditions so that adequate arrangements can be made. Smart grids distribute energy efficiently and consume less energy. Smart appliances adjust their energy usage, and smart buildings regulate heating and cooling. The IoT, a network of sensors, will be able to monitor in real-time many environmental conditions so that dangers such as wildfires, floods, and other climate calamities can be detected early and action taken immediately. GANs are a powerful tool with interesting applications in the realm of environmental changes (Bouteska, Seranto, Hajek, et al., 2023; Creswell, White, Dumoulin, Arulkumaran, Sengupta, & Bharath, 2018).

Machine learning (ML) comprises statistical techniques that autonomously learn from data to create a detection, estimation, or classification model that minimizes the disparity between training and prediction datasets without explicit programming (Sagan, Peterson, Maimaitijiang, Sidike, Sloan, Greeling, & Adams, 2020). The goal of ML is to enable semi-automatic or automated detection, classification, or pattern recognition by feeding data into a computer that may be refined using known or predetermined features or objects. Recent advances in algorithm development, processing capacity, sensor technology, and data accessibility have led to a rise in the use of machine learning for water quality monitoring via remote sensing (Peterson, Sagan, Sidike, Cox, & Martinez, 2018). Deep learning, the most potent variant of machine learning, is a hierarchically organized neural network capable of extracting valuable insights and generating predictions from unprocessed data. Deep learning has demonstrated exceptional benefits for water quality mapping (Peterson, Sagan, & Sloan, 2020).

Rising temperatures, glacier retreats, the progressive melting of ice, more extreme weather events (floods, droughts, frosts) increasing variability, destruction of ecosystems in mountains, soil degradation, and water shortage are some of the climate changes that endanger our planet (Stengel, Glaws, Hettinger, & King, 2020; Karamouz, Alipour, Roohinia, & Fereshtehpour, 2022). Even understanding the processes as such makes it hard to assess the extent of such climate change on both socio-economic and natural ecosystems. This makes it impossible to accurately predict, prepare, and/or adapt to whatever changes are bound to happen in the future. In this chapter, we attempt to present a useful framework for energy reform development as well as for the construction, comprehension, and application of climate change projections and further progress in the field of climate and environment science. The discussion of the issue of how this support for the development of innovative technologies for climate and environmental science with the help of cyberinfrastructure will be continued is also concentrated.

2. LITERATURE REVIEW

This section pinpoints GAN's applications in monitoring and mitigating environmental changes. It examines a fundamental and current study that substantiates the chapter's necessity. This section concludes with an emphasis on research gaps and the chapter's importance.

Jyoti S. Shukla et.al. (2023), have proposed a new enhancement to drought prediction techniques based on the Pix2Pix-GAN model, which effectively delineates short-term land cover alterations within the designated region of interest using historical remote sensing data. A distinctive Bidirectional Convolutional Pix2Pix integrated model for drought estimate is developed. The Pix2Pix model is combined with the Bidirectional Convolutional long short-term memory module, using the advantages of generative and temporal feature learning. It aids in the assessment of vegetative dryness. To enhance diversity in the feature set utilized for model training, the Vegetation Health Index and rainfall features are simultaneously incorporated into the dataset. The author's knowledge represents a novel combination for examining the impact of these two significant features in Vegetation Health Index-based drought prediction.

Andreas Tsatsaris et al. (2021) have carried out a thorough evaluation of geo-information technologies to aid in the monitoring of environmental hazards in the context of climate change. In light of recent developments in the field of environmental monitoring and the use of geospatial technologies, this review seeks to catalog the most devastating geo-hazards. It delineates the primary difficulty of this new period, which is the illusory utilization of information generated by environmental monitoring to facilitate the implementation of policies aimed at achieving a sustainable future. The assessment emphasizes the potential and growing significance of geographic information in facilitating environmental monitoring under climate change. The expansion of geographic information appears to be swiftly advancing due to ongoing technological and scientific advancements that are expected to progress exponentially in the future. However, this review emphasizes that ongoing environmental monitoring requires an interdisciplinary approach and encompasses measures addressing both the evolution of natural occurrences and their catastrophic consequences, primarily attributed to climate change.

Khurram Riaz et. al. (2023) have conducted a review on the management of climate resilience using 3D modeling and Digital Twin Technology. This paper asserts that the bidirectional data flow between a digital model and the physical world is a novel notion for improving climate resilience. Nonetheless, the research predominantly exists in the realm of theoretical concepts and discourse, with major research gaps persisting concerning the implementation and utilization of bidirectional data flow in an authentic digital twin. Current new research initiatives are investigating the potential of digital twin technology to tackle the issues encountered by communities in vulnerable regions, to develop practical solutions to improve climate resilience shortly.

A unique digital twin has been created (by Sakdirat Kaewunruen et. al.,) with building information modeling integration via a life cycle analysis (Kaewunruen, AbdelHadi, Kongpuang, Pansuk, & Remennikov, 2022). The Minnamurra railway bridge in Australia has been selected as a practical example to illustrate the advanced application of building information modeling to optimize operation, maintenance, and asset management, thereby enhancing the sustainability and resilience of the railway bridge. Furthermore, the digital twin has been utilized to assess greenhouse gas emissions and cost consumption via the integration of building information modeling. This research illustrates the viability of digital twin technology for railway maintenance and resilience enhancement. It facilitates virtual cooperation for co-simulations and co-creation of value among stakeholders involved in building, operation, and maintenance, hence promoting cost reduction and decreasing greenhouse gas emissions.

Ehsan Foroumandi et. al. (2024) utilizes remotely sensed images to create the first GAN for detecting flash droughts. The network has two deep learning models: the generator and the discriminator. A primary advantage of the constructed network, differentiating it from traditional deep learning models, is its adaptive loss function that adjusts according to the network's performance. As a result, overall network performance is improved and there's no longer any need to create specialized loss functions for individual problems or domains. Adversarial loss is combined with other typical loss functions in the loss function. While modeling, the network is forced to take into account interdependencies between nearby regions due to the inclusion of adversarial loss. The adversarial training process in GAN functions as a sort of implicit regularization, which is especially crucial when managing restricted or noisy datasets. This regularization enhances the model's generalization to novel data, which is essential for monitoring flash droughts. The proposed network is trained and evaluated using the Standardized Soil Moisture Index across the Contiguous United States for real-time flash drought surveillance. The model's efficacy is subsequently contrasted with that of a U-Net model and a Naive model.

Alex Bihlo (2021) employed a conditional deep convolutional generative adversarial network to forecast the geopotential height at the 500 hPa pressure level, the two-meter temperature, and total precipitation over the subsequent 24 hours across Europe. The suggested models are trained on four years of ERA5 reanalysis data to predict the corresponding meteorological fields. The forecasts demonstrate a strong qualitative and quantitative correlation with the actual reanalysis data for geopotential height and two-meter temperature, but not for total precipitation, suggesting that data-driven weather forecasts may be feasible for certain meteorological parameters.

Data from satellites is of critical significance for monitoring and protecting the ocean ecosystem. On the other hand, the quality of satellite data is significantly diminished because of the absence of values in satellite data, which is brought about by a variety of force majeure events such as cloud cover, poor weather, and sensor failure. This makes it difficult to put satellite data to use in practical situations. In light of this, numerous approaches have been suggested for doing missing data imputation for satellite data to enhance the quality of the data. On the other hand, these algorithms are not able to learn the short-term temporal dependency and the dynamic spatial dependence in satellite data very well. Consequently, the imputation performance is poor when the data-missing rate is high. In order to solve this problem, Shuyu Wang et. al. (2022) presented the spatio-temporal attention-GAN for the imputation of missing values in satellite data.

The robust generative network known as the Generative Adversarial Network, which was proposed by Goodfellow et al. (2014), makes use of an architecture that is based on machine learning and is rather adaptable. GAN is implemented such that patterns in the input domain can be automatically learned and then used to generate outputs based on those particular patterns. In other words, the network is in charge of producing target data and replicating the spread of data. On the other hand, some ML models aim to learn how to train a function that can take an input and turn it into a target. In comparison to more traditional deep learning models, GAN offers a clear advantage. Bear in mind that GANs are equipped with a system that can alter its loss function based on the target data. Because of this change, it is now much simpler to tell the difference between real and fake results.

2.1 Summary of Foundational Studies

An issue with dataset noise, spare, and incomplete has been resolved using GAN. Goodfellow et al. (2014) showed a significant breakthrough in data-driven environmental research. It paved a strong platform for creating realistic datasets with GAN. Creswell et al. (2018) proved how GANs can be used in various environmental domains. This research helps in acquiring complex data distributions without direct supervision (Creswell, White, Dumoulin, Arulkumaran, Sengupta, & Bharath, 2018). Temporal dynamic has more impact on environmental data studies. Shukla et al. (2023) demonstrated how to acquire temporal information by combining Pix2Pix-GAN models. The model efficacy has been proved by drought prediction, and vegetative aridity (Shukla & Pandya, 2023). A similar study by Wang et al. (2022), addresses problems with data continuity and quality. A spatial-temporal attention-GAN has been developed for imputing absent values in satellite data.

2.2 Summary of Recent Studies

Recent studies have expanded the use of GANs in the field of environmental real-time monitoring. For real-time drought detection, Foroumandi et al. (2024) employed GAN, stressing adaptive loss functions as a means of significantly enhancing model performance. Tsatsaris et al. (2021) emphasized the importance of combining geospatial technologies with GANs to monitor environmental hazards. On the other hand, Bihlo (2021) utilized conditional deep convolutional GANs for meteorological forecasting. This work demonstrates the applicability in situations that are related to weather (Bihlo, 2021). A summary of recent studies of GAN is included in Table 1.

Table 1. Summary of recent studies

References	Use case	Methodology	Identified Research gap
(Wang, Li, Hou, Guan, & Yao, 2022)	Ocean environment monitoring	Spatio-Temporal Attention GAN	Data quality and availability
(Shukla & Pandya, 2023)	Environmental hazards monitoring	-	Generalization Across regions
(Foroumandi, Gavahi, & Moradkhani, 2024)	Drought monitoring	GAN	Computational Costs
(Creswell, White, Dumoulin, Arulkumaran, Sengupta, & Bharath, 2018)	Air quality modelling	GAN	Interpretability
(Riaz, McAfee, & Gharbia, 2023)	Deforestation prevention	Improved Conditional-GAN	Robustness to input variations.
(Tsatsaris, Kalogeropoulos, Stathopoulos, Louka, Tsanakas, Tsesmelis, et al., 2021)	Weather forecast	GAN	Temporal dynamics handling.
(Shukla & Pandya, 2023)	Drought prediction	cGAN	Spatial resolution

This book chapter can offer an extensive road map for developing advanced GANs across multiple areas to address the mentioned research gaps. It can provide information on how to improve generalization, data efficiency, and GAN topologies, which will increase the technology's scalability, interpretability, and suitability for real-world problems. Researchers, practitioners, and business experts can benefit greatly from the conversation, which can also highlight new developments, trends, and future directions.

3. IMPACT OF ENVIRONMENTAL CHANGES

According to the Natural Resources Defence Council, the most significant threat to the entire existence of our planet is climate change. If we do not take measures to curb the production of greenhouse gases resulting from the combustion of fossil fuels, the implications of rising global temperatures will include the collapse of large crop and fishery systems, the extinction of hundreds of thousands of species, and the inaccessibility of entire communities to human habitation. The effects of climate change are already causing suffering and death, even if it is possible that these outcomes can still be avoided. The compounding impacts of it may be felt today, even outside our windows, in the form of blazing wildfires and storms that have been supercharged. In addition to causing damage to ecosystems and infrastructure, flooding can also result in the spread of illness and death. The human health problems that arise as a consequence of drought, flooding, and other meteorological conditions lead to an increase in the death rate, a shift in the availability of food, and a restriction on the amount of work that a person can accomplish, which ultimately leads to a decrease in the productivity of our economy. The different aspects and effects of climate change is illustrated in figure 1.

Figure 1. Effects of climate change

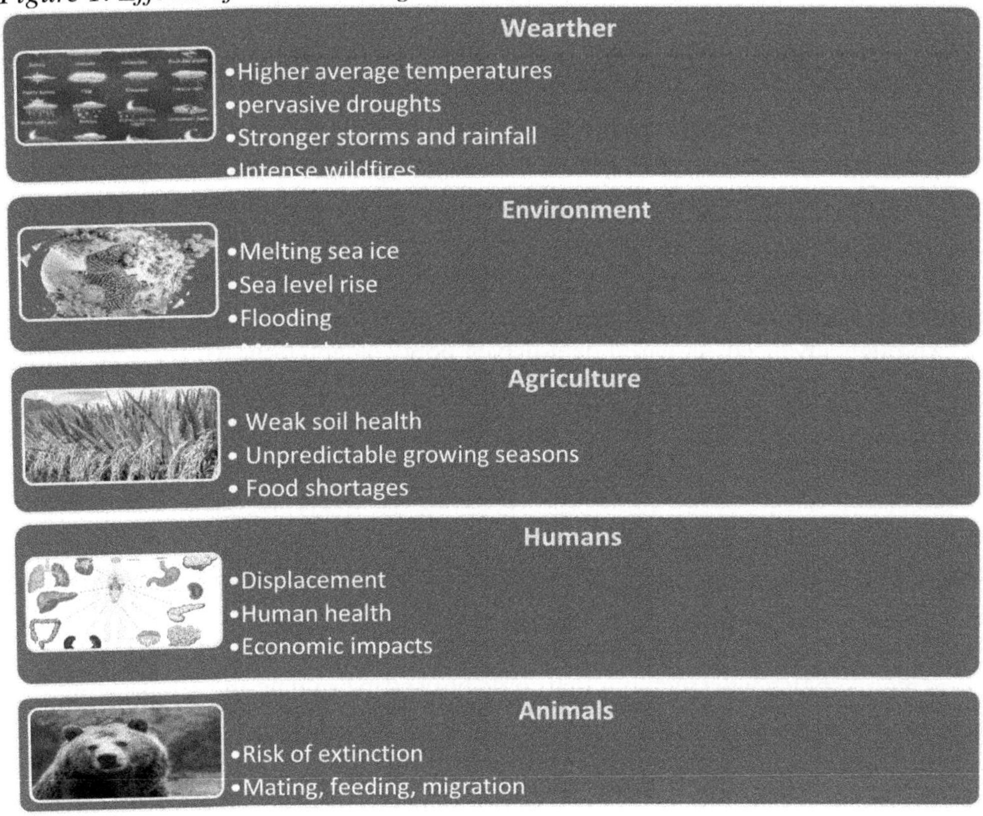

3.1 Urgency of Environmental Issues and the Role of GANs

The most severe global challenge is environmental change, predominantly climate change, and loss of biodiversity. It demands high attention and policy changes globally. The integration of advanced technologies such as GAN, AI, IoT, and remote sensing could help in addressing serious environmental issues. This section highlights the urgency of environmental issues and the role of GAN in solving them.

- **Increase in climate change risks-** the impression of climate change being sensed worldwide. life-threatening weather events, temperature, and changes in climate patterns are the symptoms of increasing climate risks. Currently, GAN helps in providing localized climate projections. It directs the policy-makers to be prepared with action plans. However, GAN adaption in a wide range requires more effort to meet climate disasters.

- **Accelerated biodiversity loss-** It affects ecosystem services with severe consequences on human health and species loss. Even though GAN helps in monitoring wildlife populations, deforestation, and ecosystem health, the challenge remains limited to global conservation policies.
- **Data-driven environmental governance-** Recently, governments and NGOs have relied on data-driven decisions to make environmental policies. For instance, the urban planning body uses GAN to simulate air pollution patterns and evaluate mitigation plans. However, the regulatory framework stays restricted due to weak collaboration between policymakers and AI researchers.

Through advanced AI-based data-driven policy adaptation, we are able to better prepare ourselves for some of the expected repercussions of climate change by minimizing the negative effects that these effects will have on ecosystems and the well-being of humans.

4. IMPRESSION OF GAN IN ENVIRONMENTAL MONITORING

GAN can be applied in environmental monitoring because of the need to implement more sophisticated methods that could be used to evaluate and predict complex environmental phenomena, as well as provide a simulation to the changes. The process monitors the environment by collecting massive volumes of data obtained from remotely sensed sources through satellites, sensors, and technologies that utilize remote sensing. Before they are actually analysed, data based on traditional models often prove noisy, incomplete, or sparse. GAN creates possibilities for overcoming these challenges by taking advantage of its capability to generate good data, thereby enhancing the model of prediction. Its main drivers are seen in Figure 2 below.

Figure 2. Motivation of GAN

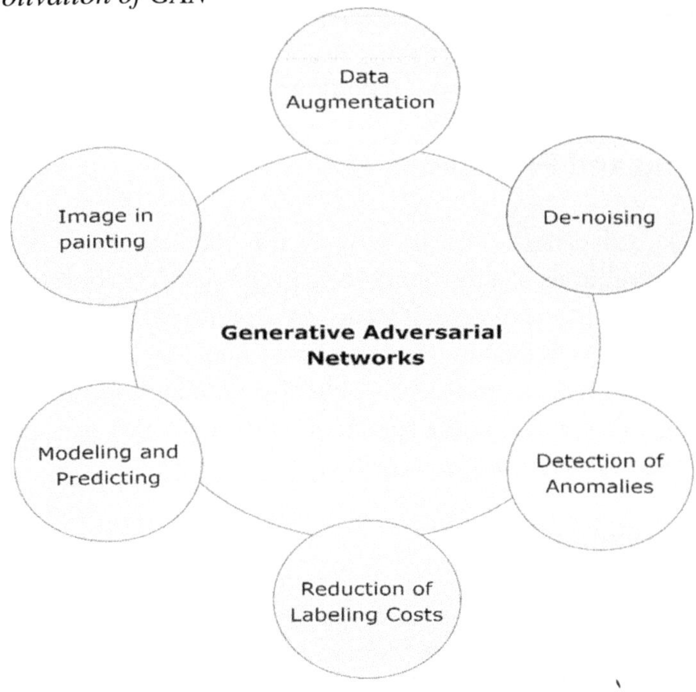

4.1 Data Augmentation

Environmental datasets are normally missing, poor quality, or imbalanced. This will particularly be the case when such events or changes are rare in nature, such as in unusual climatic events or long-term environmental modifications. GAN may produce artificially synthesized data that replicates real-world observations. This gives machine learning models enough ability and further enhances their performance in classification as well as prediction tasks. It involves the creation of synthetic samples for rare events, like extreme weather conditions, that are essential in proper modeling but are usually not well represented in the data that is observed (Yun, Kim, Kim, Song, & Woo, 2023).

4.2 De-Noising

Using remote sensing technology to get some source of environmental data, such as a satellite; however, there will be noise in data due to cloud cover, and malfunctioning sensors, among other factors. Using GAN, missing data in satellite images can be made easy to erase and will be supplemented with believable and high-resolution images wherein there is a lack of data. Higher-resolution images can be developed from low resolutions from which greater surveillance of environmental factors like the health condition of flora, water bodies, and land can be done with the aid of GAN (Luleci, Catbas, & Avci, 2022).

4.3 Modelling and Prediction

A complicated dynamics process is involved in the forecasting of variations in environmental conditions like temperature variation, land deterioration, or varying levels of pollution. GAN can predict future environmental conditions by learning the underlying patterns in the historical data and simulating probable future scenarios. Development of realistic models capable of simulating aspects such as climate change, pollutant dispersion, or the spread of wildfires so that researchers can test multiple theories and intervention measures

4.4 Anomaly Detection

At times, the environmental monitoring system may have to identify anomalies such as a change in land use or weather patterns that are not typical. With the generation of the predicted outputs and their subsequent comparison with the actual observations, GAN can be used to detect anomalous changes in the environment. By training the discriminator on the GAN to distinguish normal patterns from abnormal patterns, one can find anomalies in the data relating to the environment.

4.5 Reduction in Labelling Cost

Supervised learning models, which need a huge amount of labeled data to function appropriately for tasks like image classification, cannot be applied properly because environmental data-manual classification, especially remote sensing-are time-consuming and expensive. This can be reduced with the help of GAN which can produce synthetic labelled data and thus large dataset requirements for manual annotation could be avoided.

The motivation for using GAN in environmental monitoring arises from its potential to generate data that closely resembles real information, filling the gaps in missing information, and enhancing predictive modeling. Some of the issues that environmental researchers and monitoring systems face- inadequate dataset sizes, unpredictable changes in the environment, and demands for extensive and accurate simulations be effectively overcome when they make use of GAN. The various usage aspect of GAN is visually represented in Figure 3.

Figure 3. Usage aspects of GAN

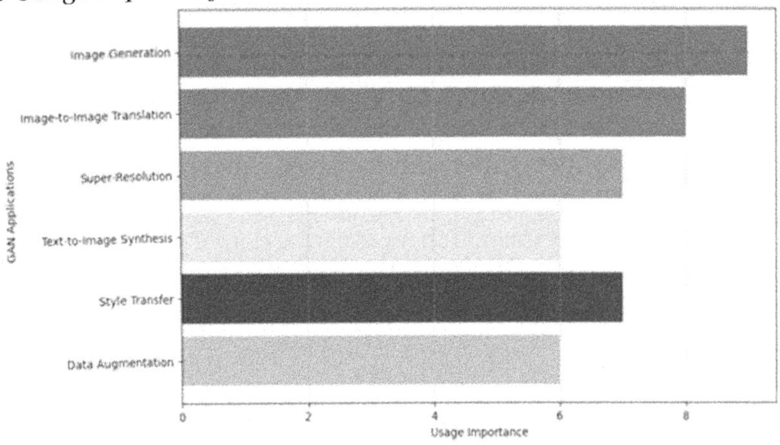

The significance of GAN in environmental monitoring is substantial. In comparison to other contemporary technologies, GAN demonstrates significant advantages due to its versatility in various applications (Koochali, Bakhshipour, Bakhshizadeh, Habermehl, Dilly, Dittmer, et al., 2024; Spyrou, Tsoulos, & Stylios, 2022; Ullo & Sinha, 2020; Afaq & Manocha, 2021; Shafique, Cao, Khan, Asad, & Aslam, 2022). Various studies indicate that GANs demonstrate superior accuracy and efficiency compared to traditional machine learning, remote sensing, IoT, and statistical models. Figure 4 presents a comparison of GAN with other technologies.

Figure 4. Comparison of GAN with other technologies

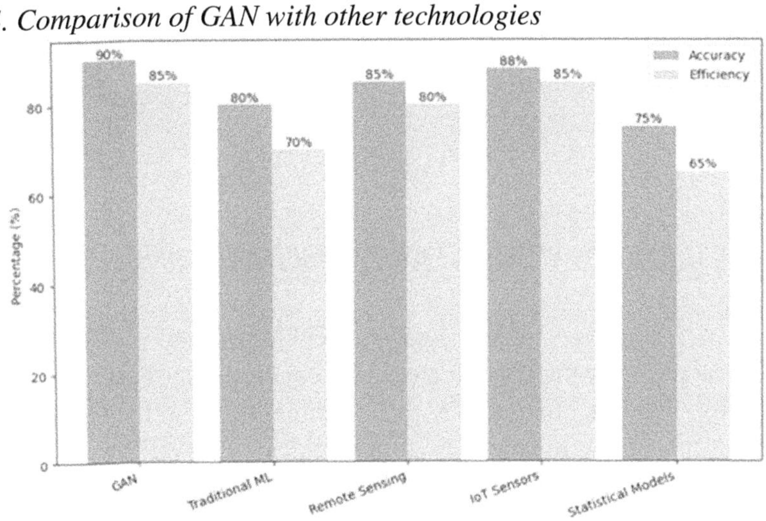

5. APPLICATIONS OF GAN IN ENVIRONMENTAL MONITORING

The applicability of Generative Adversarial Networks has shown wonderful potential in a lot of application domains, including environmental science itself (Guo, Jia, Dong, Ye, Chen, Zhang, & Zheng, 2020; Boroujeni & Razi, 2024; Hassan, Ali, Trigui, Savaria, & Sawan, 2019; Mansourifar, Moskovitz, Klingensmith, Mintas, & Simske, 2022; Wang, Sun, Chehri, & Song, 2023; Boroujeni & Razi, 2024). Their implementation in the environmental changes can have a very particularly significant influence in a variety of ways. Table 2 demonstrates the application scenarios of GAN in environment monitoring.

Table 2. Applications of GAN in environment monitoring

Generic Applications	Specific Use Cases
Modeling and Prediction	Data augmentation
	Simulating Future Climate Scenarios
Image Analysis	High-resolution of Satellite Images
	Land Cover Classification
Environmental Monitoring and Restoration	Deforestation Monitoring
	Wildlife Habitat Mapping
	Water Resource Management
	Ocean Monitoring
Pollution Monitoring and Control	Air and Water Quality Monitoring
	Soil Health Monitoring
Disaster Management	Flood Simulation and Prediction
	Biodiversity Observing
	Post-Disaster Assessment (Damage)

6. CASE STUDY: DEFORESTATION MONITORING

Deforestation continues to be a significant environmental issue; it results in the loss of biodiversity, alteration of climate, and ecosystem disturbances. Sophisticated technology like satellite photography and machine learning, including Generative Adversarial Networks, introduces new opportunities for surveillance and analysis of deforestation. In the olden days, evaluation of deforestation was carried out based on the interpretation of satellite images by human experts and traditional image processing techniques. All such approaches are usually not sophisticated enough to execute in real-time. GANs have transformed the landscape of image processing. Generating high-resolution images from low-resolution inputs helps to improve readability of satellite data.

6.1 Data Collection

Acquire high-resolution images of forest areas using satellite images. Satellites may use remote sensing technologies, such as those by Sentinel and Landsat, to obtain the images. The datasets provide multiple time points to track change across space in time. Key variables to track include forest cover density, tree canopy, and topography.

6.2 Pre-Processing

Raw remote sensing images may contain noise or distortion or lack necessary information. Pre-processing includes cleaning, georeferencing to a consistent geographic scale, and pixel normalization so that pixel brightness and contrast are consistent. This pre-processing step makes the data ready for GAN models and algorithms in deep learning.

6.3 GAN

Use the GANs to generate high-resolution truthful images of remote sensing satellite technology capturing deforestation patterns. GANs consist of two neural networks, namely a generator and a discriminator, that work together, as shown in Figure 5. The generator creates artificial images mimicking the real deforestation trends and the discriminator tries to differ between the real and the fake images. With time, the GAN becomes an expert at producing natural forest change and deforestation events representations. The GAN model is trained on a set of satellite photographs showing areas that are both clear-cut and undamaged.

Figure 5. GAN Generator and Discriminators models

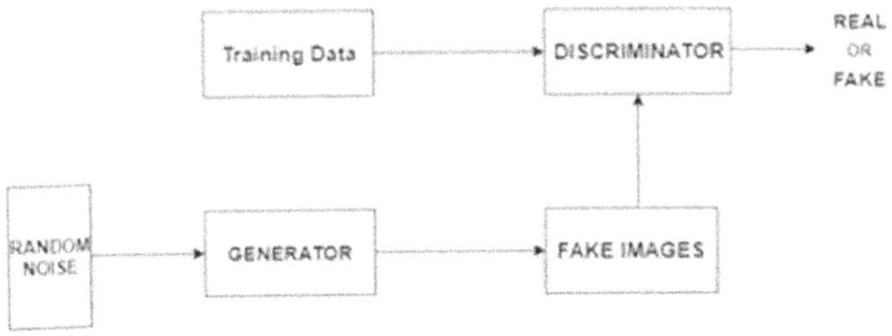

6.4 Detection and Segmentation

Mark and isolate the cleared zones from the satellite image. It then would be that after producing super high-resolution imagery, the machine learning methods comprising convolutional neural networks besides algorithms applied in image segmentation would automatically detect the areas earmarked for forest clearing. These algorithms detect and contrast changes in the level of vegetation and topography to identify deforestation and forested regions. This would thus be applied in detecting the scale of deforestation.

6.5 Monitoring and Analysis

It involves constant monitor of the forests to track changes over time. The cleared lands are then monitored for years by viewing pictures taken at periodic times. This time-series analysis will be able to find present activities such as deforestation, logging patterns, or reforestation projects. Automatic systems track changes and highlight areas of concern, thus providing a dynamic view of forest conditions.

6.6 Reporting and Decision-Making

Deliberate actionable information to policymakers and environmentalists. Analytical results are displayed on interactive maps, indicating the degree of deforestation and potential threatened regions. Reports, graphs, and maps are relayed to the policymakers, forest management departments, and environmental organizations. It is essential information in the implementation of conservation policies, development of deforestation prevention plans, and making prompt decisions about logging activities or unlawful deforestation.

6.7 Monitoring Deforestation Using GAN-Boosted Satellite Images

In the observation of deforestation, GANs greatly helped to increase the resolution and quality of satellite images observing tiny movements in the forest cover.

Traditional satellite images might not carry the resolution required to interpret small-scale operations that characterize many deforestation acts. Deep forest canopies are particularly problematic in such applications, and GANs solve this problem by delivering high-resolution images from lower-quality inputs, thus offering more accurate ways to detect patterns in deforestation. This study in the Amazonian rainforest used GAN-boosted images to track illegal logging activities over time. Such an approach led to early warnings of deforestation and helped foresters begin suitable actions in time. The GAN model had been trained based on historical satellite data relating to deforested and undamaged parts of the region. From the dataset, it learned how to generate realistic forest changes. The identification of spots where deforestation occurred was achieved with more accuracy by using GAN-generated images compared to more conventional methods.

6.8 Infrastructure and Technologies Used in the Deforestation Monitoring

6.8.1 Sources of Satellite Imagery

- **Landsat Program (NASA/USGS):** Landsat is a free, public satellite image of Earth that dates as far back as four decades. The imagery is commonly used for monitoring the long-term changes in forest ecosystems.
- **Sentinel (ESA):** Sentinel-1 and Sentinel-2: ESA satellites offer high-resolution, multi-spectral imagery that is valuable for determining subtle differences in forest cover.
- **MODIS** (Moderate Resolution Imaging Spectroradiometer): This sensor, onboard NASA's Terra and Aqua satellites, offers near-daily imagery for the detection of regional deforestation patterns.

6.8.2 GAN for Satellite Image Implementation Tools

GAN consists of two neural networks, the generator and the discriminator, which function in tandem to generate highly realistic artificial images. While the discriminator evaluates the accuracy of the forest region images generated, the generator produces images that closely resemble actual satellite photographs. Consequently, GANs possess the capability to produce highly detailed images that can be employed for deforestation detection.

- **PyTorch:** A leading library for machine learning and facilitates the easy running of GAN models. The framework can be used for training GAN models using forest imagery datasets.
- **TensorFlow:** A deep learning framework that is widely used and provides comprehensive support for GANs, which have applications in environmental monitoring.

6.8.3 Remote Sensing Software

- **Google Earth Engine**: It is a complex cloud computing remote sensing technology that deals with huge satellite data. This allows scholars to implement GAN models in real-time satellite images on deforestation detection. Through the APIs provided by Google Earth Engine, users can access and evaluate environmental data. That means Google Earth Engine is an excellent tool for GAN-augmented deforestation detection.
- **QGIS:** A free and open-source Geographic Information System (GIS) application for rendering satellite imagery, creating maps, and processing geospatial information relevant to forest clearance.
- **ArcGIS:** Commercial GIS software that enables users to view and interpret satellite imagery. It can be used with GAN images to fully analyze and predict the frequency of deforestation activities.

6.8.4 Pre-processing Toolbox and Image Analysis Tools

- **OpenCV:** Generally, an open-source library, this is primarily image processing. For the intents and purposes of monitoring deforestation, OpenCV offers data preprocessing with cleaning, alignment, and enhancement of the data before presenting it to GAN models.
- **Scikit-learn:** A toolset used in developing machine learning algorithms for Python that are especially important when making ready satellite data to input into a GAN for training or predicting purposes.

6.8.5 Cloud Computing and Storage

- **Amazon Web Services:** It supports the use of cloud services like SageMaker, in which to host GAN models and the storage of large satellite datasets. It provides the needed computing power to train GANs on high-resolution satellite images.
- **Google Cloud:** Google Cloud Machine Learning Engine and Cloud Storage enable the storage, preprocessing, and analysis of satellite data and have a robust infrastructure to carry out large environmental monitoring programs.

6.9 Comparative analysis of GAN in deforestation monitoring

While GANs offer significant advantages, GANs present other challenges-such as computation cost and complex training. A trade-off model for implementing GANs for deforestation monitoring is illustrated in Figure 6. Thus, stakeholders should assess both the merits and demerits of using GANs for their specific monitoring needs. Adopting this view guarantees that informed decisions are made and helps maximize the benefits of GAN applications toward combating deforestation. The subsequent section considers GANs from an environmental context in general.

Figure 6. Analysis of GAN in deforestation monitoring

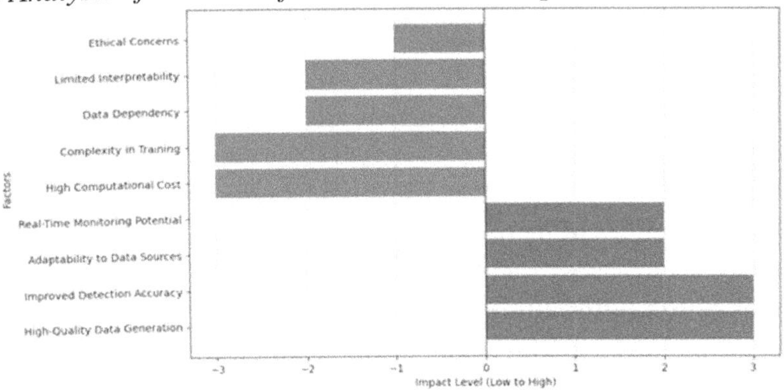

7. BALANCED PERSPECTIVE ON GANs IN ENVIRONMENTAL CONTEXT

Aside from the potential of GANs in environmental applications, numerous technical and operational difficulties impede their efficacy. A balanced perspective of GAN's successes and shortcomings in real-world applications is presented in this section.

- **Scarcity of quality data**

The effective function of GAN demands high-resolution and diverse datasets in the environmental aspect. However, the availability of high-resolution data is limited due to various external factors. The utilization of satellite data may cause cloud occlusion, unreliable historical data, and sensor malfunctions, all of which can impose constraints on data quality, hence impacting the efficacy of the GANs. Nonetheless, challenges related to data scarcity can be addressed by synthetic data augmentation and data fusion methodologies of GAN.

- **Generalisation Across Various Geographic Regions**

Attribute variations can significantly differ across areas, complicating the generalization of GANs trained on unique datasets in many locations. Consequently, a model designed to predict forest fires in one nation may not yield the same accuracy in another due to variations in plant life, temperature, and land use. Through the implementation of domain adaptation and transfer learning, the model can be enhanced to exhibit greater versatility across various environmental contexts.

- **Temporal modeling challenges**

Conventional GANs struggle to capture complex dynamics and correlations between data over time. This aspect restricts GANs from modeling long-term environmental patterns. Integration of recurrent neural networks and transformers may the temporal predictions of these models.

- **Real-time computational complexity**

Real-time environmental scanning, including wildfire monitoring and extreme weather forecasting, is advantageous; yet, it necessitates substantial computer resources, resulting in significant costs. Such strategies exhibit limited usefulness in domains where rapid decision-making is important. However, these issues could be addressed by the advancement of lightweight GANs and the efficient utilization of cloud AI.

8. ETHICAL ISSUES

While GANs offer promise in environmental monitoring, ethics raises several issues, for example, data exploitation hazards and biased results that arise from poor or biased training data. Also, relevant legislation to govern AI-based environment solution deployment is also scanty. These factors might limit the application of GANs in sensitive or critical contexts especially if policy decisions depend on the outputs of such models. The ethical concerns related to GAN are data privacy, data exploitation, bias, and black-box nature.

- **Data exploitation:** Ethical lapses may result from misuse of the capabilities of the GAN to create realistic synthetic data. It could be used to disrupt privacy by replicating sensitive information from the training dataset, which is used to train GAN.
- **Privacy breach:** Satellite image datasets are widely used in training environmental-based GAN models. In such cases, privacy breaches may exist

as a result of capturing images of restricted or private places. Even the unintentional inclusion of such images in a database may lead to ethical lapses.

- **Bias-Skewed Datasets:** GAN models trained using geographically restricted or incomplete datasets likely experience high bias. For instance, A GAN forecast model trained on deforestation data from one area, would not be able to replicate the same degree of accuracy for another area.
- **Black-Box Nature:** the decision-making procedure of the GAN model is hard to understand. Its black-box nature diminishes the confidence of the government, public, and policymakers.

The ethical concerns mentioned above can be a roadmap for building sustainable GAN models in an environmental context.

8.1 Suggestions for Policymakers and Researchers

- **Risk Assessment** - The advanced predictive models allow the government to address climate risks proactively.
- **Cost-Effective Solutions** - GAN-generated data minimizes extensive field research, making environmental monitoring affordable.
- **Public Awareness and Engagement** - Practitioners can use GAN-generated visualization to educate the public and promote sustainable practices.
- **Environmental Policy** - Environmental policymakers would be able to derive realizable environmental policies based on data-driven AI models like GAN. It delivers evidence-based knowledge that contributes to refinements of environmental policies.
- **Real-world implementation**—Strong collaboration among environmental agencies, practitioners, and the government could actually facilitate more real-world implementation rather than simulation.

9. FUTURE DIRECTION

To maximize the effectiveness of GAN, researchers, and practitioners should focus on the
following limitations.

- **Data acquisition:** The acquisition and improvement of the availability and quality of environmental data shall be advanced through satellite technology, sensor networks, and programs for the sharing of data. Pre-treatment data techniques, such as noise removal and data augmentation can improve the power of GAN models for predictive purposes.

- **Domain Adaptation:** Developments in transfer learning techniques shall enable the trained GAN models on one domain to adapt well to many other geographical and environmental conditions. This may be accomplished through the fine-tuning of pre-trained GAN models on localized datasets that need minimal retraining.
- **Hybrid Models:** Hybrid models combining GANs with other machine learning and physics-based models can improve the accuracy and robustness of predictions. Hybrid models combining GANs with physical climate models will likely deliver far better forecasts of climate change by infusing data-driven insights with known scientific knowledge.
- **Computational Efficiency:** Optimizing GAN architectures for faster inference and real-time environmental applications.
- **Explainable AI (XAI):** Improving XAI methods tailored for GANs can increase even greater confidence in such models. Such techniques as saliency maps and mechanisms for attention would help researchers better understand what the GAN is predicting. Using this approach, subsequent application of GANs on edge computing platforms eases up the computational complexity associated with real-time environmental monitoring tasks. The ability to analyze satellite images and sensor data will be closer to the source of data of GAN models. Decreased latency and relieve burdens on the central servers will be achieved as a result. Therefore, artificial neural networks must be learned to provide for the need for a superior aspect of climatic variations such as temperature, humidity, wind characteristics, and ocean currents for complex environmental events like hurricanes, droughts, and wildfires.
- **Regulatory Framework:** There is an overriding need for explicit ethics criteria and regulatory frameworks in place for the application of GAN in environmental monitoring. These would encompass the safety of the data, justice, and accountability of the models towards the people, with a minimum risk to vulnerable people and ecosystems.
- **Policymaking:** Collaborative, open-source platforms should be developed for academics, governments, and organizations to share GAN models, datasets, and best practices in order to expedite development and usage.

By tackling these challenges, GAN can emerge as a significant asset for sustainable environmental management, disaster preparedness, and climate resilience.

10. CONCLUSION

The implementation of Generative Adversarial Networks in environmental monitoring and research is on the brink of revolutionizing how we monitor, comprehend, and respond to environmental changes. GANs are a potent instrument for addressing critical issues such as deforestation, climate change, and water management due to their capacity to generate accurate and realistic data, fill gaps in datasets, and model complex environmental processes. Such capacities would arm policymakers and researchers with tools for strategic planning and proactive intervention through anomaly detection in environmental patterns and scenario simulation forward in time. Another opportunity that will likely remain ever-expanding is the use of GANs for environmental research, creating new doors to real-time monitoring for disaster preparation and mitigation. This case study illustrates how the transformative power of GANs can be used for environmental sciences and the necessity of keeping research and development alive in this discipline. As GAN tools become advanced and intelligent, they are bound to become major instruments in formulating policies and actions to address the urgent environmental issue of our times. To conclude, the GAN offers a window where informed decisions may be data-driven and decisions made in favor of long-lasting support for environmental sustainability.

REFERENCES

Afaq, Y., & Manocha, A. (2021). Analysis on change detection techniques for remote sensing applications: A review. *Ecological Informatics*, *63*, 101310. DOI: 10.1016/j.ecoinf.2021.101310

Bihlo, A. (2021). A generative adversarial network approach to (ensemble) weather prediction. *Neural Networks*, *139*, 1–16. DOI: 10.1016/j.neunet.2021.02.003 PMID: 33662648

Boroujeni, S. P. H., & Razi, A. (2024). Ic-gan: An improved conditional generative adversarial network for rgb-to-ir image translation with applications to forest fire monitoring. *Expert Systems with Applications*, *238*, 121962. DOI: 10.1016/j.eswa.2023.121962

Boroujeni, S. P. H., & Razi, A. (2024). Ic-gan: An improved conditional generative adversarial network for rgb-to-ir image translation with applications to forest fire monitoring. *Expert Systems with Applications*, *238*, 121962. DOI: 10.1016/j.eswa.2023.121962

Bouteska, A., Seranto, M. L., Hajek, P., & Abedin, M. Z. (2023). Data-driven decadal climate forecasting using Wasserstein time-series generative adversarial networks. *Annals of Operations Research*. Advance online publication. DOI: 10.1007/s10479-023-05722-7

Creswell, A., White, T., Dumoulin, V., Arulkumaran, K., Sengupta, B., & Bharath, A. A. (2018). Generative adversarial networks: An overview. *IEEE Signal Processing Magazine*, *35*(1), 53–65. DOI: 10.1109/MSP.2017.2765202

Foroumandi, E., Gavahi, K., & Moradkhani, H. (2024). Generative adversarial network for real-time flash drought monitoring: A deep learning study. *Water Resources Research*, *60*(5). Advance online publication. DOI: 10.1029/2023WR035600

Goodfellow, I., Pouget-Abadie, J., Mirza, M., Xu, B., Warde-Farley, D., Ozair, S., & Bengio, Y. (2014). Generative adversarial nets. *Advances in Neural Information Processing Systems*, 27.

Guo, H., Jia, X., Dong, Y., Ye, J., Chen, D., Zhang, R., & Zheng, Y. (2020). Applications of AlGaN/GaN high electron mobility transistor-based sensors in water quality monitoring. *Semiconductor Science and Technology*, *35*(12), 123001. DOI: 10.1088/1361-6641/abb8fb

Hassan, A., Ali, M., Trigui, A., Savaria, Y., & Sawan, M. (2019). A GaN-based wireless monitoring system for high-temperature applications. *Sensors (Basel)*, *19*(8), 1785. DOI: 10.3390/s19081785 PMID: 31013978

Kaewunruen, S., AbdelHadi, M., Kongpuang, M., Pansuk, W., & Remennikov, A. M. (2022). Digital twins for managing railway bridge maintenance, resilience, and climate change adaptation. *Sensors (Basel)*, *23*(1), 252. DOI: 10.3390/s23010252 PMID: 36616851

Karamouz, M., Alipour, R. S., Roohinia, M., & Fereshtehpour, M. (2022). A remote sensing driven soil moisture estimator: Uncertain downscaling with geostatistically based use of ancillary data. *Water Resources Research*, *58*(10), e2022WR031946. DOI: 10.1029/2022WR031946

Koochali, A., Bakhshipour, A. E., Bakhshizadeh, M., Habermehl, R., Dilly, T. C., Dittmer, U., Ahmed, S., Haghighi, A., & Dengel, A. (2024). Water depth prediction in combined sewer networks, application of generative adversarial networks. *Discover Applied Sciences*, *6*(3), 123. DOI: 10.1007/s42452-024-05787-4

Luleci, F., Catbas, F. N., & Avci, O. (2022). Generative adversarial networks for data generation in structural health monitoring. *Frontiers in Built Environment*, *8*, 816644. DOI: 10.3389/fbuil.2022.816644

Mansourifar, H., Moskovitz, A., Klingensmith, B., Mintas, D., & Simske, S. J. (2022). GAN-based satellite imaging: A survey on techniques and applications. *IEEE Access : Practical Innovations, Open Solutions*, *10*, 118123–118140. DOI: 10.1109/ACCESS.2022.3221123

Peterson, K. T., Sagan, V., Sidike, P., Cox, A., & Martinez, M. (2018). Suspended sediment concentration estimation from land sat imagery along the lower missouri and middle mississippi rivers using an extreme learning machine. *Remote Sensing (Basel)*, *10*(10), 1503. DOI: 10.3390/rs10101503

Peterson, K. T., Sagan, V., & Sloan, J. J. (2020). Deep learning-based water quality estimation and anomaly detection using Landsat-8/Sentinel-2 virtual constellation and cloud computing. *GIScience & Remote Sensing*, *57*(4), 510–525. DOI: 10.1080/15481603.2020.1738061

Riaz, K., McAfee, M., & Gharbia, S. S. (2023). Management of climate resilience: Exploring the potential of digital twin technology, 3D city modelling, and early warning systems. *Sensors (Basel)*, *23*(5), 2659. DOI: 10.3390/s23052659 PMID: 36904867

Sagan, V., Peterson, K. T., Maimaitijiang, M., Sidike, P., Sloan, J., Greeling, B. A., & Adams, C. (2020). Monitoring inland water quality using remote sensing: Potential and limitations of spectral indices, bio-optical simulations, machine learning, and cloud computing. *Earth-Science Reviews*, *205*, 103187. DOI: 10.1016/j.earscirev.2020.103187

Shafique, A., Cao, G., Khan, Z., Asad, M., & Aslam, M. (2022). Deep learning-based change detection in remote sensing images: A review. *Remote Sensing (Basel)*, *14*(4), 871. DOI: 10.3390/rs14040871

Shukla, J. S., & Pandya, R. J. (2023). Deep learning-oriented c-GAN models for vegetative drought prediction on peninsular india. *IEEE Journal of Selected Topics in Applied Earth Observations and Remote Sensing*.

Spyrou, E. D., Tsoulos, I., & Stylios, C. (2022). Applying and comparing LSTM and ARIMA to predict CO levels for a time-series measurements in a port area. *Signals*, *3*(2), 235–248. DOI: 10.3390/signals3020015

Stengel, K., Glaws, A., Hettinger, D., & King, R. N. (2020). Adversarial super-resolution of climatological wind and solar data. *Proceedings of the National Academy of Sciences of the United States of America*, *476*(2240), 20190493. DOI: 10.1073/pnas.1918964117 PMID: 32631993

Tsatsaris, A., Kalogeropoulos, K., Stathopoulos, N., Louka, P., Tsanakas, K., Tsesmelis, D. E., Krassanakis, V., Petropoulos, G. P., Pappas, V., & Chalkias, C. (2021). Geoinformation technologies in support of environmental hazards monitoring under climate change: An extensive review. *ISPRS International Journal of Geo-Information*, *10*(2), 94. DOI: 10.3390/ijgi10020094

Ullo, S. L., & Sinha, G. R. (2020). Advances in smart environment monitoring systems using IoT and sensors. *Sensors (Basel)*, *20*(11), 3113. DOI: 10.3390/s20113113 PMID: 32486411

Wang, S., Li, W., Hou, S., Guan, J., & Yao, J. (2022). STA-GAN: A spatio-temporal attention generative adversarial network for missing value imputation in satellite data. *Remote Sensing (Basel)*, *15*(1), 88. DOI: 10.3390/rs15010088

Wang, X., Sun, L., Chehri, A., & Song, Y. (2023). A review of GAN-based super-resolution reconstruction for optical remote sensing images. *Remote Sensing (Basel)*, *15*(20), 5062. DOI: 10.3390/rs15205062

Yun, J., Kim, D., Kim, D. M., Song, T., & Woo, J. (2023). GAN-based sensor data augmentation: Application for counting moving people and detecting directions using PIR sensors. *Engineering Applications of Artificial Intelligence*, *117*, 105508. DOI: 10.1016/j.engappai.2022.105508

Chapter 9

Precision Agriculture and Artificial Intelligence: Towards Sustainable Resource Management in the Face of Climate Challenges

Bikie Gerald Anicet
https://orcid.org/0009-0003-2068-3494
National Advanced School of Publics Works, Cameroon

Amol D. Vibhute
Symbiosis Institute of Computer Studies and Research, Symbiosis International University (Deemed), Pune, India

Imane Aitouhanni
https://orcid.org/0009-0004-1400-8027
Mohammed V University, Morocco

Yassine Mouniane
https://orcid.org/0000-0002-6037-5301
Ibn Tofail University, Morocco

Issam El-Khadir
Ibn Tofail University, Morocco

Ahmed Chriqui
Ibn Tofail University, Morocco

Sakina Mehdioui
Ibn Tofail University, Morocco

Driss Hmouni
https://orcid.org/0000-0001-7598-6204
Ibn Tofail University, Morocco

ABSTRACT

Precision agriculture, combined with advances in artificial intelligence (AI), plays a decisive role in the sustainable management of agricultural resources in the face of climate challenges. This field aims to optimize farming practices by using technologies such as sensors, drones, and massive data analysis to monitor, predict and respond to specific crop needs. AI tools, such as machine learning and neural net-

DOI: 10.4018/979-8-3693-6900-5.ch009

works, make it possible to anticipate climatic impacts, efficiently manage irrigation and fertilizer use, while reducing the waste of resources. This marriage between technology and agriculture aims to strengthen the resilience of agricultural systems and promote sustainable production that meets the challenges of food security and environmental impact.

1. INTRODUCTION

Agriculture, as the oldest source of mankind's life on the planet, has undergone various processes of dynamic change, starting from the production and distribution of food, through the protection against different harmful factors, digestion, and food transformation, to the discovery of new types of animals and plants. However, classic farming techniques are not enough to eliminate all the existing challenges associated with their time-consuming, labor-intensive, and low-efficiency methods. In order to meet the increased demand for food, the concept of precision farming has been introduced (Zhang et al., 2024). It is a production management system that aims to use data on the variability of different agricultural objects, combining innovative technologies with services that optimize productivity, quality, and environmental performance. The decision-making process to achieve these goals is automated and fully supported by the AI algorithm. It is guided by specialists who post models of different situations or products, and experiments are performed to clarify issues such as the accuracy of native features and the representativeness of processed data. The result of the AI algorithm is in line with the objective to reduce the cost of products and the subsequent increase in their competitiveness in the market (Araújo et al., 2021; Monteiro et al., 2021; Abbasi et al., 2022).

The research related to the utilization of AI algorithms in the field of agriculture is vivid, especially in the last twenty years. The subject of the research is abundant, from solving the problem of predicting the yield of agricultural crops through rainfall prediction, predicting soil moisture, and plant growth, to the concept of automation of various agricultural instruments and technologies as a significant contribution to reducing the total operational costs per unit product. In general, the paper will focus on presenting the utilization of AI models in the field of precision agriculture. AI algorithms will be critically analyzed and presented as an idea based on the Internet of Things. Their application is a significant contribution to the development and improvement of the concept of precision agriculture by providing detailed information related to supporting resource management. Its application is an intelligent, feasible solution that can be used through sensors located on different parts of the agricultural field in order to define the particular objective by step-by-step achieving self-control, self-response, and smart advice. In the end, the overall approach can ensure optimal management of all resources, aiming for the sustainable realization of food security in the context of the greatest climate challenges faced (Sishodia et al., 2020; Bhat and Huang, 2021; Fuentes-Peñailillo et al., 2024).

Recent technologies such as artificial intelligence, big data, and the Internet of Things have been used in different scientific areas to solve many daily challenges. In agriculture, AI has been in the spotlight for the last few decades, mainly to overcome climate challenges. With the help of many new IoT devices, including UAVs, precision agriculture is possible.

This survey paper evaluates the use of AI technology in different kinds of agricultural research, highlighting the challenges as well as the benefits. With the help of AI, sustainable resource management is possible to preserve the earth's soil and water quality, enable crop management, and provide high yields and food security. With the help of precision agriculture technologies, monitoring and automation can prevent crop loss and increase yield. AI technologies are powerful tools to provide scientists with the best solutions for the very complex challenges in agriculture. Agriculture is an important area, as it is the main resource for food production. In the last few years, new challenges and concerns have arisen about land and water usage, decreasing biodiversity, heavy use of chemical applications, land depletion, and greenhouse gas emissions from irrigated lands, in addition to the use of energy and fuel. All these challenges are intensified by climate change, rising temperatures, and extreme weather conditions that affect global food production. Humans increase their demand for non-renewable resources, and precision agriculture has the potential to play a significant role in food security with its ability to transform traditional agricultural management systems, plantation, and fieldwork management. Lastly, this paper evaluates the differences between conventional agriculture and precision agriculture, pointing out AI utilization and its application in the development of these modern technologies (Javaid et al., 2023; Akintuyi, 2024; Ben Ayed and Hanana, 2021).

In this context of adapting the agricultural sector to climate change, the search for cutting-edge techniques to reduce the impact of agricultural production on the environment is a significant challenge. Precision agriculture helps achieve a crucial objective: to produce more and better, with less. This text highlights how artificial intelligence technologies can fit into this theme; they help improve the management of resources while increasing productivity and the quality of crops (Monteiro et al., 2021).

This chapter describes the concept of precision agriculture, before presenting the application methods of quasi-sensor techniques for implementing precision agriculture methods. This is followed by a presentation of how artificial intelligence technologies can be used to help solve specific problems, such as resource management to combat climate change.

2. PRECISION AGRICULTURE

2.1. Definition of Precision Agriculture

Precision agriculture has entered the realm of ownership of many rural landowners, for many different products. Precision farming offers farmers three main advantages: increased productivity, reduced production costs and environmental protection. It can also reduce phytosanitary product consumption by over 30%. This technology enables them to practice a more specific form of agriculture, taking into account specific parameters for the organization of the harvest, and not a priori, based on a cereal and oilseed logic (Petrović et al., 2024).

2.2. Using Sensors, Data and Technologies

Precision agriculture is the use of sensors, data and contemporary technologies to help farmers achieve better yields in their fields. Sensors from drones or other aerial vehicles are used to assess crop condition, while satellites are used for climatological and soil mapping. Once this information has been established, the data is organized in geographic information systems and specific software to help farmers precisely apply seeds, fertilizers and phytosanitary products to areas with specific needs. This sharpens yields, reduces input consumption and saves farmers valuable time.

In the context of intensification of climate risks, precision agriculture is considered to be a fast-track option to overcome potential challenges. Precision agriculture can contribute significantly to addressing a variety of structural challenges prevailing in the agriculture sector, specifically food security and growing disparities between the haves and have-nots. The concept of precision agriculture typically refers to the generation, collection, processing, and accurate utilization of information to make real-time, site-specific decisions for more profitable, sustainable, and competitive inclusive farming. Artificial intelligence acts as a catalyst in driving the movement towards precision agriculture. It comprises a group of related innovations that include a range of intelligent solutions to address site-specific management and decision-making support, shattering inherent thresholds associated with information-driven and farmer-oriented pragmatic high-impact decision-making (Sanders et al., 2021; Potts et al., 2023).

In light of the above, the role of artificial intelligence, specifically machine learning and deep learning algorithms, is highly significant in providing comprehensive solutions for addressing weather variability and other challenges for driving precision agriculture in a sustainable manner, enabling the adequate management of increasingly complex modern farming operations. In arable agriculture, the availability and cost of advanced input materials for field production are often not a constraint to the implementation of new technologies. The potential high levels of farm profitability on medium to large commercial farming operations create a considerable commercial incentive to adopt and adapt the most advanced technology packages available to realize the overall business potential. The successful operational adoption of these advanced unique technology packages is based on a favorable decision-making context to move from facts to economical and sustainable implementation of innovative solutions for a supported position of larger and efficient commercial farming operations in the food supply chain (Dewitte et al., 2021; Chen et al., 2023b).

2.3. Development of Precision Agriculture

In the context of climate change, increasing global population, and decreasing number of farmers, it is necessary to develop technologies that allow for greater food production. Precision agriculture is a field of research and development aimed at producing more efficiently. Here, we propose to describe precision agriculture, reflect on the benefits of co-innovation with the agronomic actors that are farmers, and conclude on the role of precision agriculture in the short, medium, and long term in a socially and environmentally sustainable world (Sishodia et al., 2020).

Precision agriculture has recently emerged as a pathway for progress in agricultural production to meet a growing triple societal demand. The necessary production of food with the most efficient use of resources requires minimizing excess application of fertilizers. The adoption of cultural practices such as mechanical weeding, which limit the use of chemical inputs, is partially offset by the reduction in the use of herbicides. More advanced connections could even become possible through the use of sensor-robots (Khan et al., 2021). Figure 1 shows a true model of the use of connected things in agriculture.

Figure 1. Illustration of the use of connected things in agriculture

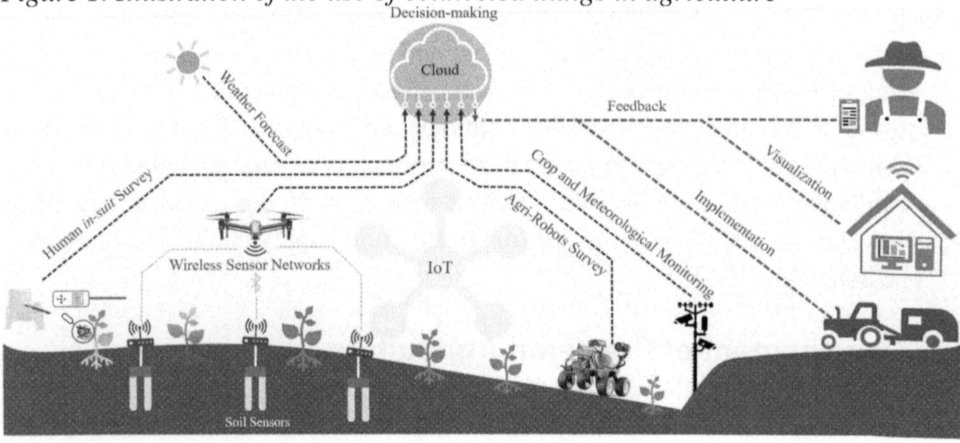

Source: Wang et al. (2025)

3. Importance of Spatial and Satellite Data in Modern Agriculture

Modern agriculture is facing new challenges, mainly linked to climate change, the growing demand for food by a growing population, major pressures on natural resources and, last but not least, strong societal expectations in terms of the environmental impact of human activities. Among all these challenges, some are more specific to agriculture itself and the way it must evolve in the face of these new pressures. The first of these is the substantial increase in crop productivity, due to the continuing growth in demand for food products, but also the prospect of the emergence of new alternatives to the fuels currently produced by petrochemicals (Gafurov et al., 2023; Segarra, 2024). The action program's premise is that space technologies are essential tools for meeting this first challenge, making a significant contribution to improving yields while reducing environmental impact. Space technologies can be applied in a number of ways. On the one hand, there are technologies for observing the Earth from space, enabling the acquisition of data for observing soils, crops and the biosphere in general, at different spatial and temporal scales. On the other hand, these technologies generate potential innovations for non-space sectors, whether in terms of image processing techniques, sensor design, cartography, remote sensing or scientific instrument design. They thus represent a major technological source of innovation for the specific field of technical assistance to farmers (Mahanta et al., 2024).

Figure 2. Intelligent sensors in agricultural use

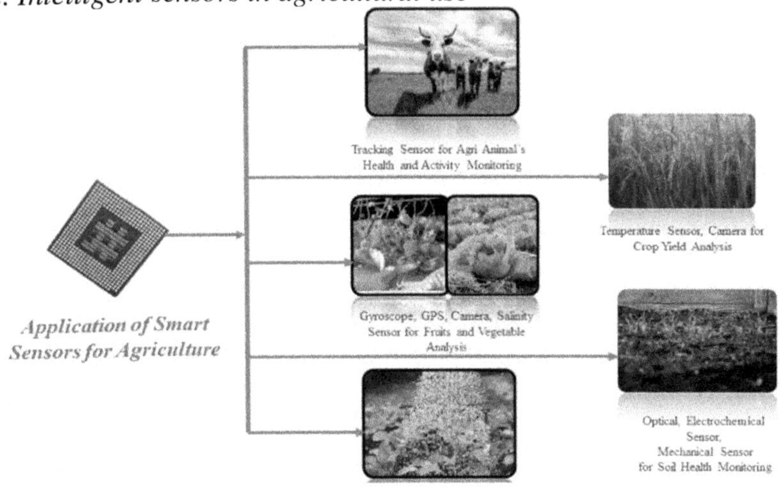

Source: Ullo and Sinha (2021)

3.1. Current Technologies: Satellite Images, Drones, Ground-Based Sensors

Today, techniques for estimating crop biophysical parameters are based on satellite images, drones and ground sensors. Satellite images provide information at regional scales and are frequently used. The advantages of these data over other remote sensing solutions for agriculture are their wide regional coverage, possessing channels in the range of human vision, and they are available at high frequencies and low cost. UAV technology also offers other advantages for quantifying biotic crop parameters. Finally, ground-based sensors can be used to obtain high-frequency, real-time measurements. Nevertheless, despite their good results, these technologies are still far from meeting all requirements. As they are too dependent on weather conditions, they struggle to provide measurements at predefined times. The field of autonomous robots and drones offers a promising answer to this problem. Despite its potential, the search for robots and drones capable of processing data on highly spatial and temporal scales, in conjunction with predictive models, remains a very active challenge. What's more, the results given are not easy to interpret for an untrained farmer. This can be circumvented by having algorithms with neural networks perform the estimation, but expert parameterization of the models to create a network can be an obstacle for them (Sishodia et al., 2020).

3.2. Satellite Images and Their Role in Agriculture

The use of satellite data in agriculture and how these data can solve problems encountered in the agricultural field, with a focus on the target region: Africa south of the Sahara. The challenge is to identify agricultural areas in Africa south of the Sahara in these satellite images, in order to boost agricultural productivity. This is expected to be achieved by using image segmentation and deep neural network techniques to identify fields and the different crops grown in them, as this information is crucial for people working in agriculture, as well as for decision-makers, economists and other stakeholders with an interest in farming communities. A better knowledge of these areas can help identify best practices in these fields and promote equal and sustainable agricultural development (Abd El-Ghany et al., 2020).

It is well known that countries and communities with strong agricultural economies are often more resilient when faced with economic crises and/or development challenges. In sub-Saharan Africa, agriculture is the source of food for most people in school. It is therefore important to identify best practices and create programs with sustainable impact. For this, knowledge of existing crop distribution and productivity in the surrounding area is essential, especially in countries without extensive agricultural data resources. Model formation using computer-aided methods is highly advantageous, as it enables a coarse representation of crop distribution over large tracts of land to be obtained in a short space of time **(Vellingiri et al. 2025**).

Precision agriculture offers the possibility for farmers to increase their input management according to the variability of the cultivated area, in order to produce more, more sustainably and with less drudgery for farmers. On this basis, the main objective of this research is to see how the development of artificial intelligence and imaging technologies can contribute to better resource management in the context of climate change. Artificial intelligence would enable specific work systems to achieve better phytosanitary applications and, above all, to work with autonomous vehicles. It also enables better monitoring of weed control and fertilization. Satellite imagery makes it possible to monitor crop development in both field and vegetable crops. Drones are also increasingly used to monitor crops but are also used in livestock farming to monitor herds and detect sick animals. The development of artificial intelligence technologies, which enable increasingly accurate prediction of production on the basis of image time series, could make it easier to decide on changes in practices to adapt digging rescue methods and irrigation techniques, as well as for the choice of inter-season cover and feed in the fight against drought **(Mouniane et al., 2024**).

To achieve a sustainable development aimed at the well-being of future generations, it is necessary to improve knowledge of the pressures exerted on the environment and ecosystems in the context of global climate challenges. In environmental risk management and agriculture, Earth observation satellite data have played a structuring and growing role since the 1990s. The goal of these research works is to increase the use of hyperspectral satellite imaging for key environmental variables detection. In this analysis, we examine the current state of the art in crop monitoring and the possibilities that hyperspectral satellite data represent. To do so, we compare supervised/human-assisted models based on field measurements taken in the agricultural zone investigated by a set of 95 flight lines over the years between 2018 and 2020. The tabulated database is then used in artificial intelligence models to establish statistical links between incident spectral images and measured parameters. After assessing the learning results of AI algorithms on large datasets, we propose suggestions and applications aimed at developing the satellite hyperspectral offering **(Bodrud-Doza et al., 2023)**.

3.3. Agricultural Applications

AI applications in agriculture are becoming increasingly numerous. They enable large quantities of data to be analyzed in real time to detect useful information for decision-making. For farming, AI applications can use algorithms to both monitor and predict greenhouse temperatures, plant growth levels, crop quality and the use of fertilizers or chemicals, positively influencing crop yields. On the administrative side, public institutions in charge of agriculture and livestock can remotely assist farmers by providing more accurate and efficient market recommendations and forecasts (Caparros-Santiago et al., 2021).

In many developing countries, particularly in Africa, agriculture accounts for 43% of total employment and 61% of the working population. The potential of these technologies deployed on a small scale in these countries has the potential to greatly increase the yield of family farms, thereby reducing farmer poverty. In addition, the growing use of these innovative technologies makes it possible to capture and analyze vast quantities of data on the agricultural sector, and to improve decision-making on both the buyer and consumer sides. The aim of this note is to provide a broad overview of what artificial intelligence systems are, the fields of application in agriculture, the benefits of their use, the obstacles to their deployment in the agricultural sector, and a literature review presenting cases of AI deployment in the agricultural sector in Africa (Giovos et al., 2021; Lanucara et al., 2024).

3.4. Monitoring Crop Growth

Crop growth monitoring is an essential form of agricultural issue monitoring. It can be a costly and messy process for growers. Resources, including money, need to be devoted. Alternatively, optical information can be collected using micro-drones, satellite data or cheaper aircraft. High-performance cameras can be attached to these devices to capture crop interaction at different wavelengths. Vegetation content at these wavelengths can be converted into information that can be useful to growers. This analysis uses optical information to help farmers quickly solve problems in their crops. In addition, the observation of fields contaminated by radicals can lead to agricultural practices that promote sustainable development and reduce the effects of storms. Vegetation surveys can be useful after storms to measure the extent of crop damage. The content of the optical information gathered at these times can be converted into grids that show stress points. These grids can then be used to investigate further agricultural issues. Sensor-equipped drones can be used to map crops at high resolution. Aerial images taken before and after an event, based on the algorithm's calculation, can be compared to obtain information on the destruction inflicted on crops. Growers would then be in a better position to make more informed decisions about the measures received by their crops (Garg et al., 2024; Saini et al., 2025).

3.5. Detection of Disease and Water Stress

AI is a powerful means of detecting and diagnosing biotic and abiotic stresses in plants. Abiotic stresses include water stress, which affects all agricultural systems. The work highlighted the crucial role of drones for measuring soil moisture and NDVI, hyperspectral imagery for identifying water stress and estimating crop losses, and deep learning models for estimating sap flow in plants. This work has proposed a frontier between machine learning and hyperspectral imagery for the characterization and management of water stress in greenhouses. This work thus contributes to the management of water stress by characterizing water stress to guide the automatic piloting of watering systems (Misra and Mall, 2024).

With regard to biotic stresses, in a challenge entitled Plant Pathology, a plant disease prediction record has been set on the basis of machine learning and deep learning models. Others have demonstrated the effectiveness of machine learning models for visually determining the level of biotic stress in plants. However, this work focuses mainly on cropping images from the aerial part of plants, without taking into account the detection of disease and water stress by tracking water and nutrients in the soil and in the plant (Pantelidakis et al., 2022; Bishshash et al., 2024).

3.6. Soil and Water Resources Mapping

Mapping soil and water resources is an essential aspect of precision agriculture. Not as good as on-station mapping, soil water analysis tools, such as capacitive probes or indirect methods based on optical, electromagnetic sensors, or near- and mid-infrared chemical imaging technologies, rapidly produce low-cost moisture and salinity maps (Kumar et al., 2022). Recent advances in artificial intelligence, which can easily process massive and diverse volumes of data from on-board sensors mounted on unmanned aerial vehicles, satellite sensors or ground monitoring stations to produce climate-focused information for managers and decision-makers, can identify and quantify phenomena and changes. In particular, artificial intelligence can provide very useful information on extreme phenomena that are at the root of natural climate disasters such as droughts and floods. This is the case with deep learning, which we use to identify and quantify extremes in temperature and precipitation, which nevertheless have biases that can be corrected a posteriori (Cui et al., 2023; Vellingiri et al. 2025).

3.7. Yield Prediction

Agricultural yields are linked to numerous regional factors of the plot, some of which cannot be predicted by the farmer or modeled scientifically before the growing season, outside of theoretical parameters of crops modeled by parameters derived from satellite measurements that strongly depend on weather conditions. For these main reasons, and because crops do not all face the same hazards, nor at the same time during the seasons, yields are not predicted with precision, and reliable ones such as statistical yields are published several months after the end of the season, while the market needs more information and proactive forecasting policies. To better assist farmers in their preparation for yields, yield forecasting programs are implemented each year, depending on the means to achieve this. Models and vegetation measurement tools by drone or ground measurements can be used to more quickly trigger the information "yield" in relation to the plant's stage, or from early estimates of a constant satellite observation that characterizes the field, the sowing date, and the first measurements of village statistics. These conditions allow for mapping multi-year stages of homogeneity under satellite sensor measurements within 10 days from mid-June, to feed spectral yield forecasting models based on meteorological peripheral data or the farmer's daily water balance **(Craswell, 2021;Qiao et al., 2022)**.

3.8. Cost-Benefit Analysis for Small Farmers

The adoption of AI-based technologies involves significant initial investment in hardware, software and training. However, the potential benefits, such as improved yields, reduced resource use and better disease detection, need to be weighed against the costs. In-depth cost-benefit analysis studies are needed to demonstrate the profitability of these technologies, particularly for small-scale farmers who often operate on narrow margins.

Figure 3, on the other hand, shows a simplified diagram of the integration of various sensors and feedback mechanisms for the implementation of intelligent agriculture, illustrating the interactions and information flows that are essential for optimised farm management.

Figure 3. Diagram illustrating the integration of diverse sensors and feedback systems for the implementation of smart agriculture

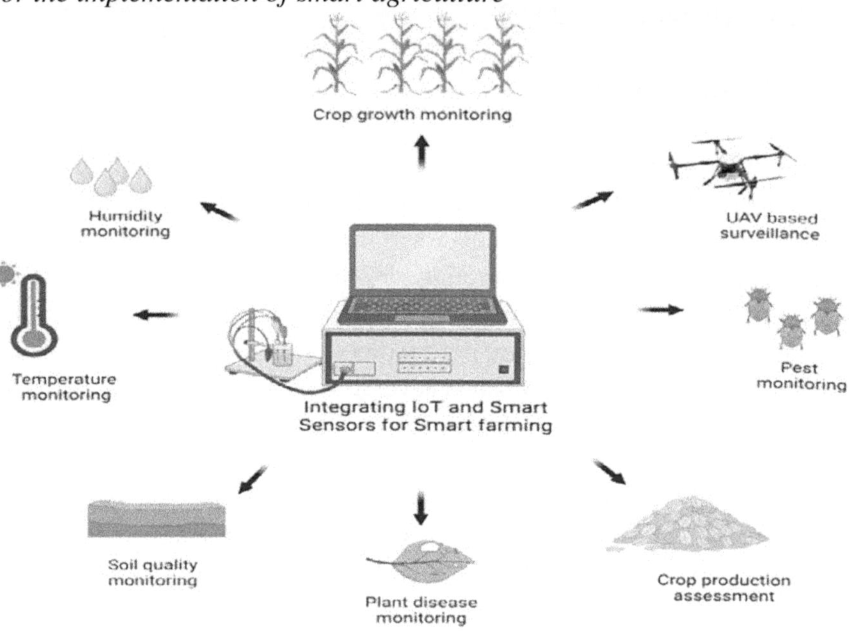

Source: Taseer and Han (2024)

4. ADVANTAGES OF SATELLITE IMAGES AND VEGETATIVE INDEXES

Optical sensor images have the advantage of covering large areas, while providing highly detailed data for mapping the surface features we need for different applications. While we can collect such types of data with drones, satellites offer such coverage of every land location on a regular scale. The additional advantages of satellites lie in the fact that images can be acquired at regular frequencies, as well as the multi-annual image databases scaled by today's satellites. These attributes of satellite images make them attractive candidates for meeting the requirements of a variety of precision farming applications. Satellite trajectory and integration lines currently exist in geostationary orbit that could provide near-time services for crop assessment and other resource-limiting applications, which we will explain below. While these constellations offer a new option for regular imaging, their cost of image acquisition is prohibitive for many agricultural users (Gao et al., 2020; Gong et al., 2024).

The recent advancements related to artificial intelligence have generated the potential of precision agriculture in order to achieve the optimized successful implementation of various beneficial operations associated with plant growth stages. The purpose of this paper is to illustrate the numerous possibilities and applications involving this field. In the first episode, this module presents a review of the scientific study predicting plant growth and health through an optimized classification approach. The distinguished properties of the NDVI will be summarized. They help in the precise detection and estimation of the volumetric content in water or in chlorophyll (vital components for the sustainability of agriculture) in the vegetative parts of the plant. The values of the NDVI give precise information about the phenologic state of the plant in order to carry out optimal inputs for the benefit of the field (De Natale et al., 2023; Mouniane et al., 2024). Indeed, the use of the NDVI to estimate the content in water and chlorophyll is very important from the point of view of the limitation of irrigation water and also the limitation of nitrogen fertilizer. This concerns the management and conservation of soils, nutrients, and water for their social, economic, and environmental sustainability. Subsequently, we exposed in the second part an original approach for the preprocessing and dimension reduction stages, which are essential for modeling relevant features of the scientific study. This approach emerges as a technology capable of capturing and evaluating the simplified patterns of the learning database found in the agro-meteorological fields. The main objective is to minimize the prediction error in order to ensure optimized saved costs. Indeed, the concept of size reduction associated with the LDA is based on the fact that by reducing the number of dimensions, the algorithm speed is faster, the error of overfitting is reduced, it is easier to visualize the database, and the most discerning feature coefficients and therefore attributes can be used for better agricultural monitoring. This approach captures and evaluates the simple patterns of the original signal available in the agrometeorology data. The application of this approach aims to optimize the features of NDVI indexes; the objective of all below is to analyze the agrometeorological behavior in order to estimate the chlorophyll and water content in the vegetative part of the apple tree (Huang et al., 2021; Ebrahimzadeh et al., 2023; Huo et al., 2024; Vellingiri et al. 2025).

5. ARTIFICIAL INTELLIGENCE (AI) FOR AGRICULTURE

Modeling, analysis and quality of agri-food systems, remote sensing and agriculture drones are part of digital agriculture and make a significant contribution to the modeling, analysis and quality of agri-food systems. By means of aerial surveys, observation drones provide considerable help in collecting data in the field, unlike conventional remote sensing using spacecraft. This time-consuming work can be put to good use for tasks such as inventorying phytosanitary and crop data, or supplying areas affected by natural disasters (Goswami et al., 2022). The main advantage is that photos can be taken at high spatial and temporal resolution. Numerous methods exist to extract information regarding soil coverage, plant growth, parcel classification, disease detection, automated segmentation of plants, parts of plants, objects, and the estimation of agronomic parameters, among others. Deep Learning algorithms are often used for their adaptive, nonlinear, noise-tolerant, and evolutionary capabilities. The advantages of deep neural networks offer interest for real-time missions. Drones capture computerized models that allow for monitoring collected data, conducting follow-ups, and helping to protect vegetation. They enable the quantification and localization of the specificities of agricultural parcels throughout the production cycle (Dhanush et al., 2023).

5.1. AI in Agricultural Data Analysis

From the mechanical to the digital age, from the organic to the artificial. The peasant farmers of the earlier period overthrew the important role of the administrative workers of the modern age. Yet, in turn, the new types of workers erected artificial intelligence on their shoulders, which became more and more necessary and required more and more precision. The reasons for this interference were many. The ever-increasing global population, climate change and the need for environmental conservation are among the best-known charges. New objects of study for the method are technologies in the agricultural sector. Artificial intelligence methods are used in a wide range of fields that support agriculture, such as crop protection and seed selection (Chen et al., 2023a).

With the evolution of technologies, data, increasingly colossal, has led to challenges in their processing and storage. In the scientific and industrial context, the evaluation of this data is valuable for supporting decisions and forecasting future events. This is why the field of machine learning is sought after in order to model this data, while respecting criteria such as prediction quality, speed, and simplicity of implementing a system. However, the AI industry has become a lever for development and innovation in the community; meeting these criteria is a real challenge since the use of particular cases becomes complex. In this document, we aim to estimate wheat stocks and yield before harvest using multispectral and hyperspectral optical data to estimate the water content of plants. We have spectra of wheat collected every hour for 17 days. We present a selection of data set partitioning, implement algorithms frequently used to successfully predict a measurement made by a multiparametric wheat probe. The prediction relies on three elements: distribution across different indicators and learning methods, aggregation of results, and transfer of the learned model (Backman et al., 2023).

Precision agriculture has been implemented to limit the impact related to production. In the current context of climate challenges, it proves to be even more relevant by providing not only an improvement in resource management but also by increasing production in at-risk regions. Indeed, it exploits the significant volume of multispectral data from digital agriculture to make an objective prediction. In our study, we test different learning methods for predicting water content based on hyperspectral data collected from five wheat plots under construction. Observing crops over time to adapt resource management plans (Sachithra and Subhashini, 2023).

5.2. Predictive Analysis to Optimize Farming Practices

Taking climate change into account in agricultural decisions and practices is vital if we are to keep the world's growing food needs within an acceptable envelope of environmental sustainability. Irrigation and fertilization are costly for crops and have environmental impacts through changes in the availability and quality of water and soil resources. Artificial intelligence offers new prospects for exploiting the full potential of soil-plant-atmosphere phenomena and guiding these practices to maximize production and minimize costs for the farmer and impact on the environment. The effectiveness of this approach has been demonstrated on irrigated annual crops with constrained water regimes (Vinod Chandra et al., 2024).

5.3. Automatic Detection of Anomalies

Automatic detection of plant anomalies such as water stress through plant energy, pigmentation alteration and leaf diseases brings intelligence to the watering process, photosynthesis regulation and accelerates decision making to combat disease. These anomalies are measured by indices extracted from thermal and hyperspectral images. To obtain the economic value of these indices, a combination of supervised learning techniques such as linear regression can be used to visualize key indicators of yield, efficiency and quality, which the grower can then develop with a view to their optimal adaptation. For the detection of water stress, indices such as Stress Water Index / Transpiration on Leaf Temperature and Temperature Gradient expressed by thermal imagery are often used, as well as hyperspectral imagery in the near-infrared range by a Soil-adjusted Vegetation Index (Shendryk et al., 2021; Prabha et al., 2024). Hyperspectral analysis is an opportunity for early expression of polyphenols in grapes and proanthocyanidins in grapes, as part of genetic understanding and grape selection in viticulture, for which the system monitored must be capable of performing all three operations: acquisition, pre-processing and construction from calibration data. New imaging technologies, such as long-wave imaging and deep learning with convolution networks, offer great promise for the classification and detection of plant anomalies in maize. A predictive model based on different wavelengths of the electromagnetic spectrum incorporates the effect of the thermo-optical behavior of plants in general, which can be further developed to increase the intelligence associated with decision-making. Low-cost thermal imaging is likely to bring dynamism and efficiency to behavioral monitoring of crop intruders (Sharma and Shivandu, 2024).

5.4. Applied AI Algorithms

The estimation of fertilization needs based on features related to the plant and its environment involves a set of measurements and observations, also called phenoscaping. The use of different categories of AI algorithms, each appropriate for specific applications in precision agriculture in the context of climate challenges, is essential. Indeed, these approaches are positioned in the generation and transformation of information for each feature to finally deduce the required doses of fertilization. The nature of the features used during the phenoscaping steps in precision agriculture can be very different (Javaid et al., 2023). It can include observed parameters of corn diagnostics, images acquired by UAV, measurements deduced from models, datasets derived from farmers' practices, and databases. These steps play an important role in estimating and optimizing plant responses to efficient management of resources and are the focus of the recent use of large modern wave AI algorithms (Bhat et al., 2024).

5.5. Regression Algorithms for Yield Prediction

Precision farming aims to optimize crop yields while minimizing environmental hazards. Drone monitoring and aerial imagery can help track multiple agronomic parameters, enabling targeted interventions for more sustainable management. That said, prediction in raster for very large yield series is often an intractable problem at lower cost. Another avenue for yield prediction is the modeling of yield time series, using adaptive aggregation concepts to search for efficient algorithms. Promising results for algorithms are obtained in the analysis of image time series or multi-temporal index ratio time series. Explicit modeling of spatio-temporal aberrations (caused by random or systematic influences) can improve the non-sampling cost. A new regression model for yield prediction is introduced. The aim is to economically model a complete raster image for yield prediction based on spatialized relationships of temporal variables and many original non-spatialized predictor variables. The latter are implemented by integrating a temporal model for each raster pixel into a dimension reduction model. The model consists of combining estimated and aggregated spatial parameters of the relationship on each non-spatialized variable in relation to a residual of the stationary normal distribution. Efficient conditional calculations are obtained using identifiability between different variables or residuals and using advanced minimization algorithms. Simulation experiments in low and high forecast situations demonstrate the model's performance. Acceptable economic results are obtained in an application for high-resolution prediction at a high cost condition (Kouadio et al., 2018; Balaji Prabhu et al., 2024).

5.6. Challenges of AI Implementation in Precision Agriculture

Although precision farming has promising potential for improving the efficiency and sustainability of farms, there are still a number of challenges to its implementation. Among these, the high cost of technologies and limited access to digital infrastructure in rural areas represent major obstacles for small-scale farmers. Adapting AI technologies to resource-limited contexts requires more affordable solutions tailored to local needs. It is essential to develop sustainable business models and subsidies to encourage widespread adoption (Linaza et al., 2021).

5.7. Technical Details on AI Algorithms and Models

Machine learning algorithms, such as random forests, convolutional neural networks (CNNs) and support vector machines (SVMs), play a key role in precision farming applications. Each algorithm has its own advantages and limitations, depending on the type of data and specific objectives. For example, CNNs are particularly effective for disease detection from leaf images, while SVM models are often used for crop classification. However, the selection and adjustment of hyperparameters, such as the learning rate and the size of the hidden layers, are critical for optimising model performance (Wang et al., 2024; Khan et al., 2024).

6. SUSTAINABLE AGRICULTURE AND RESILIENCE TO CLIMATE CHANGE

Agriculture plays an important role in building resilience and reducing the impact of climate change. Technologies such as biotechnology, advanced genomics, metagenomic engineering, synthetic biology and nanotechnology can be used to increase production yields with limited resources and address environmental and climate challenges. Nanotechnology is also playing a role in increasing agricultural productivity through the optimization of water use and the development of biosensors for environmental monitoring. Spatial imagery plays an important role in environmental monitoring, providing information on species diversity, ecological processes and land use trends and their evolution in time, space and spatial/temporal scale (Neethirajan, 2024). The development of geographic information systems and their integration with remote sensing and other emerging technologies contributes to ecosystem monitoring and the elaboration of recommendations for sustainable development and integrated management of water resources in rural and urban areas, while taking into account people's land rights. Precision agriculture plays an important role in managing natural resources, monitoring climate change and improving the resilience and adaptation of farmers and agricultural ecosystems (Mgendi, 2024; Bhat et al., 2024).

6.1. The impact of Technology on Agricultural Sustainability

Agriculture must provide more food and more diversified food that meets the needs of populations in terms of nutrition and food security. Within the context of the adaptation of agricultural systems to the challenge of climate change, major changes in agricultural practices are needed while limiting greenhouse gas emissions. The aim of sustainably managing resources in a changing climate is to reduce the negative impacts of agriculture on the environment and to provide safe and healthy products to consumers. This implies reducing the dependence on inputs. Unquestionably, new technologies, smart systems, robots, artificial intelligence, and big data analysis tools will play a prominent role in agriculture of the future. In the first part, some of the challenges facing agriculture in terms of durability will be considered. The effects of new technologies in the light of these goals will then be addressed. In conclusion, it will be argued that sustainable agriculture can indeed emerge if the implications of these new systems for the major challenges are considered and if societal choices are made with total transparency (Khan et al., 2021; Gamage et al., 2024).

6.2. Reducing the Environmental Footprint

In various studies, geographic information systems, global positioning systems, remote sensing, and yield monitors have been exploited to develop site-specific technologies at the field scale. The advent of artificial intelligence, combined with several information technologies, may enable the prediction of crop yields at field scales or at scales measuring at least 10 m. As a strategic approach to precision farming, which is compared to non-precise management techniques, AI technologies, such as machine learning, enable a rapid increase in automation levels and the use of big data for decision-support tools, channeling an increase in experience areas. These customizations should ultimately lead to increased food production (Cesco et al., 2023). As such, these systems are considered not only functional but indeed precise. CIOs may choose sustainable models for their commercial initiatives to mitigate ecological endangerments. The adoption of information technology tools has been proposed as an alternative that can support the management of extreme climate events. Among the benefits is the opportunity for timely measures to support crop production in adverse climates. Precision farming, with the use of AI models, may be useful in this context (Saleem et al., 2023). It must be emphasized that AI tools make it possible, at the same time, to promote not only the increase in food and energy production and improved economic performance but also to reduce the environmental repercussions linked to agriculture. AI combined with IoT may enhance monitoring automation levels at regional levels. In combination with big data-linked AI technological tools, they can also drive decision-making processes in relation to near-real-time weather forecasts or alert messaging. The increase in sensor sensitivity is crucial, ultimately promoting widespread sustainable development of resource usage in interconnected scenarios like interrelated crop fields. This study is based on widely reviewed published studies and aims to inform and guide the optimization of posthuman agency organizational initiatives for a reduction in the commercial and therefore corporate socio-environmental applications of the AI-based precision farming method (Javaid et al., 2023).

7. CHALLENGES AND PROSPECTS

Given the complexity, variability, uncertainties and multidimensional nature of agricultural problems and issues in the context of climate change, the integration of digital technologies, particularly artificial intelligence, raises a number of challenges and offers promising prospects. Among these challenges are those linked to the low connectivity of rural areas and, consequently, the need to optimize the costs of installing and operating digital systems. One problem linked to connectivity in rural areas remains the duplication of costs for equipping agricultural or transport machinery and general road vehicles (Piao et al., 2019; Tang et al., 2021). Given this situation, the principle of the field guide represents an approach that some researchers consider relatively recent. An agricultural harvest guide for a group of machines and tractors is a service offered on the model of carpooling. The presence or absence of a machine or tractor for a task is indicated on the application by a map, and a countdown guides the user. The second challenge is the importance of uncertainty-related issues and the emergence of expert systems capable of reasoning not only in normal mode, but also in degraded mode in the event of loss of connection or loss of information quality. Game-theoretic parameters will be integrated to enable adaptive decision-making in a context of cooperation or competition, in agrosystems and agroecosystems at plot, farm and territorial scales. These general-purpose systems will also be flexible enough to handle the aggregation and disaggregation of human activity by means of mobile or stationary agricultural robot systems (Karunathilake et al., 2023; Shekhar et al., 2024).

The creation of more robust, efficient, and trustworthy AI models, capable of managing and fully exploiting multimodal and multisource data, as well as the definition and construction of dynamic communication channels and human-machine interfaces of DSSs intrinsic to Agro-ICT architecture, are some of the differentiating factors and improvements that could contribute to making PA even more effective, rapid, and proactive in managing and controlling the variability of crops (Vibhute et al., 2024). The digitalization of agriculture, enhancing the precision of management of natural resources, minimizing the impact and risks in the environment, in terms of conservation and protection, increasing resilience to change, increasing the efficiency of distribution products in the food supply chain, and promoting sustainable rural development are very important contributors to agriculture 4.0, food security, improvement of quality of life and welfare, with long-term positive impacts from an economic and environmental-ecosystem perspective. Articulating knowledge of agronomy, economics, ecosystem management, mechanical engineering, informatics, mathematics, and computer science is only possible if there is active collaboration between researchers and experts in the field (Fuentes-Peñailillo et al., 2024).

7.1 Ethical Considerations and Socio-Economic Impacts

The massive collection of agricultural data raises crucial questions about data confidentiality and the ethical use of AI. Farmers need to be aware of data management practices and the potential implications. In addition, increased automation could affect employment in the agricultural sector, requiring proactive planning to avoid negative impacts on rural communities. An inclusive approach, taking into account social and ethical aspects, is essential to ensure the responsible deployment of technologies.

7.2. Importance of Data Heterogeneity

The heterogeneity of agricultural data, which comes from a variety of sources such as satellite images, IoT sensors and information provided by farmers, poses integration and analysis challenges. The use of robust algorithms capable of handling diverse datasets is essential to guarantee accurate predictions. Data pre-processing techniques, such as normalisation and weighted aggregation, can improve the consistency and reliability of results.

8. CONCLUSION

Precision agriculture relies on artificial intelligence (AI) technologies to solve critical productivity and sustainability challenges. To improve the clarity of our chapter, we have detailed several specific applications. For example, convolutional neural networks (CNNs) are widely used for crop disease detection, enabling the precise identification of visual symptoms from leaf images. For yield prediction, supervised regression algorithms analyse historical and climatic datasets to estimate crop performance. In addition, resource management, such as irrigation optimisation, uses computer vision models to adjust water volumes according to the specific needs of plants.

To illustrate the impact of these technologies, we have added regional case studies. In India, drones equipped with multispectral cameras monitor crop health, offering an effective solution on a large scale. In Africa, AI-based mobile applications are helping farmers diagnose diseases from photos taken with smartphones, reducing agricultural losses and increasing income. These examples demonstrate the effectiveness of AI even in resource-constrained environments.

We have also delved deeper into the analysis of economic, political and connectivity challenges to provide a global perspective. Initial investment costs are a major barrier for small-scale farmers, holding back the adoption of new technologies. On the political front, the lack of appropriate regulations complicates the integration of AI in the agricultural sector. Finally, digital infrastructures and Internet connectivity are often insufficient in rural areas, limiting the effectiveness of AI-based solutions.

The literature review has been enriched with recent references. We have included recent work on computer vision applied to disease detection and thermal analysis techniques for monitoring plant water stress. This research offers insights into the rapid development of technologies in the agricultural sector.

Finally, the chapter structure has been improved with clear sub-sections such as 'AI Technologies in Precision Agriculture' and 'Regional Challenges and Solutions for AI Adoption'. Visual aids, including diagrams showing the flow of data in AI-based systems and graphs illustrating the impact of technologies on crop yield, have been incorporated to enhance the reader's visual engagement.

REFERENCES

Abbasi, R., Martinez, P., & Ahmad, R. (2022). The digitization of agricultural industry–a systematic literature review on agriculture 4.0. *Smart Agricultural Technology, 2*, 100042. DOI: 10.1016/j.atech.2022.100042

Abd El-Ghany, N. M., Abd El-Aziz, S. E., & Marei, S. S. (2020). A review: Application of remote sensing as a promising strategy for insect pests and diseases management. *Environmental Science and Pollution Research International, 27*(27), 33503–33515. DOI: 10.1007/s11356-020-09517-2 PMID: 32564316

Akintuyi, O. B. (2024). Adaptive AI in precision agriculture: a review: investigating the use of self-learning algorithms in optimizing farm operations based on real-time data. *Research Journal of Multidisciplinary Studies., 7*(2), 16–30. DOI: 10.53022/oarjms.2024.7.2.0023

Araújo, S. O., Peres, R. S., Barata, J., Lidon, F., & Ramalho, J. C. (2021). Characterising the agriculture 4.0 landscape—Emerging trends, challenges and opportunities. *Agronomy (Basel), 11*(4), 667. DOI: 10.3390/agronomy11040667

Backman, J., Koistinen, M., & Ronkainen, A. (2023). Agricultural process data as a source for knowledge: Perspective on artificial intelligence. *Smart Agricultural Technology, 5*, 100254. DOI: 10.1016/j.atech.2023.100254

Balaji Prabhu, B. V., Shashank, R., Shreyas, B., & Narsipura, O. S. J. (2024). ARIA: Augmented Reality and Artificial Intelligence enabled mobile application for Yield and grade prediction of tomato crops. *Procedia Computer Science, 235*, 2693–2702. DOI: 10.1016/j.procs.2024.04.254

. Ben Ayed, R., Hanana, M. 2021. Artificial intelligence to improve the food and agriculture sector. Journal of Food Quality. Article ID 5584754, 7 pages. https://doi.org/DOI: 10.1155/2021/5584754

Bhat, S. A., & Huang, N. F. (2021). Big data and ai revolution in precision agriculture: Survey and challenges. *IEEE Access : Practical Innovations, Open Solutions, 9*, 110209–110222. DOI: 10.1109/ACCESS.2021.3102227

Bhat, S. A., Qadri, S. A. A., Dubbey, V., Sofi, I. B., & Huang, N. F. (2024). Impact of crop management practices on maize yield: Insights from farming in tropical regions and predictive modeling using machine learning. *Journal of Agriculture and Food Research, 18*, 101392. DOI: 10.1016/j.jafr.2024.101392

Bishshash, P., Nirob, A. S., Shikder, H., Sarower, A. H., Bhuiyan, T., & Noori, S. R. H. (2024). A comprehensive cotton leaf disease dataset for enhanced detection and classification. *Data in Brief, 57*, 110913. DOI: 10.1016/j.dib.2024.110913 PMID: 39328970

Bodrud-Doza, M., Yang, W., de Queiroga Miranda, R., Martin, A., DeVries, B., & Fraser, E. D. G. (2023). Towards implementing precision conservation practices in agricultural watersheds: A review of the use and prospects of spatial decision support systems and tools. *The Science of the Total Environment, 905*, 167118. DOI: 10.1016/j.scitotenv.2023.167118 PMID: 37717782

Caparros-Santiago, J. A., Rodriguez-Galiano, V., & Dash, J. (2021). Land surface phenology as indicator of global terrestrial ecosystem dynamics: A systematic review. *ISPRS Journal of Photogrammetry and Remote Sensing, 171*, 330–347. DOI: 10.1016/j.isprsjprs.2020.11.019

Cesco, S., Sambo, P., Borin, M., Basso, B., Orzes, G., & Mazzetto, F. (2023). Smart agriculture and digital twins: Applications and challenges in a vision of sustainability. *European Journal of Agronomy, 146*, 126809. DOI: 10.1016/j.eja.2023.126809

Chen, H. Y., Sharma, K., Sharma, C., & Sharma, S. (2023a). Integrating explainable artificial intelligence and blockchain to smart agriculture: Research prospects for decision making and improved security. *Smart Agricultural Technology, 6*, 100350. DOI: 10.1016/j.atech.2023.100350

Chen, L., Chen, Z., Zhang, Y., Liu, Y., Osman, A. I., Farghali, M., Hua, J., Al-Fatesh, A., Ihara, I., Rooney, D. W., & Yap, P. (2023b). Artificial intelligence-based solutions for climate change: A review. *Environmental Chemistry Letters, 21*(5), 2525–2557. DOI: 10.1007/s10311-023-01617-y

Craswell, E. (2021). Fertilizers and nitrate pollution of surface and ground water: An increasingly pervasive global problem. *SN Applied Sciences, 3*(4), 518. DOI: 10.1007/s42452-021-04521-8

Cui, X., Han, W., Zhang, H., Dong, Y., Ma, W., Zhai, X., Zhang, L., & Li, G. (2023). Estimating and mapping the dynamics of soil salinity under different crop types using Sentinel-2 satellite imagery. *Geoderma, 440*, 116738. DOI: 10.1016/j.geoderma.2023.116738

De Natale, F., Alilla, R., Parisse, B., & Nardi, P. (2023). A bibliometric analysis on drought and heat indices in agriculture. *Agricultural and Forest Meteorology, 341*, 109626. DOI: 10.1016/j.agrformet.2023.109626

Dewitte, S., Cornelis, J. P., Müller, R., & Munteanu, A. (2021). Artificial intelligence revolutionises weather forecast, climate monitoring and decadal prediction. *Remote Sensing (Basel), 13*(16), 3209. DOI: 10.3390/rs13163209

Dhanush, G., Khatri, N., Kumar, S., & Shukla, P. K. (2023). A comprehensive review of machine vision systems and artificial intelligence algorithms for the detection and harvesting of agricultural produce. *Scientific African, 21*, e01798. DOI: 10.1016/j.sciaf.2023.e01798

Ebrahimzadeh, G., Mahabadi, N. Y., & Bayat, H. (2023). Estimating pre-compression stress in agricultural Soils: Integrating spectral indices and soil properties through machine learning. *Computers and Electronics in Agriculture, 215*, 108393. DOI: 10.1016/j.compag.2023.108393

Fuentes-Peñailillo, F., Gutter, K., Vega, R., & Silva, G. C. (2024). Transformative technologies in digital agriculture: Leveraging Internet of Things, remote sensing, and artificial intelligence for smart crop management. *Journal of Sensor and Actuator Networks., 13*(4), 39. DOI: 10.3390/jsan13040039

Gafurov, A., Mukharamova, S., Saveliev, A., & Yermolaev, O. (2023). Advancing Agricultural Crop Recognition: The Application of LSTM Networks and Spatial Generalization in Satellite Data Analysis. *Agriculture, 13*(9), 1672. DOI: 10.3390/agriculture13091672

Gamage, A., Gangahagedara, R., Subasinghe, S., Gamage, J., Guruge, C., Senaratne, S., Randika, T., Rathnayake, C., Hameed, Z., Madhujith, T., & Merah, O. (2024). Advancing sustainability: The impact of emerging technologies in agriculture. *Current Plant Biology, 40*, 100420. DOI: 10.1016/j.cpb.2024.100420

Gao, L., Wang, X., Johnson, B. A., Tian, Q., Wang, Y., Verrelst, J., Mu, X., & Gu, X. (2020). Remote sensing algorithms for estimation of fractional vegetation cover using pure vegetation index values: A review. *ISPRS Journal of Photogrammetry and Remote Sensing, 159*, 364–377. DOI: 10.1016/j.isprsjprs.2019.11.018 PMID: 36082112

Garg, S., Rumjit, N. P., & Roy, S. (2024). Smart agriculture and nanotechnology: Technology, challenges, and new perspective. *Advanced Agrochem., 3*(2), 115–125. DOI: 10.1016/j.aac.2023.11.001

Giovos, R., Tassopoulos, D., Kalivas, D., Lougkos, N., & Priovolou, A. (2021). Remote Sensing Vegetation Indices in Viticulture: A Critical. *Agriculture, 11*(5), 457. DOI: 10.3390/agriculture11050457

Gong, Z., Ge, W., Guo, J., & Liu, J. (2024). Satellite remote sensing of vegetation phenology: Progress, challenges, and opportunities. *ISPRS Journal of Photogrammetry and Remote Sensing*, *217*, 149–164. DOI: 10.1016/j.isprsjprs.2024.08.011

Goswami, A., Sharma, D., Mathuku, H., Gangadharan, S. M. P., Yadav, C. S., Sahu, S. K., Pradhan, M. K., Singh, J., & Imran, H. (2022). Change detection in remote sensing image data comparing algebraic and machine learning methods. *Electronics (Basel)*, *11*(3), 431. DOI: 10.3390/electronics11030431

Huang, S., Tang, L., Hupy, J. P., Wang, Y., & Shao, G. (2021). A commentary review on the use of normalized difference vegetation index (NDVI) in the era of popular remote sensing. *Journal of Forestry Research*, *32*(1), 1–6. DOI: 10.1007/s11676-020-01155-1

Huo, D., Malik, A. W., Ravana, S. D., Rahman, A. U., & Ahmedy, I. (2024). Mapping smart farming: Addressing agricultural challenges in data-driven era. *Renewable & Sustainable Energy Reviews*, *189*, 113858. DOI: 10.1016/j.rser.2023.113858

Javaid, M., Haleem, A., Khan, I. H., & Suman, R. (2023). Understanding the potential applications of Artificial Intelligence in Agriculture Sector. *Advanced Agrochem.*, *2*(1), 15–30. DOI: 10.1016/j.aac.2022.10.001

Javaid, M., Haleem, A., Singh, R. P., & Suman, R. (2022). Enhancing smart farming through the applications of Agriculture 4.0 technologies. *International Journal of Intelligent Networks.*, *3*, 150–164. DOI: 10.1016/j.ijin.2022.09.004

Karunathilake, E. M. B. M., Le, A. T., Heo, S., Chung, Y. S., & Mansoor, S. (2023). Chung, Y.S.; Mansoor, S. The Path to Smart Farming: Innovations and Opportunities in Precision Agriculture. *Agriculture*, *13*(8), 1593. DOI: 10.3390/agriculture13081593

Khan, A., Patil, C. H., Vibhute, A. D., & Mali, S. (2024). A U-Net Based Approach for High-Accuracy Land Use Land Cover Classification in Hyperspectral Remote Sensing. In Patel, K. K., Santosh, K., Patel, A., & Ghosh, A. (Eds.), *Soft Computing and Its Engineering Applications. icSoftComp 2023. Communications in Computer and Information Science* (Vol. 2031). Springer., DOI: 10.1007/978-3-031-53728-8_8

Khan, N., Ray, R. L., Sargani, G. R., Ihtisham, M., Khayyam, M., & Ismail, S. (2021). Current progress and future prospects of agriculture technology: Gateway to sustainable agriculture. *Sustainability (Basel)*, *13*(9), 4883. DOI: 10.3390/su13094883

Kouadio, L., Deo, R. C., Byrareddy, V., Adamowski, J. F., Mushtaq, S., & Phuong Nguyen, V. (2018). Artificial intelligence approach for the prediction of Robusta coffee yield using soil fertility properties. *Computers and Electronics in Agriculture*, *155*, 324–338. DOI: 10.1016/j.compag.2018.10.014

. Kumar, S., Meena, R.S., Sheoran, S., Jangir, C.K., Jhariya, M.K., Banerjee, A., Raj, A. 2022. Remote sensing for agriculture and resource management. Natural Resources Conservation and Advances for Sustainability. 91-135. https://doi.org/ DOI: 10.1016/B978-0-12-822976-7.00012-0

Lanucara, S., Praticò, S., Pioggia, G., Di Fazio, S., & Modica, G. (2024). Web-based spatial decision support system for precision agriculture: A tool for delineating dynamic management unit zones (MUZs). *Smart Agricultural Technology*, *8*, 100444. DOI: 10.1016/j.atech.2024.100444

Linaza, M. T., Posada, J., Bund, J., Eisert, P., Quartulli, M., Döllner, J., Pagani, A. G., Olaizola, I., Barriguinha, A., Moysiadis, T., & Lucat, L. (2021). Data-Driven Artificial Intelligence Applications for Sustainable Precision Agriculture. *Agronomy (Basel)*, *11*, 1227. DOI: 10.3390/agronomy11061227

Mahanta, D. K., Bhoi, T. K., Komal, J., Samal, I., & Mastinu, A. (2024). Spatial, spectral and temporal insights: Harnessing high-resolution satellite remote sensing and artificial intelligence for early monitoring of wood boring pests in forests. *Plant Stress*, *11*, 100381. DOI: 10.1016/j.stress.2024.100381

Mgendi, G. (2024). Unlocking the potential of precision agriculture for sustainable farming. *Discov Agric.*, *2*(1), 87. DOI: 10.1007/s44279-024-00078-3

Misra, V., & Mall, A. K. (2024). Harnessing image processing for precision disease diagnosis in sugar beet agriculture. *Crop Design.*, *3*(4), 100075. DOI: 10.1016/j.cropd.2024.100075

Monteiro, A., Santos, S., & Gonçalves, P. (2021). Precision agriculture for crop and livestock farming—Brief review. *Animals (Basel)*, *11*(8), 2345. DOI: 10.3390/ani11082345 PMID: 34438802

Mouniane, Y., Chriqui, A., El-Khadir, I., Hbyaj, K., El-Moustaqim, K., Mezouara, M., Zirari, M., Mabrouki, J., & Hmouni, D. (2024). Impact of Remote Sensing on Irrigation Management in Precision Agriculture. In Mabrouki, J., & Mourade, A. (Eds.), *Technical and Technological Solutions Towards a Sustainable Society and Circular Economy. World Sustainability Series*. Springer., DOI: 10.1007/978-3-031-56292-1_19

Neethirajan, S. (2024). Net Zero Dairy Farming—Advancing Climate Goals with Big Data and Artificial Intelligence. *Climate (Basel)*, *12*(2), 15. DOI: 10.3390/cli12020015

Pantelidakis, M., Panagopoulos, A. A., Mykoniatis, K., Ashkan, S., Eravi, R. C., Pamula, V., Verduzco, E. C., Babich, O., Panagopoulos, O. P., & Chalkiadakis, G. (2022). Identifying sunlit leaves using Convolutional Neural Networks: An expert system for measuring the crop water stress index of pistachio trees. *Expert Systems with Applications*, *209*, 118326. DOI: 10.1016/j.eswa.2022.118326

Petrović, B., Bumbálek, R., Zoubek, T., Kuneš, R., Smutný, L., & Bartoš, P. (2024). Application of precision agriculture technologies in Central Europe-review. *Journal of Agriculture and Food Research*, *15*, 101048. DOI: 10.1016/j.jafr.2024.101048

Piao, S., Liu, Q., Chen, A., Janssens, I. A., Fu, Y., Dai, J., Liu, L., Lian, X., Shen, M., & Zhu, X. (2019). Plant phenology and global climate change: Current progresses and challenges. *Global Change Biology*, *25*(6), 1922–1940. DOI: 10.1111/gcb.14619 PMID: 30884039

Potts, J., Jangra, S., Michael, V. N., & Wu, X. (2023). Speed breeding for crop improvement and food security. *Crops (Basel)*, *3*(4), 276–291. DOI: 10.3390/crops3040025

Prabha, D., Subramanian, R. S., Dinesh, M. G., & Girija, P. (2024). Sustainable Farming Through AI-Enabled Precision Agriculture. In *Artificial Intelligence for Precision Agriculture* (pp. 159–182). Auerbach Publications. DOI: 10.1201/9781003504900-8

Qiao, L., Wang, X., Smith, P., Fan, J., Lu, Y., Emmett, B., Li, R., Dorling, S., Chen, H., Liu, S., Benton, T. G., Wang, Y., Ma, Y., Jiang, R., Zhang, F., Piao, S., Müller, C., Yang, H., Hao, Y., & Fan, M. (2022). Soil quality both increases crop production and improves resilience to climate change. *Nature Climate Change*, *12*(6), 574–580. DOI: 10.1038/s41558-022-01376-8

Sachithra, V., & Subhashini, L. D. C. S. (2023). How artificial intelligence uses to achieve the agriculture sustainability: Systematic review. *Artificial Intelligence in Agriculture.*, *8*, 46–59. DOI: 10.1016/j.aiia.2023.04.002

Saini, A. K., Yadav, A. K., & Dhiraj, . (2025). A Comprehensive review on technological breakthroughs in precision agriculture: IoT and emerging data analytics. *European Journal of Agronomy*, *163*, 127440. DOI: 10.1016/j.eja.2024.127440

Saleem, S. R., Levison, J., & Haroon, Z. (2023). Environment: Role of precision agriculture technologies. *Precision Agriculture*, •••, 211–229. DOI: 10.1016/B978-0-443-18953-1.00012-X

Sanders, C. E., Mayfield-Smith, K. A., & Lamm, A. J. (2021). Exploring Twitter discourse around the use of artificial intelligence to advance agricultural sustainability. *Sustainability (Basel)*, *13*(21), 12033. DOI: 10.3390/su132112033

Segarra, J. (2024). Satellite Imagery in Precision Agriculture. In Priyadarshan, P. M., Jain, S. M., Penna, S., & Al-Khayri, J. M. (Eds.), *Digital Agriculture*. Springer., DOI: 10.1007/978-3-031-43548-5_10

Sharma, K., & Shivandu, S. K. (2024). Integrating artificial intelligence and Internet of Things (IoT) for enhanced crop monitoring and management in precision agriculture. *Sensors International*, *5*, 100292. DOI: 10.1016/j.sintl.2024.100292

Shekhar, S., Durgam, M., Khose, S. B., Pohshna, C., & Bhalekar, D. G. (2024). Advancement and Challenges of Implementing Artificial Intelligence of Things in Precision Agriculture. In Chouhan, S. S., Saxena, A., Singh, U. P., & Jain, S. (Eds.), *Artificial Intelligence Techniques in Smart Agriculture*. Springer., DOI: 10.1007/978-981-97-5878-4_13

Shendryk, Y., Davy, R., & Thorburn, P. (2021). Integrating Satellite Imagery and Environmental Data to Predict Field-Level Cane and Sugar Yields in Australia Using Machine Learning. *Field Crops Research*, *260*, 107984. DOI: 10.1016/j.fcr.2020.107984

Sishodia, R. P., Ray, R. L., & Singh, S. K. (2020). Applications of remote sensing in precision agriculture: A review. *Remote Sensing (Basel)*, *12*(19), 3136. DOI: 10.3390/rs12193136

Tang, Y., Dananjayan, S., Hou, C., Guo, Q., Luo, S., & He, Y. (2021). A survey on the 5G network and its impact on agriculture: Challenges and opportunities. *Computers and Electronics in Agriculture*, *180*, 105895. DOI: 10.1016/j.compag.2020.105895

Taseer, A., & Han, X. (2024). Advancements in variable rate spraying for precise spray requirements in precision agriculture using Unmanned aerial spraying Systems: A review. *Computers and Electronics in Agriculture*, *219*, 108841. DOI: 10.1016/j.compag.2024.108841

Ullo, S. L., & Sinha, G. R. (2021). Advances in IoT and Smart Sensors for Remote Sensing and Agriculture Applications. *Remote Sensing (Basel)*, *13*(13), 2585. DOI: 10.3390/rs13132585

Vellingiri, A., Kokila, R., Nisha, P., Kumar, M., Chinnusamy, S., & Boopathi, S. (2025). Harnessing GPS, Sensors, and Drones to Minimize Environmental Impact: Precision Agriculture. In Designing Sustainable Internet of Things Solutions for Smart Industries (pp. 77-108). IGI Global.

Vibhute, A. D., Kale, K. V., & Gaikwad, S. V. (2024). Machine learning-enabled soil classification for precision agriculture: A study on spectral analysis and soil property determination. *Appl Geomat*, *16*(1), 181–190. DOI: 10.1007/s12518-023-00546-3

Vinod Chandra, S. S., Hareendran, A., & Albaaji, G. F. (2024). Precision farming for sustainability: An agricultural intelligence model. *Computers and Electronics in Agriculture*, *226*, 109386. DOI: 10.1016/j.compag.2024.109386

Wang, D., Zhao, M., Li, Z., Xu, S., Wu, X., Ma, X., & Liu, X. (2025). A survey of unmanned aerial vehicles and deep learning in precision agriculture. *European Journal of Agronomy*, *164*, 127477. DOI: 10.1016/j.eja.2024.127477

Wang, J., Wang, Y., Li, G., & Qi, Z. (2024). Integration of Remote Sensing and Machine Learning for Precision Agriculture: A Comprehensive Perspective on Applications. *Agronomy (Basel)*, *14*(9), 1975. DOI: 10.3390/agronomy14091975

Zhang, X., Feng, G., & Sun, X. (2024). Advanced technologies of soil moisture monitoring in precision agriculture: A Review. *Journal of Agriculture and Food Research*, *18*, 101473. DOI: 10.1016/j.jafr.2024.101473

Chapter 10
Generative Adversarial Networks in Time Series Analysis and Change Detections Using Remote Sensing

Rufai Yusuf Zakari
https://orcid.org/0000-0002-4645-6412
Skyline University, Nigeria

Wasswa Shafik
https://orcid.org/0000-0002-9320-3186
Universiti Brunei Darassalam, Brunei

ABSTRACT

This chapter explores the application of generative adversarial networks (GANs) in time series analysis and change detection using remote sensing imagery. It provides an overview of GANs, covering their architecture, training, and applications, before discussing the importance of time series analysis and change detection for monitoring environmental changes like deforestation and urban expansion. The chapter demonstrates how GANs can be adapted for tasks such as data augmentation, anomaly detection, and predictive modeling, addressing challenges like data scarcity. It also examines how integrating GANs with remote sensing imagery enhances the detection of subtle temporal changes. Practical aspects, including data preprocessing, model selection, and performance evaluation, are discussed, along with ethical concerns like privacy and bias. The chapter concludes by highlighting GANs' potential to transform time series analysis and change detection, proposing

DOI: 10.4018/979-8-3693-6900-5.ch010

future research directions.

1 INTRODUCTION

Generative adversarial networks[1] (GANs) are cross-sections that have emerged as a transformative deep-learning approach for generating high-quality and diverse data in the current technological paradigm. Since its inception, a generator network has demonstrated computational abilities in producing data, while a discriminator network assesses the authenticity of the generated data (Kloprie & Schwung, 2024). Throughout an adversarial mechanism, the discriminator learns to differentiate between real and fake data, while the generator aims to produce data that is indistinguishable from real data. These networks take the vantage of a dual-network structure to accomplish cutting-edge performance in generative tasks (Ahmad et al., 2024). The core structure of a GAN entails two neural networks: the generator and the discriminator, both of which participate in a zero-sum video game during the training process. The generator's purpose is to synthesize data that carefully appears, such as the training dataset, successfully trying to "dupe" the discriminator (Ankan et al., 2024). Conversely, the discriminator is trained to distinguish between actual data from the training collection and the synthetic data produced by the generator. A basic GAN structure is presented in Figure 1. This adversarial arrangement causes a vibrant, iterative procedure where both networks enhance simultaneously, and the generator ends up being extra skilled at creating sensible information while the discriminator develops its capability to detect counterfeits (Chakraborty et al., 2024).

GANs operate under an adversarial framework that leads to a lively and iterative learning process where both the generator and discriminator progressively adapt (Oh et al., 2019; Pan et al., 2019). This labels the competition between the generator and discriminator during training, with one component trying to perfect its work while the other tries to take it down. What this means is that the generator gets better at producing more realistic fake images as it learns from past errors (i.e., if ever the discriminator correctly identifies a fake from a real one, the generator learns from that) (Rewhelet al., 2023). The discriminator, on the other hand, is constantly learning to better distinguish between actual data and the fake examples produced by the generator. Essentially, it is this push-pull action that is the key to maintaining the emergence of the practicality of the GAN model, for the two model networks force each other to maintain a constantly escalating emergence of human-level performance (Pengqiang et al., 2020). This adversarial training process continues iteratively, with the generator getting better at generating data that resembles the real distribution while the discriminator becomes better at detecting slight differences. Over the years, this process leads to the generated outputs of the generator being indistinguishable from real data, pushing the boundaries in data generation and synthesis (Chakraborty et al., 2024). The outcomes reflect a general and performant gencrative model that can generate high-quality, diverse datasets for multiple usage scenarios such as medical and synthetic biology or other research fields (Paola et al., 2023). Its more extensive usage has been in medical or healthcare by means of remote sensing, and the training process of a GAN can be tutorially understood via the minimax optimization approach where the generator attempts to be instrumental in minimizing the ability of the discriminator to recognize fictional information accurately, and the discriminator doing everything in its power to magnify it (Kuntalp & Düzyel, 2024). Nonetheless, due to the non-convex nature of this feature, training GANs can be incredibly unpredictable. Mostly identified GAN challenges such as setting collapse, where the generator produces minimal range in its results, and disappearing slopes, specifically during the initial stages of training, are common difficulties that researchers have looked to minimize via numerous architectural and mathematical advancements (Zhang et al., 2024). A critical facet of GANs is the option of architecture and loss features. In contrast, the original GAN solution uses a binary cross-entropy loss, various modifications, such as the Wasserstein GAN (WGAN), present alternate loss functions that aim to resolve some of the instabilities in training. WGANs, for example, change the standard discriminator with a "doubter" that estimates the Wasserstein distance between genuine and produced data distributions (Bi et al., 2024). The Wasserstein distance $W(p_r, p_g)$ between the real data distribution p_r and the generated data distribution p_g is defined as:

$$W(p_r, p_g) = \inf_{\gamma \in \Pi(p_r, p_g)} \mathbb{E}_{(x,y)\sim\gamma}\left[\| x - y \| \right] \quad (1)$$

where $\Pi(p_r, p_g)$ denotes the set of all joint distributions with marginals p_r and p_g. This metric offers a better measure of distributional differences than Jensen-Shannon divergence, which is used in vanilla GANs.

To approximate the Wasserstein distance, WGAN uses a critic f to compute:

$$\mathbb{E}_{x\sim p_r}[f(x)] - \mathbb{E}_{x\sim p_g}[f(x)] \quad (2)$$

subject to the Lipschitz constraint $\| \nabla f \| \leq 1$. This ensures a smooth gradient for the generator, addressing the vanishing gradient problem common in vanilla GANs.

There is an urgency for research that merges deep learning techniques with traditional geospatial data analysis. The intersection has given a significant jump in the performance of the computed models that are capable of transforming new tools into methods that are robust enough to provide reliable scientific results. In this context, tackling time series data using a data-driven approach has the potential to blend traditional time series analysis with convolutional neural networks, particularly in an adversarial setting since systematic noise is a critical factor of the time series when it approaches remote sensing images (Ahmad et al., 2024). Generative adversarial networks, which were originally proposed for synthetic data creation, are now used for a wide range of applications, including natural images, synthetic music orchestration, and photos to simulate human movement. There are also a few studies in time series analysis that have seen success with GANs in their approach for noise-smoothing time series data representation. This will be a revolution in monitoring environmental changes in an increasingly unstable dynamic world data (Kloprie & Schwung, 2024). This piece is dedicated to the state-of-the-art adoption of GANs for time series analysis and detecting changes in remote sensing imagery.

1.1 Background and Motivation

Time series are data sequences ordered either in time or in terms of some other appropriate index. In general, time series data hold richer information than static data, as dynamic variations are recorded. A wide range of sensors probe time series data owing to technological advancements. Conventional analysis techniques have certain limitations, such as being data dependent, lacking generalization, and measurements being low-rank subspace with systematic errors that are challenging to model. In addition, two issues are driven by the advancements in technology: data scarcity and reasons at times might be due to computational resources (Oh et al., 2019; Pan et al., 2019). Consequently, new opportunities knock on the door of researchers to explore the current trends in artificial intelligence. In this context, Generative Adversarial Networks have been portrayed to discover novel ways of generating, processing, and understanding the patterns processed by large voluminous datasets that include time series data.

GAN is a principle of maximum entropy that attempts to bridge the gap between generative and discriminative models of time series data. For the first time, in an unsupervised manner, this model has provided visible state-of-the-art results for anomaly detection and image generation tasks. GANs have been portrayed to generate synthetic images or even music that is similar to artistic creations or from artistic influences, human identities, and realistic data distributions. These advances necessitate the incorporation of GANs into a proof-of-principle framework of time series and change detection studies using available remote sensing imagery data, which is complex and multi-dimensional (Rewhelet al., 2023). Remote sensing data is inherently time-sensitive, complex, and often not interpretable as dynamics are captured in it. Several sensor advancements have led to the era of big data, which exhibits potential and motivation to increase the timeliness, data points, and variability of the data when compared to static datasets. Nowadays, data acquisition becomes easier as every individual carries mobiles that have built-in remote sensing technology. Furthermore, with extensive research and a drastic reduction in their launch costs, along with the availability of hundreds of micro and nano satellites hovering in orbit around the globe, commercial global coverage using remote sensing networks is also possible. Model adaptations are needed to generalize the proposed techniques in such a dimensional space. Some of the associated challenges are the increased number and complexity of data availability, the performance of the methods degrades as the dimensionality increases, computational time increases, and more information is needed with an increase in variables (Ahmad et al., 2024). Furthermore, change detection with a single feature across a large number of samples is very difficult.

1.2 Research Objectives

Our methodology involves the following research objectives. Exploration: We aim to investigate the potential of GANs to analyze and represent time series data extracted from remote sensing imagery. Evaluation: Another critical objective of this study is to critically evaluate and benchmark the qualitative performance and quantitative accuracy of the proposed GAN-based change detection models with those of traditional convolutional, recurrent, and hybrid time-series models. Future exploration: GANs are generally used to improve the generation of an image. It is anticipated that, in the future, these networks will be used to study change detection using remote sensing. Our goal is to provide a direction for developing appropriate GAN algorithms. We also aim to examine the existing benchmark dataset to discover the general GAN architecture that can diagnose change detection, which is a dynamic dataset. Considering these gaps in the literature, we formulated our research objectives to design an appropriate architecture of GAN for change detection in remote sensing, as introduced below. According to the above discussion, the key objectives of this study are to explore the potential application of GANs to analyze and represent time series data of remote sensing imagery. Furthermore, the study aims to use and benchmark change detection techniques based on GANs using synthetic datasets. The dataset is made up of remote sensing time series. It includes a wide variety of compositions of landscapes, climate, seasons, and sensing systems, as well as a change contingency for synthetic datasets. In summary, the key objective of this study is to develop and design the GAN architecture for use in remote sensing change detection.

2 FUNDAMENTALS OF GENERATIVE ADVERSARIAL NETWORKS (GANS)

Generative Adversarial Networks (GANs) have been successful in various domains like image generation, anomaly detection, imbalanced learning, data augmentation, and others. It is very important to learn the basics of GANs, which will be helpful for readers to understand the proposed model. A GAN consists of two interconnected networks constructed as twin components, the generator and the discriminator, which are trained at the same time (Bi et al., 2024). The generator learns to map a random latent space vector to data space, while the discriminator learns to distinguish between real data samples from the training data distribution and fake samples generated by the generator. GANs are trained in an adversarial manner, competing against each other until a minimax equilibrium is reached, where the discriminator cannot differentiate between real and synthetic samples. At this equilibrium, the generator produces samples that are indistinguishable from true data samples. The generator, in turn, produces better fake samples that can fool the discriminator. This process is repeated over a number of iterations. The training of the networks proceeds iteratively and is optimized to minimize both value functions iteratively to reach Nash equilibrium (Paola et al., 2023). Value functions are free to be defined during the learning process.

There are two main use cases of GANs: generating new data samples and enhancing the quality of existing samples. The new GANs can also preserve attributes of the training data. For time series data, they operate on raw point-to-point values and thus function on temporal sequences. A variant of that, called conditional GAN, provides a way of customizing data characteristics by considering other variables as conditions. It is also possible to condition GANs on a specific class or label. GAN methods have been applied to text, time series, and network traffic datasets but have been proven robust for small- or zero-shot learning despite method adaptations (Oh et al., 2019; Pan et al., 2019). Generative models have been getting much attention for representation learning in recent years, with the latest research focusing on GANs with a large number of samples for training big models with big data. Generative Adversarial Networks (GANs) have been used in remote sensing for super-resolution of satellite imagery, augmentations of synthetic data, deblurring, classification, and generation of urban street views using aerial photos. In super-resolution applications, the generator learns to increase the resolution of input images, whereas the discriminator learns to differentiate between original and upscaled images (Chakraborty et al., 2024). GANs have also been applied to generative methods of SAR data simulation to solve the problem of limited labeled data for hyperspectral remote sensing data analysis. Due to GAN's ability to generate fake responses in a computationally efficient manner, improving the quality and interpretability of the synthetic test datasets could lead to more insightful results, likely validating the critical value of the anomaly score being used in subsequent works (Liu et al., 2021).

Figure 1. A systematic diagram of generative adversarial network architecture

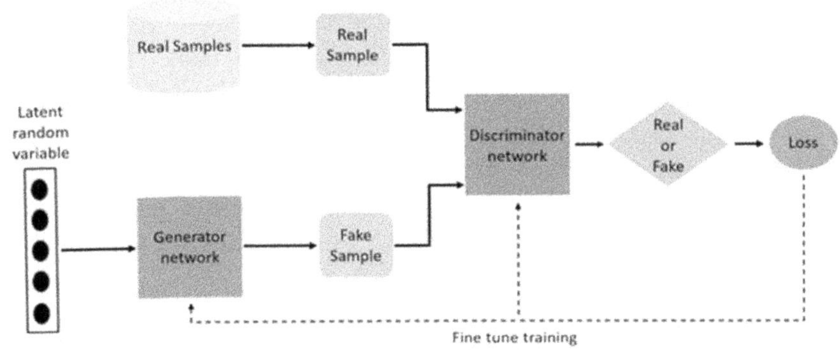

2.1 Basic Concepts and Components of GANs

In GANs, two networks are pitted against each other during training: the generator, which fabricates new data instances, and the discriminator, which ascertains whether the data it receives are generated or belong to the authentic training set. GAN improvements strive to reach a stage wherein the generator can generate counterfeit data instances that are indistinguishable from the authentic data, as computed by a series of conditions (Rewhelet al., 2023). Class components, primarily loss functions, and optimizers, are applied to GANs, but the prevailing consensus requires robust, adversarial training and networking. This paper introduces GAN in a concrete environment and in time series analysis. GANs are powerful parametric models that are efficient owing to a non-Markovian, training-based generator and zero-order generator. As a result, utilizing GANs in remote sensing converter systems is appropriate. To understand the GAN mechanism, let us contemplate an example. A clip distributor might put watermarks on countless internet video clips by artificially imitating the natural distribution of the true near-duplicate video snippet (Liu et al., 2021). The campaign would seek to make other watermark detectors' functions inadequately distinguishable from authentic videos that have no watermarks. As a result, the clip distributor with which he works becomes a GAN creator, the actual videos without watermarks are the GAN tools, and the watermark detectors become fakes. The creator tries to use their generative network to outsmart the detectors, while the detectors fight back by continuously developing new deterrence procedures – with the end goal being a quickly perceived change in performance. Most GAN implementations, for a comparative examination, utilize at least one classifier network and more.

2.2 Types of GAN Architectures

There exist several types of GANs, which can be distinguished by their architectural design. Within these architectural categorizations, an even greater number of variations on these designs exist. In experimentally evaluating these architectures, it is important to consider the particular strengths of specific architectures and attempt variations to these core designs to evaluate the performance gains in generative capacities. The base list of works that follow delineates various successful design strategies for GAN architectures. The simplest GAN variant is to design a GAN using a feed-forward neural network to map noise onto a particular dataset, allowing for a range of noise to be used to create similar data to those learned in the original set. Conditional GANs allow for the noise to be used to select a target class from the data distribution one wishes to create (Chakraborty et al., 2024). Deep Convolutional GANs improve on this concept by utilizing convolution in the generator – a feature that could also be added independently to both the discriminator and the generator in vanilla network architectures. To increase the stability of training, Coulomb-GANs add a gradient cost to the typical error function, improving training stability near monomials. Frame GANs reduce the function the generator produces to not only a sample but an entire visual stream using ideas from Video GANs and recurrent neural networks (Liu et al., 2021). To improve training stability and reduce the likelihood of the generator losing to the discriminator in an imbalanced game, teachers and students can use soft-generation techniques in modeling statistical moments of the actual data to improve generator results. Other GANs designed to bypass the usual distribution modeling techniques include the authoritarian discriminator GAN, in which the discriminator is 'in on the job' of determining percentile top-n discrimination against data samples, as well as others that build a reformulation of the GAN objective to improve better a better clustering of samples of data that are least likely to be generated from the model (Rewhelet al., 2023). While DCGANs, Conditional GANs, and feed-forward variations all have their strengths, the easiest generator to create when attempting to build a GAN that creates best-in-class data from noise is likely the LSTM-GAN, as it has an understanding of time in the output sequence (Oh et al., 2019, Pan et al., 2019). Similarly, LSTMs, on their own, function as both generative models and realistic predictive engines with long 'memories' of the past. Many topological variational autoencoder GANs are able to do similar things, but with a generative engine that creates data in untimed output.

Other specific applications might require various feedback convolution algorithms – generally, LSTMs are a strong bet when it comes to GAN speed and accuracy (Rewhelct al., 2023). Understanding these architecture types and how they might relate to each other is paramount in understanding how to design a GAN for your specific needs in a given application.

3 TIME SERIES ANALYSIS IN REMOTE SENSING

The description of the time series incorporates the time of acquisition, date, or any other time-related information. Time series analysis is essential because we can interpret the change observed in the data over time and analyze its patterns, trends, cycles, and peculiarities. Evolution can provide useful information in a wide range of fields, including economics, biology, environment, and geography. The quantity can indicate future trends and conditions and provide generic predictions (Guo & Mounjid, 2024). With the help of remote sensing technology, serial and/or sequential observations taken instantaneously or at regular time intervals at any site recorded over a period have supported change detection analysis. This analysis turns into valuable information for various fields, including environmental monitoring, climate change, agriculture, hydrology, forestry, healthcare, geophysics, soil science, and biodiversity indices (Zakari et al., 2024).

The remotely sensed time series data derived from the Earth Observation system using satellite and aerial sensors have been serving us with multiresolution over space and time. Many sensors in the current era are able to recapture the data at multiple and various temporal and spatial scales, and the fine spatio-temporal resolution provides the most accurate results in time series. These data acquired various temporal ranges, for example, 15 to 16 days, 10 days, 7 days, 2-3 days, and so on, in payloads such as high-resolution and hypertime of Europe's Sentinel-2. The next important feature of the payload is the revisit time, for example, 50 days, 16 days, etc (Chakraborty et al., 2024). Last but not least, the average lifespan of the satellite varies between 3 weeks for a CubeSat in low Earth orbit and more than 20 years for a GOES-R. Consider this real-time multispectral sensor as a snapshot.

3.1 Definition and Importance of Time Series Analysis

Time series analysis is a practical approach to track data points collected at regular intervals over time. The most significant feature of time series data is that it is sequential and depends on its position. This sequence of data helps us to understand the trend, seasonality, fluctuation, random patterns, or anomalies in the dataset. Time series analysis is beneficial in the field of remote sensing because it provides an understanding of the dynamics that change over time to resolve useful information from data and trends (Wi et al., 2024). Various remote sensing datasets are being received through satellite servers and open APIs. These datasets include environmental phenomena such as changes in land cover, landslides, surface moisture, and natural events. Time series data provide an opportunity to retrieve data over a period so that patterns present at the given location can be identified. Time series data offers an option that is highly valuable when it comes to predictive modeling in predicting probabilities or other outcome orientations at future events. One aim of this study is to uncover changes in urban development found in satellite data between two time periods. This might be useful for town planners and stakeholders as evidence (Chakraborty et al., 2024). Time series data is a set of temporally ordered observations of a data range. Time series data include factors with observations taken over time intervals. These data could include dates, years, hours, days, or minutes, and they have one outcome-dependent variable and the other independent variable. Time series analysis techniques can be used to forecast future events. The time series data could be of any form, such as univariate time series or multivariate time series. Videos are one of the best timeline examples, where multiple pieces of images run one after the other. Photos of a human being could be another significant example, where every year one or more photos of life are captured. There is also geospatial time-dependent data, such as satellite imagery, vector data, raster data, etc. that can be useful for analyzing several dynamic changes (Bi et al., 2024). Temporal diagnosis has also been carried out in several experiments that support research on land use/land cover classification.

3.2 Applications of Time Series Analysis in Remote Sensing

Timely and effective interpretation and understanding of land use changes is the prerequisite for the development of strategies to guide the rational use of land resources, express how nature influences humans, and drive research in other geospatial fields. For this purpose, time series analyses have proven to be more feasible since they can offer better insight into subtle changes, community functioning, or land use pattern modifications. Concerning environmental changes, the literature supports its importance in current global change research and management since changes due to phenomena such as earthquakes, volcanic eruptions, catastrophic floods, sea-level rise, etc., are captured to display a "before" and "after" scenario. Increasing temporal and spatial resolutions impose pressing demands for handling, mining, and analyzing large data volumes with novel methodologies based on big/long time series. Time series are used in diverse applications: cutting-edge technologies such as computer vision and artificial intelligence – time series forecasting using neural networks (Wi et al., 2024).

Scientists and stakeholders have a vested interest in applying time series methodologies to earth observation data to develop solid earth science theories, refine experimental or observational products, ensure measurement quality, make forecasts, and create knowledge from existing data. Remote sensing technologies have rapidly transformed the science of earth observation by acquiring geospatial information at a range of resolutions over vast scales. The temporal aspect of land and land-cover changes is very important to many scientists (Zakari et al., 2024). For example, a study was able to define historical estimates of burned areas in North America since 1650 as a function of climate. Landscape ecologists and conservation biologists are also interested in analyses of change over time as a method to understand factors controlling species distributions and extinction, assess the efficacy of management and policy, and identify areas meriting further study. Past time series analyses of remotely sensed data have included post-event decrease studies, urban growth curves, forest change mapping, and surface hydrology time series (Liu et al., 2021). Remote sensing of temporal measurements evaluates the rate of physical, biological, and chemical activities of the surface and lower atmosphere over a specific region (Guo & Mounjid, 2024). For example, when analyzing radar remote sensing data, a time series of radar backscatter data is analyzed to determine snow melting times over a region.

4 CHANGE DETECTION IN REMOTE SENSING IMAGERY

Change detection is the process of identifying the transformation in the state of an object or phenomenon by monitoring it over different times. Change detection helps in identifying various changes such as environmental, urban, and social changes. Furthermore, it is best to observe these changes and also support decision-making in natural resource management, disaster reduction, urban monitoring, and planning (Bi et al., 2024). Widely approved applications of time series data include geological, coastal, agricultural, aquaculture, and environmental monitoring. Traditional methods of change detection can be classified as pixel-based, object-based, and semantic-based approaches. However, these methods could not produce satisfactory results and failed to capture the temporal changes effectively (Trinh et al., 2024). With the availability of high spatial resolution multi-temporal data, many deep learning methodologies are employed in change detection with improved accuracy and efficiency using deep generative models overview as illustrated in Figure 2.

Figure 2. Deep generative models overview

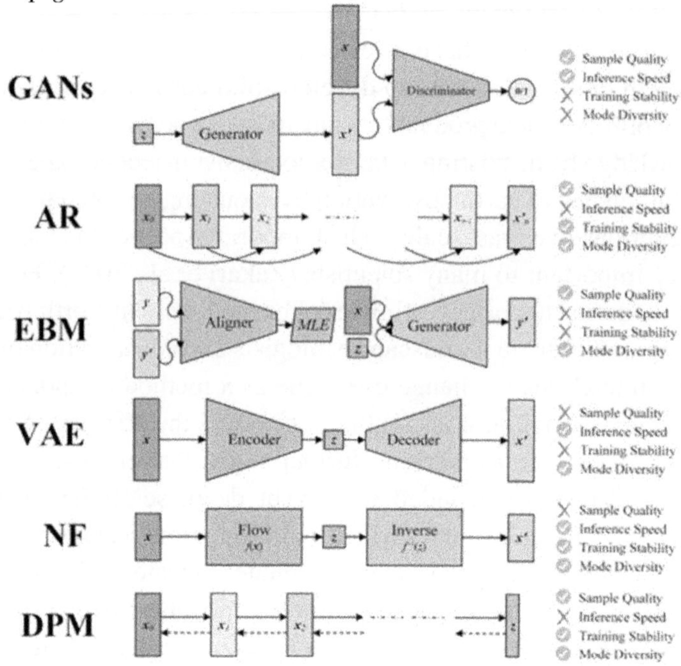

Source: Ahmad et al. (2024)

Deep learning can learn the temporal changes between the multi-temporal data to identify the changes. The accumulation of multi-temporal data in several Earth observation applications calls for efficient techniques that are capable of robustly identifying pertinent change signals over different scenes and domains. Traditional methods of change detection using remote sensing imagery have proven almost impractical in many real-world scenarios due to unrealistic assumptions such as Gaussian noise and linear behavior (Xiao et al., 2024). Henceforth, there is a continual need to develop computational techniques that are capable of capturing temporal changes exhibited in remote sensing imagery despite the complexities, including varying imaging conditions, significant changes in lighting, and factors such as shadow, pose, and occlusion. The development of advanced computational algorithms for temporal change detection in images is an emerging and active research endeavor, with significant discoveries in the employment of GANs being foreseen (Rui et al., 2021; Sara et al., 2021). In essence, change detection is a significant area where Generative Adversarial Networks can play a vital role in highlighting temporal changes.

4.1 Definition and Significance of Change Detection

Change detection is the process of identifying differences between two points in time; it is an important part of time series analysis. Time series analysis is an important aspect of a wide array of fields, including environmental science, remote sensing, and urban planning (Sharma et al., 2021). The ability to accurately determine changes that occur over some time is important for the efficient management of resources. It is important for making informed decisions to conserve resources and for shaping governmental policies. Remote sensing technologies are used in a number of different applications, so change detection is crucial for these applications (Chakraborty et al., 2024). However, change detection is a complex field because there are various factors, such as sensor or environmental noise, that cause differences in data collected at different times, which makes it difficult to differentiate between them. Change detection can help to manage resources for proper utilization and to undertake appropriate measures for development. In addition, change detection can aid in predicting disasters or planning responses in time. Change detection makes it possible to monitor and manage deforestation, soil erosion, and other harmful activities (Wi et al., 2024). In urban and rural monitoring, change detection methods can be useful to monitor urban sprawl or changes in agricultural areas. In disaster management, change detection can be used to monitor disaster impacts. Change detection is a multidisciplinary area and thus has wide applications.

4.2 Traditional Methods vs. Deep Learning Approaches

Traditional Methods. Change detection in remote sensing data is defined as the process of identifying differences in land cover, use, or characteristics, usually by comparing multiple data sets acquired at different temporal subsets. Historically, a large number of methods have been proposed to detect changes. Some of the traditional methods are visual interpretation or post-classification techniques (Zhang et al., 2024). Visual interpretation is straightforward, but it requires expertise and an unacceptably long time when a large amount of data is considered. Post-classification techniques instead use the supervised classification of multiple annual images to identify areas of change. Despite the effectiveness of these techniques during the last decades, they cannot be directly used for big data because of time-consuming multi-temporal acquisition and fine-tuning of several complex parameters (Liu et al., 2024). The main limitation is the necessity of the user to tune the threshold values of the change at multiple scales that also imply domain-specific and dataset-dependent parameters, which, in general, do not give a good knowledge of the exact change extent and impact, opening the door to change model errors. Deep Learning. Various deep learning methods have been proposed, mainly through the use of convolutional neural networks. These networks are used to solve computer vision tasks without using a system of hand-engineered features (Wang et al., 2024). Instead of manually calculating features by filtering the input images, these networks automatically learn these features from the input. Thus, deep learning methods can provide high efficiency and accuracy and can reduce human involvement and errors, and the computational costs in the classification of large-scale datasets. The use of deep learning, and especially these networks, for the detection of changes in remote sensing images, can improve the overall performance since additional information can be extracted, complementing spectral information with spatial information. Previous studies have shown significant improvement in performance with respect to hand-crafted feature-based methods (Guo & Mounjid, 2024).

5 INTEGRATION OF GANS IN TIME SERIES ANALYSIS

Integration of Generative Adversarial Networks in Time Series Analysis in Remote Sensing Observational, time-related data in the remote sensing domain are often characterized by unresolved, subtle signals embedded within considerable noise (Xiao et al., 2024). The noisy environment is a result of a variety of contributing factors such as sensor limitations, terrain topography, meteorological phenomena, and limited sampling. In addition, time series data with which we work are often of significant scale and dimension, with ground measurements or independent data to ground truth significant temporal or spatial phenomena scarce. Digital change detection methodologies intended to exploit this potentially valuable data often struggle to identify signals (events) of interest accurately and often exhibit low reliability and consistency in their outputs. The objective of the review that follows is first a study of the production of relatively high-quality synthetic dense time series imagery data which attempts to address a gap in the literature (Su et al., 2023). A review of the current status and developments of GANs in temporal and multi-temporal remote sensing and applications is then undertaken.

Admittedly, GANs' capabilities to produce high-quality, realistic, unlabelled synthetic data have peaked interest recently in multiple research domains; however, the universal adoption of these methods in time series analysis has not been as forthcoming. To the best of our knowledge, the specific challenges and potential opportunities that interest in using this class of techniques in temporal and multi-temporal remote sensing are discussed not more widely (Kloprie & Schwung, 2024). Here we highlight the importance of accurately articulating those challenges, noting complexities in spatio-temporal dynamics and also across different states of change - continuous processes such as land degradation versus discrete changes such as those following a bushfire. It is anticipated that generative methods can be developed from disparate temporal acts that generally restrict attention to those representing multi-temporal stacks (Kuntalp & Düzyel, 2024). Other recent reviews have discussed change detection in remote sensing filling the literature, reviewing specific literature and recent approaches using deep learning geospatial data, understory fire, and multi-temporal remote sensing imagery, but do not address this research gap. They also consider variations in GAN architecture complementing many GANs for time series of publications which propose architectures they apply to the time series. The most popular in previous works targeting time series of GANs have not been considered here either (Zakari et al., 2024). It is hoped that we can demonstrate the inclusion of this type of research as a significant research gap within the current literature to allow for future contributions.

5.1 Challenges and Opportunities

The nature of changing requirements and evolutions in technologies can pose critical challenges in adapting them for time series data. The rarity of occurring patterns, especially in natural systems, results in data scarcity, which is a challenge for a network that works best with a lot of data. Overfitting and high variability in time series data can further bring about challenges in fake and real sample discrimination, an issue that is intensified with small amounts of data (Zhang et al., 2024). In generative models, the underlying probability distribution of the data is implicitly learned through long-term learning over a large amount of data, which is a bottleneck for time series data applications due to instability in training. Careful consideration of the number of epochs for training is also critical. The overlearning of synthetic images in a number of epochs can also make the generator produce data samples that are very similar to the training set, regardless of the intrinsic nature of the generator. Despite these challenges, generative models have a range of opportunities in analyzing time series data (Bi et al., 2024). Generative models can be seen as a gap bridge between hospitals with a huge amount of data and those without, as it could be a fruitful way to enrich the model training of medical time series and have direct applications for predictive modeling. Deep learning and generative models can be helpful in recognizing patterns in non-static data representation for many scientific fields where simple mathematical functions are used in previous time series detection methods. Shallow models, for example, in environmental monitoring, look inconceivable to represent the system, such as remote sensing data, which changes over time and, if needed to be modeled, also depends on metadata. Generative models offer a unique mindset compared to the above models in capturing the changes in data as they occur over time, as in remote sensing images (Wi et al., 2024). They will be capable of capturing the true manifold and the existence of new data, as is, without risking adding new extreme data into an existing dataset.

6 CASE STUDIES AND APPLICATIONS

Our first case study is urban growth. The case study has proved that GANs, the cycle-consistency model, seem to outperform others. Another new contribution of using GANs in urban growth monitoring is the consideration of utilizing a sequence of class maps for training (Liu et al., 2024). The study provides a take-home message for further research by suggesting that modeling vast landscape change in low-resource areas requires a much more generic fitting. Our study focused on this connection, investigating the ability of one of today's state-of-the-art methods to estimate the growing process using data from almost two years, allowing the capabilities of GANs to be verified (Wang et al., 2024). The use of GANs in deforestation detection should focus on such an analysis, ideally aimed at detecting the illegal deforestation highlighted by purely remote-sensing-based tools. In summary, out-of-sample assessment of deep learning models is a fundamental requirement when the considered model is used to predict future trends, as their ability to forecast change over long timeframes will impact the available notice period for local stakeholders who are generally unaware (Guo & Mounjid, 2024). As such, one can take reassurance that very different timeframes in this respect yield a similar output. Inspired by this, the effort to understand the scope and scalability limits of generative adversarial networks (GANs) over such 26 years is documented. The results clearly demonstrate that the generative adversarial model that has dominated the field of image processing will likely not maintain its preeminent results when the incoming observations on the variable "Y" become multi-sourced in time, multi-possibly-sourced in the spatial dimensions and multidimensional as in our big Earth system science observational dataset (Xiao et al., 2024). Besides its application in time series analysis and change detection, there are several different domains of application, as illustrated in Figure 3.

Figure 3. Generative adversarial networks (GANs): Applications, variants and evaluation metrics

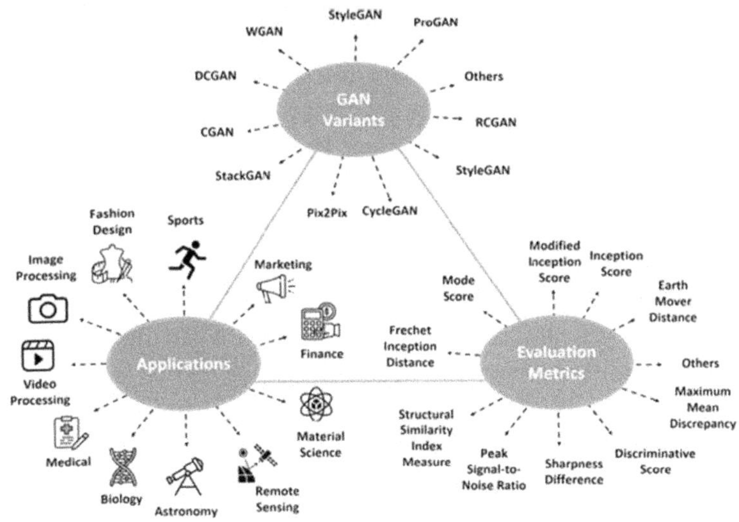

Source: Ankan et al. (2024)

6.1 Urban Growth Monitoring

For many years, urban growth has been studied through methods that comprise time series satellite image analysis for urban expansion and land use change detection. Using a stack of multi-temporal satellite images, different data mining or statistical approaches have been implemented to reflect urban growth patterns and predict urban expansion. However, urban land cover types that changed into different ones have not been properly managed or monitored (Yang et al., 2024). To address this issue, syntheses of different urban land cover types can be efficiently derived using the ability to produce outputs as real as the real samples. For the urban growth monitoring application, the fully connected adversarial network was implemented to produce results on par or better than those using PCA. As a matter of fact, the GAN more effectively detected the changes and produced an accurate land cover map. A study modeled an urban scenario using time series from remote sensing satellite images, where the results were used for urban planning and land management to obtain a sustainable environment. The method exploited GANs to study urban growth (Trinh et al., 2024). The results are quite different with potential use to model time series urban patches and have important implications for urban land planning.

As reported, GAN started as noise and became trivial at the end of the process, and our results for continuous urban patches are evidence for that, as we keep adding epochs. Urban growth elements govern the increase in the population density of urban patches. Employing GANs to monitor these changes can provide valuable results for land managers and urban planners. However, such a model also faces some difficulties, as explained in the analysis section (Zheng et al., 2024). Continuous Urban Growth Monitoring To contribute to the urban growth monitoring studies, a case of applying GANs for urban color images is presented. In this context, a method was utilized for modeling deterministic urban expansion to simulate urban sprawl. Based on the per capita estimated relationship, it can model the urban growth pattern, though, as previously mentioned, this method can be improved (Gao et al., 2024). The idea is to use GANs to generate realistic urban growth images from urban patches treated as temporal series. Standard urban studies use end-of-year land use maps because all available flooded data are registered during the dry season, which are less visible temporal land extent variations (Jafarigol et al., 202). Thus, most of the current studies do not accommodate the continuous monitoring of temporal changes, even though they might be used for modeling urban growth or changes.

6.2 Deforestation Detection

In recent years, technological advancements have led to the application of remote sensing data for environmental changes across the globe. The use of time series analysis of remote sensing data has enabled the identification of illegal logging activities and environmental degradation, causing deforestation and forest degradation. Accurate sub-pixel temporal land cover changes can provide clear insights and require unique conservation strategies (Bengesi et al., 2024). However, it is a challenge to detect land cover changes using an ordinary time series analysis approach. Various studies have focused on reviewing deforestation decoders using sub-pixel-based temporal datasets based on conventional machine learning algorithms. Generative adversarial networks can capture variations and generate unique data as required, which can effectively represent deforestation pattern changes. A technique was proposed in which scarce data are generated by the use of generative adversarial networks that lose track of changes in sub-pixel approximations. Even though deforestation detection is a significant issue, there are very few time series datasets available, especially high-resolution ones (Li et al., 2021). GANs have played a critical role in developing powerful models for remotely sensed data to offer valuable insights into forest loss.

Change detection is important for environmental policymakers to be informed about concerning activities. Detection of changes in land applications is relatively straightforward, indicating activities such as deforestation or reforestation in areas (Yin et al., 2024). Most traditional change detection methods can detect changes; however, they determine how land has changed only to limited classes, such as forest or agricultural, and not the type of forest loss. This information is more beneficial for environmental policymakers to take necessary steps. The customary change detection methods fail to reveal fine-grained deforestation patterns. A model was developed to determine two types of deforestation observations. The traditional model resulted in 37.32% accuracy, indicating countrywide predictions. However, it lost time sequence relations (Pingi et al., 2024). Studies demonstrate that convolutional LSTM presented a fusion network for several bands. The model achieved 95.74% overall countrywide accuracy. However, the proposed model did not consider sub-pixel deforestation and reforestation. The model does not account for pixel inaccuracies for accurate bidirectional models. According to conventional remote sensing change detection methods, the thresholds are often set and used to look at the differences between two orthomosaics, which mainly focus on sub-pixels (Zheng et al., 2024). A deforestation decoder focuses on present data, and any sales can offer modifications via specific spectral features through pixel-based machine learning methods.

6.3 Domain Adaptation

The procedure of transferring knowledge from one domain, where labeled data is abundant, to another correlated domain, where labeled data is scarce or differs substantially. Conventional device discovering versions repeatedly have problems with domain adaptation since they tend to throttle when related to information that differs from the training collection, such as varying noise levels, sampling rates, or underlying precedents (Zheng et al., 2024). GANs foster in finding synthetic information that bridges the gap between the resource and target domains. Particularly, a GAN can be trained to map time series data from the resource domain to a representation that looks like the target domain (Yu et al., 2024). The generator creates time series information that aligns with the target domain's features, while the discriminator finds out how to compare the adapted resource information with the actual target data. This adversarial procedure makes it possible for the generator to generate information that is tantamount to the target domain, effectively transferring understanding throughout the domain (Gao et al., 2024).

6.4 Irregular Time Series Handling

Ordinary models typically fill gaps between data points, making them much less effective when exposed to irregularly sampled data. GANs use an effective approach to address this concern by generating regularized time series information from irregular inputs or by discovering how to handle irregular intervals directly within the modeling process. Data collection continues to be hard in healthcare, where patient information such as important indicators, laboratory outcomes, and medication logs are often recorded at uneven periods, as demonstrated (Wi et al., 2024). For instance, a patient might have their high blood pressure measured several times in a single day throughout a hospital, however, only when a week after discharge. GANs can be utilized to create a regular time series from this irregular information, which can, after that, be used to educate predictive designs for patient results, such as anticipating the chance of an issue or the development of a disease. This regularization permits more consistent and reputable input to artificial intelligence models, resulting in better predictions (Liu et al., 2024).

6.5 Risk Management

Risks might have different definitions depending on the application, and context is applied to model and simulate various risk situations by producing synthetic time series data that shows potential future conditions. This capacity is especially important in economic markets, insurance policies, and operational risk monitoring, where understanding and getting ready for unusual or severe events is vital (Allen et al., 2024). Standard risk models frequently depend on historical data to forecast future threats; however, these versions can be restricted by the schedule and representativeness of past data. GANs, nevertheless, can develop a vast array of reasonable situations that consist of both usual and rare occasions, consequently improving the toughness of danger assessments (Wang et al., 2024). In monetary risk management, GANs can generate artificial time collection information for property rates, interest rates, or credit rating spreads that capture not only regular market conditions but additionally prospective extreme events such as financial situations or abrupt market collisions. These synthetic scenarios enable danger managers to stress-test portfolios and evaluate just how different methods might be carried out under damaging conditions that are not well-represented in historical data. This leads to extra thorough danger evaluations and better-informed decision-making processes (Guo & Mounjid, 2024). An insurance policy fosters replication of tragic occasions, such as natural catastrophes, by creating synthetic time collection information for variables like weather conditions, damage assessments, and claims prices.

6.6 Data Denoising

This includes eliminating noise and undesirable or unnecessary variations from time series data to enhance the top quality and dependability of the information for analysis. In lots of areas, such as finance, health care, and environmental monitoring, the gathered time collection information is usually contaminated with sound because of sensor errors, transmission errors, or exterior environmental variables (Xiao et al., 2024). Typical denoising techniques, such as relocating standards or wavelet changes, may be limited in their capability to efficiently separate sound from the underlying signal, specifically in complex or highly variable datasets (Yang et al., 2024). GANs supply an innovative technique for data denoising by learning to distinguish between the truth underlying patterns in the data and the noise. In this context, the generator network tries to produce a "clean" version of the moment series, while the discriminator network tries to distinguish between the initial loud data and the denoised outcome from the generator. Via this adversarial procedure, the generator enhances its capacity to remove sound while preserving vital attributes of the time collection. This denoising process boosts the signal quality, permitting more reputable analyses and much better scientific outcomes (Trinh et al., 2024).

6.7 Semantic Change Detection: Classifying the Nature of Changes

Semantic change detection not only identifies where the changes have taken place but also categorizes and grasps the very nature and semantics of that change. GANs are particularly useful in this respect since they allow discrimination between different classes of change and, therefore, a deeper understanding of landscape transformation (Choi et al., 2018). GANs can be trained on all types of land-use transitions—from natural habitats into agricultural uses or urban expansion into previously rural lands to the natural regrowth of forests. In this way, it could be very useful for underpinning the driving forces behind landscape alterations, be they natural processes or resulting from human activities. Furthermore, the semantic interpretation given by GANs to results from detection makes such outcomes even more accessible for further processing to the very end-user (Jozdani et al., 2022). These insights obtained with GAN are supposed to help urban planners cope with the understanding of urban growth, while environmental policymakers are able to receive an answer about the impact of expanding agriculture on natural ecosystems. Disaster response teams can take advantage of a more disaggregated analysis of these detected changes and use this to help in determining and developing priorities for providing relief and reconstruction works to affected communities. Thus, GANs integrated into the traditional approaches of remote sensing could leverage both strengths combined, which, in the final instance, improves the capability in general for detecting changes (Gao et al. 2023).

6.8 Advancing Change Detection Capabilities with GANs

GANs integrated into remote sensing workflows for change detection can dramatically enhance the quality, speed, and intuitiveness of such processes significantly. The improvement in the model's accuracy can be gained with GANs since the models reduce false-positive and false-negative cases from performing the task of changing detection (Kazeminia et al., 2018). That, in return, will make the monitoring of dynamic landscapes more accurate. To this effect, effective ecological management and decision-making cannot be complete without it. In addition, GANs can be used to synthesize training data, which is advantageous in cases of limited labeled data, in order to create better models that even perform well under challenging situations. The fast processing and analysis of volumes of data in real or near real-time allow the observation of changes in almost real-time (Li et al., 2020). The scalability of GANs further enables them to apply change detection at regional or global scales and hence serves as a critical means of global monitoring programs in tracking forest, ice sheet, and urban area changes across the world.

7 EVALUATION METRICS FOR CHANGE DETECTION

The evaluation metrics can be used to evaluate the change detection methodologies. When performing a change detection from the remote sensing imagery, it always yields two outcomes: the correct predictions and the incorrect ones. As we are able to count the number of images by direct visual inspection, quantitative measures have been introduced to assess predictability (Mohan et al., 2024). To do this, some of these measures have been adapted from the confusion matrix used in the classification task. The confusion matrix consists of a 2×2 table, which tabulates four outcomes: the true/observed labels of the negative class and true/observed labels of the positive class on the rows, and the false predictions as negative and false predictions as positive on the columns (Trinh et al., 2024). These are termed as the correctly classified, while the others are termed as the incorrectly classified. From these four statistics, we can obtain the three transparency and four binary outcomes: the total number of negative instances or total count of true negatives and false positives, the total number of events or total count of true positives and false negatives, and the total count of predictions or summation of the total number of negative instances and incorrectly classified, where total number of negative instances = true negatives + false positives, events = true positives + false negatives, and predictions = total number of negative instances + incorrectly classified (Miftahushudur et al., 2024).

Hence, we can quantify the level of accuracy, precision, and recall. The accuracy computes the proportion of the correct predictions out of the total number of predictions, which favors balanced classes. Precision and recall evaluate the degree of correctness in all instances of the positive and the appearance of all cases of actual positive categories. Thus, precision could reduce the impact of false positives leading to the wastage of resources, and recall would reduce false negatives that have the heavy cost of political, commercial, or other high risks. It should be noted that the accuracy measures overall agreement between the predicted and the actual classes (Ahsan et al, 2022). All the abovementioned evaluation measures are related to the incorrectly classified by the calculations for each of them. A change detection would decrease false positives but could have some side effects on the possibility of increasing false negatives, which derives from the trade-offs of the model. One point worth noting is that as the scenario of the same false prediction extent is applied to a different model, one could argue that due to the varying ratio between true positives and true negatives determinations, the results of model A would be either better or worse than model B as we use a different evaluation metric (Bengesi et al., 2024). The choice here would significantly impact the interpretations. Although a variety of confusion matrices and evaluation metrics are applied to disparate scenarios, to guarantee that performance could be confirmed or compared to other models, the performance must be appraised and undoubtedly should be directed by a valid evaluation metric. In all scenarios, we aim to apply the concise measurements from the three transparency measurements to determine the model's performance success (Ali-Gombe et al., 2019). A best practice is using a series of evaluations to decide the model's performance. In most circumstances or scenarios, the evaluation in change detection could also contribute to guiding research in terms of possible improvements and help to make decisions with practical value in real problematized problems or cases. Changes in uncertainty for a particular class can help to direct the bioinformatician or other responsible investigators formally and further additional unguided processes based on already obtained results (Cao et al., 2022).

7.1. Accuracy, Precision, and Recall

Within the literature, performance matrices can be calculated through: Accuracy = (TP + TN) / (TP + TN + FP + FN), Precision = TP / (TP + FP), and Recall = TP / (TP + FN), where TP is the number of true positives, FP is the number of false positives, TN is the number of true negatives, and FN is the number of false negatives (Zheng et al., 2024). The differences between precision, recall, and accuracy are summarized as follows: (1) accuracy can reflect the overall classification level, but it cannot differentiate the recognition of samples under different categories; (2) precision can reflect the proportion of actual positive examples, but it cannot identify the misidentified negative examples; (3) recall can reflect the recall rate of all examples under positive conditions; however, excessive inclusion of negative examples is ignored; (4) any combination of the three must consider the usability of change detection models in remote sensing. Through the confusion matrix, we can focus on a specific class of samples and calculate accuracy, precision, and recall between the number of correctly recognized samples and the number of samples to be recognized (Choi et al., 2018).

For example, the receiver operating characteristic curve of the false alarm rate against the true positive rate, and the precision-recall curve are common in the quantitative analysis of change detection results. Draw the precision-recall curve, in which the recall value is displayed on the x-axis and the precision value is displayed on the y-axis, or draw the receiver operating characteristic curve, in which the false positive rate (i.e., recall) is displayed on the x-axis and the true positive rate is displayed on the y-axis. For comprehensive evaluation, in the change detection hyperspectral dataset, the scatter diagram of precision-recall value, precision-accuracy value, and figure descending the average of precision, recall, and F1-score value are also displayed. In conclusion, in the process of change detection, accuracy, recall, and precision are considered equally important. It is difficult to simultaneously improve precision, recall, and accuracy, and improvements in one metric may result in the degradation of others (Gao et al. 2023). It is wrong to simply use the high-accuracy result as the best to evaluate the research model without considering the recall and precision results, and directly changing detection based on a single metric result may cause many false positives and false negatives (Jozdani et al, 2022).

8 FUTURE DIRECTIONS AND RESEARCH OPPORTUNITIES

The field of data analytics is continuously developing due to the emerging trends in technological advancements, particularly in the IoT and remote sensing domains. Researchers are, therefore, continuously required to develop new advancements and tackle fundamental technical challenges of using GANs and time series analysis in remote sensing (Kazeminia et al., 2018). Systematically sharing interdisciplinary research knowledge and connecting different expertise together is also one significant research opportunity. Furthermore, in the context of artificial intelligence and deep learning, numerous important problems remain to be tackled (Li et al., 2020). The most significant among these are the lack of change detection reports and repositories, confusion about the performance drawbacks of the existing GAN architectures for synthetic and large datasets, and the development of a need for large-scale research in this sector. One of the most demanding opportunities is future data analytics with GANs for remote sensing big-time series array data. As a result of the unorganized shared objectives and corresponding research questions, the current study provides many paths for future research to focus exclusively on GAN applications and overall technical topics (Liu, et al., 2023). We propose challenging cross-discipline research projects for technical researchers, together with our interdisciplinary futures for remote sensing time series data analytics.

(1) Developing change detection GANs for environmental and meteorological extreme event studies and remote sensing applications. There is no GAN specifically developed for environmental time series change detection outcomes to date.
(2) Due to intrinsically authored joint generative-interpretive process training applications, design novel big remote sensing time series latent space models.
(3) Innovative heterogeneous feature extraction-based GAN models used in GANs rather than just the conventional feature space learning.
(4) The collection of large labeled remote sensing time series datasets for fine-tuning pre-trained GANs generated using large standard-time data. The printable format of missing data can be filled by GANs. As a result of the acute need for continuous machine synthetic data, which is an important aspect as per the landscape encountered while dealing with real-world remote sensing datasets.
(5) Comprehensive climate and agriculture system dynamics, disaster mitigation, and utility forecasting are other important practical applications. In terms of data availability, deep learning has been utilized in this region to explore simple models of time series change detection.
(6) Identify time series monitoring challenges in remote sensing, specifically with hyper-spatial resolution images that require unique continuous data analytics.

It will also provide GAN research applications for big data analysis in biomedical and epidemiological time series and the possibility of hybrid GAN anomaly monitoring. The research opportunities present the overall untouched, which could be a significant regional trigger for the next wave in remote sensing analytics, monitoring, and precision agriculture (Liu et al., 2021). The testing of unreported research questions, which will enable the results of GAN studies to meet international conference publications, should begin based on a detailed literature review. Given monitoring requirements, this paper specifically focuses on identifying specific objectives for future remote sensing studies. It has been proposed as a change detection resolution for detecting localized critical changes in resolution-rich data. A detailed research endeavor would be very critical if we consider contemporary temporal monitoring in deep learning work and very detailed sensor research (Yu et al., 2022). We predict that objectives will arise in time domain-specific spatial feature extraction methodologies in future remote sensing studies. It will go further in remote sensing to test deep learning-based change detection.

It is clear that while GANs represent a sophisticated machine learning approach, they are not exempt from further improvements. For instance, attention mechanisms could be integrated into the generator in an attempt to extract important features from the time series (Ma et al., 2021). The highly competitive environment of GAN architectures also leaves open a plethora of research possibilities for future innovators. Another direction that is largely unexplored is the combination of GANs with other machine learning approaches, potentially leading to hybrid models that have both generative and discriminative powers (Nyberg et al., 2021). In addition, the importance of further optimizing the training process has been observed, considering that training a GAN is often challenging and solving issues such as mode collapse, training instability, and overfitting—especially in the case of small data or data with skewed distributions—will bring about tangible real-world benefits. An alternative research direction could focus on the integration of unsupervised and semi-supervised change detection settings into an end-to-end framework, potentially resulting in further benefits for scientific applications. Moreover, the optimization of the generator and discriminator architectures for specific downstream tasks could represent a promising research avenue. A last direction could integrate with emerging areas of transdisciplinary investigation, leveraging themes such as transfer learning with a focus on the development of actionable knowledge for pressing challenges in the environmental domain, e.g., migration patterns and global climate assessment (Shamsolmoali et al., 2020). By replicating this process, additional potential enhancements could be identified, contributing towards a more actionable and innovative mapping of the developments in these research areas.

9 CONCLUSION

Tremendous efforts and research have been carried out in time series analysis for a long time, ranging from traditional statistical techniques to deep learning methods. Recently, convolutional and recurrent neural networks have been applied to process and detect changes in remote sensing data, such as optical and SAR, and their temporal domain. GANs have shown effectiveness in transforming common approaches and have demonstrated the potential to strengthen these methods compared to their original versions, even though they have some limitations. Advanced machine learning-based technology such as GANs is available to enhance the capabilities of scientific revelation and technical technology in continuous monitoring and assessment of the ponderable and imponderable environment caused by human activity and natural changes. For example, GAN-based methods could offer improvements in generated digital twins of the Earth or assist in verifying target object displacement and movement in high-spatial-temporal-frequency VHR video information. Developed and improved methods with advanced technology and machine learning, as well as GANs, can contribute to the automatic detection and further understanding of environmental and security situations, and improve the work of staff responsible for geography, Earth, environmental sciences, and security. The continuous growth in machine learning in continuous observations is essential for supporting these observations and investigating the emergence of changes, such as the environment, with emphasis on abuse in fine-tuned methods that lead to operational procedures and applications. With the application of these methods, future work is proposed in applying soft thresholding for the modified loss functions. In conclusion, the research on the use of GANs has presented the potential for new studies to achieve effective monitoring and applications for analyzing and classifying changes in time series and remote sensing applications. On the dates, suggesting that the selection of built classifiers to identify the effect of changes can go beyond qualitative and leap to enhancing the use of early, raw features, and geography classifiers. This study motivated the development of this completed study to extend the investigation of GANs and time series remotely sensed in areas never curated before and application in the future, with these results.

REFERENCES

Ahmad, Z., Jaffri, Z. U. A., Chen, M., & Bao, S. (2024). Understanding GANs: Fundamentals, variants, training challenges, applications, and open problems. *Multimedia Tools and Applications*, •••, 1–77. DOI: 10.1007/s11042-024-19361-y

Ahsan, M., Ali, M., & Siddique, Z. (2022). Imbalanced Class Data Performance Evaluation and Improvement using Novel Generative Adversarial Network-based Approach: SSG and GBO. ArXiv, abs/2210.12870. https://doi.org//arXiv.2210.12870.DOI: 10.48550

Ali-Gombe, A., & Elyan, E. (2019). MFC-GAN: Class-imbalanced dataset classification using Multiple Fake Class Generative Adversarial Network. *Neurocomputing*, *361*, 212–221. DOI: 10.1016/j.neucom.2019.06.043

Allen, D. E., Mushunje, L., & Peiris, S. (2024). GANs and synthetic financial data: Calculating VaR. *Applied Economics*, •••, 1–16. DOI: 10.1080/00036846.2024.2365456

Ankan, D., Junyi, Y., & Guiling, W. (2024). *A Review of Generative Adversarial Networks (GANs) and Its Applications in a Wide Variety of Disciplines: From Medical to Remote Sensing*. IEEE., DOI: 10.1109/ACCESS.2023.3346273

Bengesi, S., El-Sayed, H., Sarker, M. K., Houkpati, Y., Irungu, J., & Oladunni, T. (2024). Advancements in Generative AI: A Comprehensive Review of GANs, GPT, Autoencoders, Diffusion Model, and Transformers. *IEEE Access*, 69812 – 69837, https://doi.org/DOI: 10.1109/ACCESS.2024.3397775

Bi, Y., Abrol, A., Sui, J., & Calhoun, V. (2024, April). Cross-Modal Synthesis of Structural MRI and Functional Connectivity Networks via Conditional ViT-GANs. In *ICASSP 2024-2024 IEEE International Conference on Acoustics, Speech and Signal Processing (ICASSP)*, Seoul, Korea, (pp. 1756-1760). IEEE. https://doi.org/DOI: 10.1109/ICASSP48485.2024.10446450

Cao, J., Luo, M., Yu, J., Yang, M., & He, R. (2022). ScoreMix: A Scalable Augmentation Strategy for Training GANs With Limited Data. *IEEE Transactions on Pattern Analysis and Machine Intelligence*, *45*, 8920–8935. DOI: 10.1109/TPAMI.2022.3231649 PMID: 37015400

Chakraborty, T., Reddy K S, U., Naik, S. M., Panja, M., & Manvitha, B. (2024). Ten years of generative adversarial nets (GANs): A survey of the state-of-the-art. *Machine Learning: Science and Technology*, *5*(1), 011001. DOI: 10.1088/2632-2153/ad1f77

Cheng, W., Li, Y., & Ma, T. (2024). S-KDGAN: Series-Knowledge Distillation with GANs for Anomaly Detection of Sensor Time-Series Data in Smart IoT. *IEEE Sensors Journal*, *24*(15), 24344–24354. DOI: 10.1109/JSEN.2024.3415390

Choi, H., Jung, D., Kim, S., & Yoon, S. (2021). Imbalanced Data Classification via Cooperative Interaction Between Classifier and Generator. *IEEE Transactions on Neural Networks and Learning Systems*, *33*(8), 3343–3356. DOI: 10.1109/TNNLS.2021.3052243 PMID: 33531305

Choi, Y., Choi, M., Kim, M., Ha, J. W., Kim, S., & Choo, J. (2018). Stargan: Unified generative adversarial networks for multi-domain image-to-image translation. In *Proceedings of the IEEE conference on computer vision and pattern recognition* (pp. 8789-8797). DOI: 10.1109/CVPR.2018.00916

Gao, G., Lai, H., & Jia, Z. (2023). Unsupervised image dedusting via a cycle-consistent generative adversarial network. *Remote Sensing (Basel)*, *15*(5), 1311. DOI: 10.3390/rs15051311

Gao, Y., Hu, Z., Shi, S., Chen, W. A., & Liu, M. (2024). Adversarial discriminative domain adaptation for solar radiation prediction: A cross-regional study for zero-label transfer learning in Japan. *Applied Energy*, *359*, 122685. DOI: 10.1016/j.apenergy.2024.122685

Guo, X., & Mounjid, O. (2024). GANs training: A game and stochastic control approach. *Mathematical Finance*, *34*(2), 522–556. DOI: 10.1111/mafi.12427

Jafarigol, E., & Trafalis, T. B. (2024). A distributed approach to meteorological predictions: Addressing data imbalance in precipitation prediction models through federated learning and GANs. *Computational Management Science*, *21*(1), 22. DOI: 10.1007/s10287-024-00504-3

Jozdani, S., Chen, D., Pouliot, D., & Johnson, B. A. (2022). A review and meta-analysis of generative adversarial networks and their applications in remote sensing. *International Journal of Applied Earth Observation and Geoinformation*, *108*, 102734. DOI: 10.1016/j.jag.2022.102734

Kazeminia, S., Baur, C., Kuijper, A., Ginneken, B., Navab, N., Albarqouni, S., & Mukhopadhyay, A. (2018). GANs for Medical Image Analysis. *Artificial Intelligence in Medicine*, *109*, 101938. DOI: 10.1016/j.artmed.2020.101938 PMID: 34756215

Klopries, H., & Schwung, A. (2024). ITF-GAN: Synthetic time series dataset generation and manipulation by interpretable features. *Knowledge-Based Systems*, *283*, 111131. DOI: 10.1016/j.knosys.2023.111131

Kuntalp, M., & Düzyel, O. (2024). A new method for GAN-based data augmentation for classes with distinct clusters. *Expert Systems with Applications*, *235*, 121199. DOI: 10.1016/j.eswa.2023.121199

Li, J., Shen, Y., & Yang, C. (2020). An adversarial generative network for crop classification from remote sensing time series images. *Remote Sensing (Basel)*, *13*(1), 65. DOI: 10.3390/rs13010065

Li, X., Yuan, Y., Liu, G., Han, Z., & Stouffs, R. (2024). A predictive model for daylight performance based on multimodal generative adversarial networks at the early design stage. *Energy and Building*, *305*, 113876. DOI: 10.1016/j.enbuild.2023.113876

Li, Z., Wu, X., Xia, B., Zhang, J., Wang, C., & Li, B. (2022). A Comprehensive Survey on Data-Efficient GANs in Image Generation. ArXiv, abs/2204.08329. https://doi.org//arXiv.2204.08329.DOI: 10.48550

Liu, J. J., Yao, J. P., Liu, J. H., Wang, Z. Y., & Huang, L. (2024). Missing data imputation and classification of small sample missing time series data based on gradient penalized adversarial multi-task learning. *Applied Intelligence*, *54*(3), 2528–2550. DOI: 10.1007/s10489-024-05314-3

Liu, L., Lei, S., Shi, Z., Zhang, N., & Zhu, X. (2021). Hyperspectral Remote Sensing Imagery Generation From RGB Images Based on Joint Discrimination. *IEEE Journal of Selected Topics in Applied Earth Observations and Remote Sensing*, *14*, 7624–7636. DOI: 10.1109/JSTARS.2021.3099242

Liu, P. (2023). A review on remote sensing data fusion with generative adversarial networks (GAN). *Authorea Preprints*.

Ma, T., Ma, J., Yu, K., Zhang, J., & Fu, W. (2021). Multispectral Remote Sensing Image Matching via Image Transfer by Regularized Conditional Generative Adversarial Networks and Local Feature. *IEEE Geoscience and Remote Sensing Letters*, *18*(2), 351–355. DOI: 10.1109/LGRS.2020.2972361

Miftahushudur, T., Grieve, B., & Yin, H. (2024). Permuted KPCA and SMOTE to Guide GAN-Based Oversampling for Imbalanced HSI Classification. *IEEE Journal of Selected Topics in Applied Earth Observations and Remote Sensing*, *17*, 489–505. DOI: 10.1109/JSTARS.2023.3326963

Mohan, S., Manke, A., Verma, S., & Baskar, K. (2024). Machine Learning at the Edge: GANs for Anomaly Detection in Wireless Sensor Networks. In *Enhancing Security in Public Spaces Through Generative Adversarial Networks (GANs)* (pp. 305-317). IGI Global. https://doi.org/DOI: 10.4018/979-8-3693-3597-0.ch021

Nyberg, D. (2021). *Exploring the Capabilities of Generative Adversarial Networks in Remote Sensing Applications*. Linkoping University Electronic Press.

Oh, J., Hong, J., & Baek, J. (2019). Oversampling method using outlier detectable generative adversarial network. *Expert Systems with Applications*, *133*, 1–8. DOI: 10.1016/j.eswa.2019.05.006

Pan, X., Yang, F., Gao, L., Chen, Z., Zhang, B., Fan, H., & Ren, J. (2019). Building extraction from high-resolution aerial imagery using a generative adversarial network with spatial and channel attention mechanisms. *Remote Sensing (Basel)*, *11*(8), 917. DOI: 10.3390/rs11080917

Paola, Z. L., Jesús, L. S., Christian, A. H., & Sonia, R. U. (2023). Correction of Banding Errors in Satellite Images With Generative Adversarial Networks (GAN). *IEEE Access : Practical Innovations, Open Solutions*, *11*, 51960–51970. DOI: 10.1109/ACCESS.2023.3279265

Pengqiang, Z. H. A. N. G., Bing, L. I. U., Xuchu, Y. U., Xiong, T. A. N., Fan, Y. A. N. G., & Zenghua, Z. H. O. U. (2020). Generative adversarial networks for hyperspectral image classification. Bulletin of Surveying and Mapping, (3), 29. http://tb .chinasmp.com/EN/10.13474/j.cnki.11-2246.2020.0073

Pingi, S. T., Zhang, D., Bashar, M. A., & Nayak, R. (2024). Joint representation learning with generative adversarial imputation network for improved classification of longitudinal data. *Data Science and Engineering*, *9*(1), 5–25. DOI: 10.1007/s41019-023-00232-9

Rewhel, E. M., Li, J., Hamed, A. A., Keshk, H. M., Mahmoud, A. S., Sayed, S. A., Samir, E., Zeyada, H. H., Mohamed, S. A., Moustafa, M. S., Nasr, A. H., & Helmy, A. K. (2023). Deep Learning Methods Used in Remote Sensing Images: A Review. *Journal of Environment and Earth Science*, *5*(1), 33–64. DOI: 10.30564/jees.v5i1.5232

Rui, X., Cao, Y., Yuan, X., Kang, Y., & Song, W. (2021). Disastergan: Generative adversarial networks for remote sensing disaster image generation. *Remote Sensing (Basel)*, *13*(21), 4284. DOI: 10.3390/rs13214284

Sara, D., Mandava, A. K., Kumar, A., Duela, S., & Jude, A. (2021). Hyperspectral and multispectral image fusion techniques for high resolution applications: A review. *Earth Science Informatics*, *14*(4), 1685–1705. DOI: 10.1007/s12145-021-00621-6

Shamsolmoali, P., Zareapoor, M., Shen, L., Sadka, A., & Yang, J. (2020). Imbalanced Data Learning by Minority Class Augmentation using Capsule Adversarial Networks. ArXiv, abs/2004.02182. https://doi.org/.DOI: 10.1016/j.neucom.2020.01.119

Sharma, A., Singh, P., & Chandra, R. (2021). SMOTified-GAN for Class Imbalanced Pattern Classification Problems. *IEEE Access : Practical Innovations, Open Solutions*, *10*, 30655–30665. DOI: 10.1109/ACCESS.2022.3158977

Sharma, S., & Namboodiri, V. (2018). No Modes left behind: Capturing the data distribution effectively using GANs. ArXiv, abs/1802.00771. https://doi.org/.DOI: 10.1609/aaai.v32i1.11790

Su, L., Sui, Y., & Yuan, Y. (2023). An Unmixing-Based Multi-Attention GAN for Unsupervised Hyperspectral and Multispectral Image Fusion. *Remote Sensing (Basel)*, *15*(4), 936. DOI: 10.3390/rs15040936

Suh, S., Lee, H., Lukowicz, P., & Lee, Y. (2020). CEGAN: Classification Enhancement Generative Adversarial Networks for unraveling data imbalance problems. Neural networks: the official journal of the International Neural Network Society, 133, 69-86 . https://doi.org/.DOI: 10.1016/j.neunet.2020.10.004

Trinh, L. T., & Hamagami, T. (2024). Latent Denoising Diffusion GAN: Faster sampling, Higher image quality. *IEEE Access : Practical Innovations, Open Solutions*, *12*, 78161–78172. DOI: 10.1109/ACCESS.2024.3406535

Wang, C., Chen, B., Zou, Z., & Shi, Z. (2023). Remote sensing image synthesis via semantic embedding generative adversarial networks. *IEEE Transactions on Geoscience and Remote Sensing*, *61*, 1–11. DOI: 10.1109/TGRS.2023.3335484

Wang, J., Wang, S., Lv, M., & Jiang, H. (2024). Forecasting VaR and ES by using deep quantile regression, GANs-based scenario generation, and heterogeneous market hypothesis. *Financial Innovation*, *10*(1), 36. DOI: 10.1186/s40854-023-00564-5 PMID: 35251896

Wang, X., Sun, L., Chehri, A., & Song, Y. (2023). A review of GAN-based super-resolution reconstruction for optical remote sensing images. *Remote Sensing (Basel)*, *15*(20), 5062. DOI: 10.3390/rs15205062

Wi, H., Shin, Y., & Park, N. (2024, March). Continuous-time Autoencoders for Regular and Irregular Time Series Imputation. In *Proceedings of the 17th ACM International Conference on Web Search and Data Mining* (pp. 826-835). https://doi.org/DOI: 10.1145/3616855.3635831

Wu, A. N., Stouffs, R., & Biljecki, F. (2022). Generative Adversarial Networks in the built environment: A comprehensive review of the application of GANs across data types and scales. *Building and Environment*, *223*, 109477. DOI: 10.1016/j.buildenv.2022.109477

Xiao, H., Wang, X., Wang, J., Cai, J. Y., Deng, J. H., Yan, J. K., & Tang, Y. D. (2024). Single image super-resolution with denoising diffusion GANs. *Scientific Reports*, *14*(1), 4272. DOI: 10.1038/s41598-024-52370-3 PMID: 38383573

Xiao, J., Li, J., Yuan, Q., Jiang, M., & Zhang, L. (2021). Physics-Based GAN With Iterative Refinement Unit for Hyperspectral and Multispectral Image Fusion. *IEEE Journal of Selected Topics in Applied Earth Observations and Remote Sensing*, *14*, 6827–6841. DOI: 10.1109/JSTARS.2021.3075727

Yang, G., Lu, H., Li, C., Zhou, G., Wu, H., & Lu, Z. (2024, April). Progressive Image Synthesis from Semantics to Details with Denoising Diffusion GAN. In *ICASSP 2024-2024 IEEE International Conference on Acoustics, Speech and Signal Processing (ICASSP)* (pp. 7495-7499). IEEE. https://doi.org/DOI: 10.1109/ICASSP48485.2024.10445885

Yin, Y., Yuan, Z., Tanvir, I. M., & Bao, X. (2024). Electronic medical records imputation by temporal Generative Adversarial Network. *BioData Mining*, *17*(1), 19. DOI: 10.1186/s13040-024-00372-2 PMID: 38926718

Yu, M., Zhang, W., Chen, X., Liu, Y., & Niu, J. (2022). An End-to-End Atrous Spatial Pyramid Pooling and Skip-Connections Generative Adversarial Segmentation Network for Building Extraction from High-Resolution Aerial Images. *Applied Sciences (Basel, Switzerland)*, *12*(10), 5151. DOI: 10.3390/app12105151

Yu, Y., Chen, W., Chen, F., Jia, W., & Lu, Q. (2024). Night-time vehicle model recognition based on domain adaptation. *Multimedia Tools and Applications*, *83*(4), 9577–9596. DOI: 10.1007/s11042-023-15447-1

Zakari, R. Y., Lawal, Z. K., Kalinaki, K., & Maiwada, U. D. (2024). Time Series Forecasting Techniques for Climate Trend Prediction. In *Reshaping Environmental Science Through Machine Learning and IoT* (pp. 282–310). IGI Global., DOI: 10.4018/979-8-3693-2351-9.ch014

Zakari, R. Y., Shafik, W., Kalinaki, K., & Iheaturu, C. J. (2024). Internet of forestry things (IoFT) technologies and applications in Forest management. In *Advanced IoT technologies and applications in the industry 4.0 digital economy* (pp. 275–295). CRC Press. DOI: 10.1201/9781003434269-15

Zhang, H., Song, Y., Han, C., & Zhang, L. (2020). Remote sensing image spatiotemporal fusion using a generative adversarial network. *IEEE Transactions on Geoscience and Remote Sensing*, *59*(5), 4273–4286. DOI: 10.1109/TGRS.2020.3010530

Zhang, N., Wang, Y., Zhang, X., Xu, D., & Wang, X. (2020). An unsupervised remote sensing single-image super-resolution method based on generative adversarial network. *IEEE Access : Practical Innovations, Open Solutions*, *8*, 29027–29039. DOI: 10.1109/ACCESS.2020.2972300

Zhang, Y., Peng, C., Wang, Q., Song, D., Li, K., & Zhou, S. K. (2024). Unified multi-modal image synthesis for missing modality imputation. *IEEE Transactions on Medical Imaging*. Advance online publication. DOI: 10.1109/TMI.2024.3424785 PMID: 38976465

Zheng, B., Zhang, R., Diao, S., Zhu, J., Yuan, Y., Cai, J., Shao, L., Li, S., & Qin, W. (2024). Dual domain distribution disruption with semantics preservation: Unsupervised domain adaptation for medical image segmentation. *Medical Image Analysis*, *103275*, 103275. Advance online publication. DOI: 10.1016/j.media.2024.103275 PMID: 39032395

Zheng, Y., Su, J., Zhang, S., Tao, M., & Wang, L. (2022). Dehaze-AGGAN: Unpaired remote sensing image dehazing using enhanced attention-guide generative adversarial networks. *IEEE Transactions on Geoscience and Remote Sensing*, *60*, 1–13. DOI: 10.1109/TGRS.2022.3225843

Zhu, C., Deng, S., Li, J., Zhang, Y., Gong, L., Gao, L., Ta, N., Chen, S., & Wu, Q. (2023). Hyperspectral and multispectral remote sensing image fusion using SwinGAN with joint adaptive spatial-spectral gradient loss function. *International Journal of Digital Earth*, *16*(1), 3580–3600. DOI: 10.1080/17538947.2023.2253206

Zhu, C., Deng, S., Zhou, Y., Deng, L., & Wu, Q. (2023). QIS-GAN: A Lightweight Adversarial Network With Quadtree Implicit Sampling for Multispectral and Hyperspectral Image Fusion. *IEEE Transactions on Geoscience and Remote Sensing*, *61*, 1–15. DOI: 10.1109/TGRS.2023.3332176

Zhu, L., Chen, Y., Ghamisi, P., & Benediktsson, J. (2018). Generative Adversarial Networks for Hyperspectral Image Classification. *IEEE Transactions on Geoscience and Remote Sensing*, *56*(9), 5046–5063. DOI: 10.1109/TGRS.2018.2805286

ENDNOTE

1 https://www.analyticsvidhya.com/blog/2021/10/an-end-to-end-introduction-to-generative-adversarial-networksgans/

Chapter 11
Case Studies on Generative Adversarial Networks in Precision Farming:
GAN for Precision Agriculture

Pradnya Awate
Dr. G.Y. Pathrikar Collge of Compute Science and IT, MGM University, India

Ajay D. Nagne
https://orcid.org/0000-0002-9652-5631
Dr. G.Y. Pathrikar Collge of Compute Science and IT, MGM University, India

ABSTRACT

The chapter reviews the applicability of Generative Adversarial Networks in precision agriculture, with an emphasis on its role in enhancing remote sensing technology. This ranges from resolution augmentation for satellite and drone images using GAN-based models like SRGAN and CycleGAN to generating synthetic data for training models that will help in crop health monitoring, soil analysis, and yield prediction. This case study demonstrates tremendous improvements in image quality and decision-making, with further reach into weather simulation, real-time UAV monitoring, and IoT integration.

DOI: 10.4018/979-8-3693-6900-5.ch011

1. INTRODUCTION

1.1. Overview of Precision Farming

Precision farming, also referred to as precision agriculture, is a major breakthrough in farm management: it deals with the optimum levels of crop farming by incorporating advanced technologies at field scales. It combines high-end geospatial techniques with remote-sensing technologies and Internet of Things devices to enhance agricultural data for making decisions (Rimpika, Anushi, Manasa, Anusha, Thakur, Shilpa, & Sood, 2023).

GPS technology has introduced in the system, has enabled precise mapping, variable rate application of inputs, like fertilizers, pesticides among many others. In addition to that, unmanned aerial vehicles equipped with multi-spectral and hyper-spectral sensors can provide very high-resolution imagery capable of monitoring crop health, assessments of biomass, and even detection of pest infestations very precisely (AWATE, 2023).

Machine learning algorithms and artificial intelligence are increasingly used for the analysis of big datasets, finding patterns, and developing predictive models for crop yield optimization and disease forecasting. Supported by internet-of-things sensors spread in the field to gather real-time data on soil moisture, nutrient levels, and climatic conditions, it feeds into an advanced DSS for decision-making. At the same time, they help to make an analysis of agronomic data spatially and create specific management zones and site-specific management practices. Advances in soil-sampling procedures and in soil-health diagnostics have been important in tailoring interventions to the particular needs of different areas within larger fields (Manikandababu, Preethi, Kanna, Vedhathiri, & Suresh Kumar, 2024).

The rise of such mechanisms guarantees that agricultural produce may be traced from the farm to the table with new technologies like blockchain. Figure 1 shows new view of precision farming. In general, these advances in precision farming have brought about greater efficiency, sustainability, and productivity in modern agriculture, which alludes to the changes in farming toward being data-driven and technology-based (Lakshmi, Brindha, Sathish, Vinay Kumar, & Puttur Santhosh-kumar, 2024).

Figure 1. Precision Farming

1.2. Introduction to Generative Adversarial Networks (GANs)

Generative Adversarial Networks (GANs), introduced by Ian Goodfellow et al. in 2014, are a new generative model paradigm used for high-fidelity synthetic data generation in the manner of corresponding real-world data. The crux of GANs is their dual-network architecture, coupling a generator with a discriminator that is trained simultaneously in a zero-sum game. A generator attempts to generate plausible samples of data, and a discriminator tries to make decisions identifying which samples are real and which are generated. This adversarial framework with continuation training forces the generator to put out increasingly more realistic-looking stuff (Ran, 2024).

The basic architecture of GANs has been innovated in a bunch of ways to combat problems such as training instability, mode collapse, and convergence issues. One of these innovations is known as Wasserstein GANs, or WGANs for short, which use a new loss metric, the Wasserstein distance, in place of the older loss metric, Jensen-Shannon divergence. WGANs address common issues such as mode collapse by enforcing a less spiky and much smoother process of gradient descent (Bobadilla, 2024).

Other refinements include Conditional GANs (cGANs), where the generator and discriminator are conditioned with respect to auxiliary information like class labels or data attributes. This enables controlled generation, allowing a synthesis of only specific types or classes of data, extending the applicability to tasks such as image-to-image translation and data augmentation (DeVries, Romero, Pineda, Taylor, & Drozdzal, 2019).

Another important breakthrough in the development of GANs is progressive growing, called ProGANs. It introduces a systematic training strategy in which the complexity level of the generator and discriminator networks is progressively increased. This approach has proved effective with applications that intend to synthesize high-resolution images by stacking layers incrementally and refining the resolution in each period of training (Han, Zhang, Zhou, & Wang, 2019).

There's an even newer one, StyleGAN, which integrates style transfer techniques within the GAN architecture, providing a more detailed way of controlling generated images, including aspects such as texture, color, and fine detail. This is implemented by untangling the latent space into multiple scales, with each corresponding to a different level of abstraction in the generated output. StyleGANs have been influential models and have spearheaded the generation of hyper-realistic human faces and art content with an unprecedented level of detail (Karras, Laine, Aittala, Hellsten, Lehtinen, & Aila, 2020).

Based on GANs, unsupervised and semi-supervised learning paradigms are being developed into more versatile applications beyond data generation: anomaly detection, domain adaptation, and data privacy preservation. Moreover, the introduction of GANs into adversarial machine learning has been pushing research in robustness and security further, trying to make these networks less vulnerable to adversarial attacks that tend to exploit an insufficiency in their design (Goodfellow et al., 2020).

Overall, GAN research has been rapidly evolving, with other key works aimed at making them more scalable, interpretable, and ethically sound, which firmly secures their place among the most powerful and important tools in the field of artificial intelligence and machine learning.

1.3. Importance of GANs in Enhancing Remote Sensing for Agriculture

Generative Adversarial Networks (GANs) have come to be one of the transformative tools for remote sensing applications in agriculture, enabling advanced capabilities of data generation, enhancement, and analysis. Remote sensing has formed a very crucial platform of precision agriculture, requiring high-resolution, accurate, and timely information about crop health, yield prediction, and resource management. However, most of such potential is hindered by problems like data scarcity, noise, and significantly low temporal resolution common to many forms of remote sensing. GANs address such problems by generating data of higher quality and utility for remote sensing (Jozdani, Chen, Pouliot, & Johnson, 2022).

One of the main impacts that GANs have on this field is data augmentation. GANs can synthesize remote sensing imagery that is indistinguishable from real data and hence lead to the expansion of training datasets. This is especially important in the field of agriculture, where annotated datasets are usually small. This data could simulate various stages of crop growth, the environmental condition, and the modality of sensors, and thus improve the model robustness and generalization. SRGANs have, in particular, been applied to increase the spatial resolution of satellite images. In realizing higher resolutions from the lower ones, SRGANs give a more detailed and exact way to monitor agricultural fields. This is important for capturing small-scale features of crop health, detecting pest infestations, and optimizing site-specific management practices. Apart from this, SRGANs maintain the fine detail in vegetation indices that become very crucial for diagnosing plant health and estimating yield (Dou, Li, Shi, & Liu, 2020).

GANs also play a significant role in cloud removal and gap filling in remote sensing images. One of the most limiting parameters in such imagery is the percentage of cloud cover that always interferes with satellite observation, yielding missing or corrupt data. We develop cGANs to reconstruct these occluded regions based on the inherent information learned in the cloud-free regions; hence, they are continuous and complete over the entire temporal series. This is a quite strong development for applying consistent monitoring and decision-making processes in agriculture (Rodríguez-Suárez, Quesada-Barriuso, & Argüello, 2022).

For multispectral and hyperspectral imaging, GANs increase the resolution of spectra to better predict any missing spectral band or to further enhance the already existing ones. This could further aid in defining the crop type, soil property evaluation, and detection of any early-stage diseases, supporting more accurate agricultural intervention. GANs thus enable domain adaptation due to the nature of data synthesis between different sensor platforms or even geographic regions. This characteristic is very important in enabling the application of models that are trained on one dataset and thus improves the usability of such remote sensing models in different agricultural environments (Kerdegari, Razaak, Argyriou, & Remagnino, 2019).

Generally, GANs improve the quality, resolution, and applicability of the remote sensing process in agriculture, and hence improve the monitoring of crops, the management of resources, and even predictive modelling in that area. Their continuous integration into workflows of remote sensing is promising further revolutionizing precision agriculture toward sustainable and efficient food production systems. This chapter addresses how GANs can be integrated with remote sensing to improve precision farming. It provides an overview of the issues and how GANs can address them, such as low-resolution images, limited labeled data, and the necessity to reliably monitor and estimate crop productivity. With a case study, the efficacy of SRGAN in image improvement and CycleGAN in spectrum translation is demonstrated to improve crop health evaluation, soil analysis, and forecasting. The chapter also discusses emerging themes including multi-spectral imaging, integration with edge computing, and climate impact simulation. The results show significant improvements in image quality and forecast accuracy, indicating a promising application for GANs in sustainable agriculture.

2. LITERATURE REVIEW

2.1. Historical Context and Development of GANs

Generative Adversarial Networks are a key breakthrough in machine learning, redesigning a whole new way in which one thought about the problem of generative modelling and artificial intelligence. Originally proposed by Ian Goodfellow, with his colleagues, in the year 2014, these networks have been made for the sole purpose of the generative modelling problem, meaning learning an underlying probability distribution of a dataset and generating new samples that look like the original data (Salimans, Goodfellow, Zaremba, Cheung, Radford, & Chen, 2016).

Underneath its surface, GANs generally work in a game-theoretic style, in which two competing neural networks play against each other. The generator network is responsible for creating new samples from the training data, while the discriminator network basically distinguishes the generated samples from the real data (Ge, Xia, Chen, Berry, & Wu, 2018). It is this set-up of adversaries which provokes both networks to continue getting better: the one gives ever more realistic samples that should fool the discriminator, while the other learns to identify the generated sample. It implies that, in this iterative process, GANs could learn the true underlying probability distribution of the training data to produce new samples that are both highly realistic and mutually dissimilar (Ge, Xia, Chen, Berry, & Wu, 2018).

2.2. Applications of GANs in Various Fields

Generative adversarial networks have taken many different domains by storm because they offer an ability to generate real data across a wide spectrum of applications. We present in this paper an overview of GAN applications in some domains, focusing on their technical contributions and challenges.

- Computer Vision

The areas in which GANs probably made the biggest impacts are the fields of computer vision, including image generation, super-resolution, and style transfer. For instance, in image creation, they act instrumentally to come up with high-fidelity synthetic images hardly distinguishable from the real ones. This capability applies to everything from creating photorealistic avatars and data augmentation of training datasets for machine learning models to the fashion industry for designing new patterns of apparel. The algorithm upscales low-resolution images into a higher one using super-resolution GANs, retaining fine details and texture. Another variant of GANs called StyleGAN is one that would disentangle the high-level features, say the structure of a face, from the lower-level features like the hair component of a face, so we could synthesize completely new images with a set of chosen stylistic features (Wang, She, & Ward, 2021).

- Natural Language Processing (NLP)

For NLP, GANs find use in text generation, machine translation, and dialog systems. While traditional language models focus on generating the next word in a sequence, GANs, which model complex language distributions, can be very effective in providing coherent and contextually relevant text. In fact, GANs have also been applied to the problem of text-to-image synthesis, such that for a given textual description, matching images are generated—a problem that thus bridges vision and language (Thirumagal & Saruladha, 2021).

- Healthcare and Bioinformatics

GANs have been widely adopted in healthcare, generating synthetic medical data for training ML models when data scarcity is significant. GANs can generate highly realistic medical images, including MRI scans, X-rays, and CT scans; therefore, they can help augment data sets for disease detection and segmentation tasks. GANs are helping apply drug discovery by generating novel molecular structures with features desirable in a compound, hence accelerating the early phase of drug design (Sharmilan & Chaminda, 2017).

- Finance

In this regard, GANs are being used in finance, mainly for the synthesis of data when the dataset is inadequate or when it is imbalanced. This is the range where synthetic stock market data can be created and then, in turn, give a source to develop strong trading algorithms and risk models. Further, GANs are applied to fraud detection by creating adversarial examples that assist in testing the strength of developed models and enhance their accuracy and reliability (Vuletić, Prenzel, & Cucuringu, 2024).

- Robotics and Autonomous Systems

Other areas of application of GANs include robotics for developing simulation environments through which training of autonomous systems is possible. Case samples of simulated environment cases can be used to train autonomous cars and drones on the course of operations under normal circumstances. GANs further help in developing synthetic data for sensor fusion tasks where different modalities of data, like LIDAR and camera inputs, are integrated to have a better perception in complex environments (Schnell, Bott, Puck, Buettner, Roennau, & Dillmann, 2022).

- Art and Creative Industries

In the domain of creativity, GANs have opened new possibilities for both artists and designers by generating new forms of art. GAN-based models are promising for generating art in the style of famous painters, new music compositions, and even whole new visual aesthetics. The capabilities of GANs are by no means confined to human-art imitation but go far beyond, toward generation of unique and never-before-seen kinds of art that would challenge conventional wisdom concerning creativity (Anantrasirichai & Bull, 2022).

- Scientific Research and Data Augmentation

GANs have found wide application in scientific research, where they are used to generate synthetic data for different types of experiments, especially when data acquisition is difficult or very costly. For example, in astronomy, GANs are employed to synthesize high-resolution cosmic microwave background radiation maps that are necessary to understand the universe's earliest stages. In climate science, GANs are simulating realistic patterns of weather, which helps in making not only forecasts but also in analysing the impact of changes in climate (Tran, Tran, Nguyen, Nguyen, & Cheung, 2021).

2.3. Previous Studies on GANs and Remote Sensing in Agriculture

Rapid development in the technology of remote sensing and the increasing availability of high-resolution satellite and aerial images have opened up ways of implementing new kinds of applications related to agriculture. Earlier revelations have shown that prospecting potential in remote sensing techniques is very high for the purposes of crop monitoring, yield estimation, and precision farming (Desai, Bhavsar, & Pandya, 2023).

The assessment of crops in continuous monitoring of growing crops is one of the prime research areas. Remote sensing is based on the techniques and principles based on electromagnetic radiation interacting with the soil and crops, thereby allowing for measurements of the applied radiation in terms of any reflected radiation quantities. The information could be used in order to assess various aspects regarding crop health and growth status like nutrition-related status, water stress, or the amount of pests and diseases (Sarmah, Jyoti Baruah, & Sarmah, 2024).

Moreover, the development of Generative Adversarial Networks has shown the new possibilities of better monitoring in agriculture and more exact, reliable decision-making. This is further reinforced by recent work in the development of suitable GAN models that do a splendid job in the generation of images, so this application in the field of remote sensing is hoped to be the one that will allow for still more accurate and reliable assessment of crop condition status (Anuradha, Kuchipudi, Ashreetha, Venkata, Naga Ramesh, & Ayadi, 2024).

This will empower farmers and agricultural researchers to have a brilliant tool through the use of remote sensing data and GAN-based techniques for precision agriculture. The basic idea is that maybe a combination of spatial information and temporal within remotely sensed data with the capacity that GANs have in synthesizing and enhancing images could constitute the basis for the development of new generation agricultural monitoring systems to enable better decision-making and resource allocation (Jia, Wang, Khandelwal, Karpatne, & Kumar, 2019).

3. CASE STUDY OVERVIEW

This is a case study on state-of-the-art hyperspectral data processing techniques that help transform volumes of information into meaningful insights for environment and precision agriculture applications. As agriculture becomes more and more informed by data-driven strategies, hyperspectral imaging provides important information about vegetation health, species identification, and soil properties that help enable more effective and sustainable agriculture (Hennessy, Clarke, & Lewis, 2021).

3.1. Background of the Agricultural Enterprise

It is focused on the fact that improved methods are needed for the conversion of vast amounts of hyperspectral data into useful information related to environmental monitoring and precision agriculture. Precision agriculture, in particular, increasingly relies on advanced data-driven methods to ensure increases in productivity while remaining sustainable. Hyperspectral imaging can cover a very wide range of wavelengths that give detailed information about vegetation health, species identification, and soil properties—very important parameters for effective agricultural management (Hennessy, Clarke, & Lewis, 2021).

3.2. Objectives of the Case Study

The main objective of the paper is to make a GAN that generates realistic hyperspectral vegetation data, preserving the dependency of different classes of vegetation. All this goes towards having more training data available for machine learning classifiers on farm applications. The authors also examined the efficiency of the synthetic data in enhancing classification accuracy for various kinds of vegetation, hence firmly establishing the potentials of GANs in agricultural data analysis (Hennessy, Clarke, & Lewis, 2021).

3.3. Methodology and Data Collection

In this work, they train a GAN to approximate the distribution of two vegetation hyperspectral datasets using the Cramer distance in such a way that it is capable of generating very realistic synthetic spectra. Data collection involves two vegetation hyperspectral datasets, from which a GAN will be trained. Thereafter, statistical properties of the synthetic data are compared against those of the original datasets.

- Types of Remote Sensing Data Used

The underlying data used in this study is hyperspectral in nature, capturing all wavelengths beyond the visible spectrum. It is very useful in agriculture in assessing plant health, recognizing species, and monitoring environmental conditions. This paper checks the degree to which generated synthetic hyperspectral data retains statistical properties from real spectra, so as to enable its effective use in training classifiers for improved accuracy in vegetation classification. The work thus brings out the creative application of GANs in generating synthetic hyperspectral data, which can greatly improve the potentials of remote sensing in agriculture (Hennessy, Clarke, & Lewis, 2021).

4. IMPLEMENTATION OF GANS IN PRECISION FARMING

GANs have emerged as one of the transformative technologies in precision farming, helping to enhance and synthesize agricultural data. In this respect, satellite and drone image resolution can be improved with the use of GANs. Further, precision farming systems can be installed with the use of GANs to augment training datasets with synthetic remote sensing data that helps convert images for advanced vegetation analysis, which in turn enables the optimization of crop monitoring and management processes ("Generative Adversarial...", 2022). Figure 2 outlines the overall process of implementing GANs in precision farming, outlining critical processes from data gathering to evaluation. It shows how GAN models use AI and real-time decision assistance to improve the quality of remote sensing data for crop health monitoring, soil analysis, and yield prediction in sustainable agriculture.

Figure 2. Described Methodology

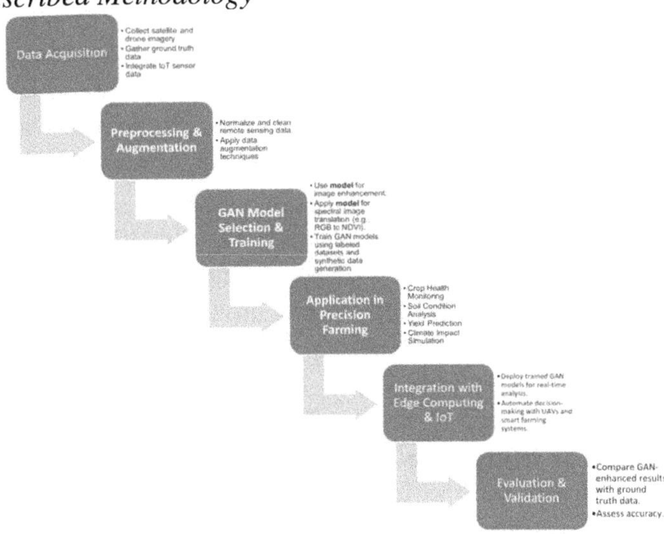

4.1. Super-Resolution GAN (SRGAN)

- Enhancing Satellite and Drone Image Resolution

SRGANs apply deep learning to super-resolve satellite and drone images whose spatial resolution is normally limited due to constraints on imaging hardware and atmospheric conditions. In this case, a neural network can be trained to learn the mapping from low-resolution to high-resolution imagery, and thus SRGANs reconstruct fine details lost by standard image processing methods. This becomes of particular interest in precision farming, whereby high-resolution imaging will allow the detection of very small changes in crop health, the early stage discovery of disease or pest invasion, and a more precise estimation of soil quality. In this respect, improved imagery enables the execution of more precise interventions related to targeted fertilization, irrigation, and pest control, all aiming to ensure better harvests and more efficient use of resources (Maqsood, Mumtaz, Haq, Shafi, Zaidi, & Hafeez, 2021).

4.2. Data Augmentation with GANs

- Generating Synthetic Remote Sensing Data for Training Models

In this regard, high-quality, labelled remote sensing data may be limited in availability within the domain of precision agriculture, which impacts the performance of the machine learning model. On the other hand, GANs can mitigate this problem by generating synthetic data very similar to real remote-sensing imagery. Further, by simulating a wide array of conditions—including different weather patterns, lighting conditions, and crop growth stages—GANs could create variety-enriched training datasets. Such synthetic data augmentation makes the model generalize better to become more robust against real-world condition variations. This kind of synthetic data augmentation enhances the model's robustness to real-world condition variations. Therefore, models trained from augmented datasets are better placed in the analysis and interpretation of satellite and drone images, hence able to make accurate predictions and recommendations in precision farming (Fawakherji, Potena, Prevedello, Pretto, Bloisi, & Nardi, 2020).

4.3. CycleGAN for Image-to-Image Translation

- Converting RGB to NDVI Images for Vegetation Analysis

These CycleGANs offer a general framework in the translation of images into images, whereby the approach converts standard RGB images into NDVI images without requiring paired training data. NDVI is rather very commonly applied in agriculture and as such captures something meaningful about vegetation health by highlighting differences in plant biomass and chlorophyll content. In other terms, using CycleGANs, RGB images captured from drones or even satellites can be converted into NDVI images, hence making valuable and actionable information regarding crop health, vigor, and stress levels available to farmers and agronomists. This value is typically useful in areas where the existence of dedicated NDVI sensors is not available, hence guaranteeing constant and cost-effective vegetation monitoring. It enables one to create NDVI images from standard RGB data and helps with irrigation, fertilization, and pest management decisions in crops (Cheng et al., 2024).

5. REMOTE SENSING APPLICATIONS IN PRECISION FARMING

5.1. Crop Health Monitoring

- Early Detection of Pests and Diseases

The work applies Time-Series Generative Adversarial Networks in the synthesis of data for agricultural sensing to forecast pests, so as to raise an early warning on sustainable agriculture by generative models. This research focuses on the multivariate synthesis of agricultural sensing data with the aim of improving pest incidence forecasting in sustainable agriculture. This is important for timely, appropriate pest control, particularly in view of the lack of data availability in most of these applications due to challenges of long-term agricultural data collection. The authors used a Time Series Generative Adversarial Network to generate synthesized agricultural sensing data. This will enable the generation of real time-series data to be used in training neural network models such as RNN, LSTM, and GRU. Other activities conducted during this research work on the data were at the level of data preprocessing, which involved cleaning the data, processing the missing values, resampling, and fusing data to get it ready for effective model training.

Other advantages that have been noted by the author in this research are, among many others, data augmentation. It effectively tackles the challenge of missing data by generating synthetic datasets very similar to real agricultural sensing data. Such augmentation can considerably improve the performance of predictive models. Accurate Predictions: Models built using synthetic data have shown to perform almost like models built using actual data, suggesting that synthetic data can be an excellent stand-in whenever real data are limited. Well-timed Pest Control: Since this research makes it possible to predict pest populations at the right time, this work will aid in timely applications for the management of pests. This is important if agriculture is to remain sustainable.

Quality of Synthetic Data: Though the data generated is useful, concerns as to the quality and representativeness of synthetic data with respect to the real-world scenario cannot be completely ruled out. If the synthetic data doesn't hold the nuances of the actual data, it would predict not as accurately. Implementation Complexity: Computational resources and expertise to implement TimeGAN and associated preprocessing techniques could be very high, thus standing in the way for some agricultural practitioners. That is, this study aims to enhance pest incidence forecasting by using innovative data synthesis techniques that offer a lot of advantages but also have their own share of limitations (Tai, Wang, & Huang, 2023).

5.2. Soil Condition Analysis

- Mapping Soil Quality Using Enhanced Imagery

The overall objective of the study is to create soil electrical conductivity maps of greenspace in the GTA area. This would be for increasing precision regenerative agriculture with advanced techniques in remote sensing and artificial intelligence.

The author used a conditional generative adversarial network, which is a method for deep learning. In this approach, one neural network is trained on another to accurately predict the soil's conductivity. Notably, this work mentions a future possibility where other data types, such as infrared spectrum and LiDAR data, can be added to increase the model's accuracy. Study outcome: The analysis found that the CGAN model was able to predict the conductivity of the soil with an accuracy of 34.6%. That is to say, it will be medium, then, in effectiveness, which means that the model demonstrates some potential and is open for improvement. Emphasis was further put on more comprehensive data inputs for improvement in the prediction accuracy as part of the study.

Advantages: Novel Approach: CGAN stands as the most futuristic approach in soil analysis through the integration of remote sensing with artificial intelligence. Automation Potential: This study visionally automates the process by using a microdrone equipped with computer-vision-enabled sensors for soil sample analysis, enhancing efficiency and decreasing labor requirements. Commercially Available Data: All data cited in this research, such as electromagnetic bands and LiDAR, are commercially available and thus can be used in the analysis, so this method can be applied to real life.

Study limitations: Moderate Accuracy: The accuracy of the model prediction at 34.6 percent simply indicates the unreliability of the model and further refinement and validation that needs to be done. Concealed Soil Problems: The results show that, in areas with soil coverage by foliage, efficiency seems to be lower, thus further limiting its applicability in those environments. Need for Additional Data: The additional changes the author has suggested to the model are the inclusion of more electromagnetic bands and three-dimensional imagery, hence clearly specifying that there could be a deficiency in terms of the current data inputs.

In summary, this research opens an avenue for the application of AI and remote sensing for soil analysis, with apparent benefits but serious limitations that should be eliminated in order to ensure better accuracy and applicability (Kemp, 2024).

5.3. Yield Prediction

- Utilizing High Resolution and Synthetic Data for Accurate Forecasting

The goal of the research is to present an effective approach for agricultural yield forecasting enhancement by integrating advanced deep learning techniques with high-resolution satellite imagery. The correct approach to achieve food security is against the background of the rapidly increasing global population and environmental changes. The authors employ Deep Convolutional Generative Adversarial Networks to generate synthetic satellite images representing real agricultural conditions. This is used to augment the dataset of a Convolutional Neural Network tasked with yield estimation. This combination of DCGAN and CNN will confer improved feature extraction and generalization across a range of environmental conditions. Outcomes of the study reveals a high degree of precision—namely, the precision of agricultural yield forecasting is around 98.6 percent, which is an improvement of 3.62 percent compared to traditional CNN-based methods. In this regard, it decisively acknowledges the complex trends that correlate with the growth phases, health, and yield potential of crops. Interestingly, the inclusion of synthetic images enhances the model's robustness and adaptability in various geographical locations and changing climatic conditions.

Advantages being Data Augmentation: the utilization of DCGANs will further be enabled to generate synthetic images, which will make the training dataset rich and improve the generalization properties of the model. High Accuracy: The proposed approach yields better accuracy in predicting yield compared to the conventional model. Interpretability: The study also elaborates on the interpretability of the CNN model, shedding light on features learned and their relation with yield-related factors. This supports sustainable farming: The model underpins sustainable farming practices with better yield predictions and informed decisions to ensure food security.

The limitations observed by author are: This is in line with the fact that there have been problems with traditional agricultural yield prediction models in extracting their features from satellite data and generalizing during different environmental conditions. Such problems include an inhomogeneous pixel intensity distribution within those images, influenced by different variables, for example shadows and lighting.

To sum up, this study proposed a novel approach to forecasting agricultural yields, taking deep advantage of recent advances in deep learning to make sure that prediction accuracy is not limited by existing limitations in the literature (Anuradha, Kuchipudi, Ashreetha, Venkata, Naga Ramesh, & Ayadi, 2024).

6. RECENT TRENDS IN GANS FOR PRECISION FARMING

Generative Adversarial Networks are increasingly becoming a part and parcel of precision farming, integrating their tools with Agri-Tech to innovate in the realms of imaging, real-time data processing, and predictive modelling. The emerging GAN applications in these areas include enhancing efficiencies around resources and shaping the technologies for precision agriculture to ensure better crop monitoring in response to environmental challenges.

6.1. Multi-Spectral and Hyper-Spectral Imaging

Multispectral and hyperspectral imaging technologies are essential for capturing data at multiple wavelengths, which is required to identify plant health, soil composition, and water content. In such spectral bands, the quality synthetic images generated by GANs compensate for gaps real-world data faces due to limitations in sensors or environment conditions. In this regard, GANs can be used to increase the accuracy of vegetation indices and other crucial measurements by synthetically generating data in specific spectral bands that could be missing or corrupted. This contributes to more accurate crop condition monitoring, early identification of nutrient deficiency supply, and appropriate intervention to reach optimal crop yields (Kerdegari, Razaak, Argyriou, & Remagnino, 2019).

6.2. Integration with Edge Computing for Real-Time Analysis

One of the major trends in precision farming is represented by the integration of GANs with edge computing devices, such as drones, sensors, and IoT nodes. Edge computing enables processing at source locations without having to move the large datasets to some central server for analysis. Running GANs on edge devices enables real-time image enhancement, anomaly detection, and predictive analytics in the field. For example, a drone that includes GANs could process and enhance imagery in-flight to give farmers real-time information on the health status of crops or the existence of pests. This ability to do things in real time is very important for time-critical agricultural decisions, since fast responses reduce crop losses and bring about more efficient use of resources (Qi, Gao, Chen, Shu, & Pearson, 2022).

6.3. Use of GANs for Climate Impact Simulation

The application of GANs is becoming increasingly relevant in view of climate change, which is having continuous impacts on agricultural systems through simulation and prediction. These models can simulate changes in temperature patterns, precipitation levels, and extreme weather events by generating synthetic climatic data to show their impact on crop growth, soil health, and water availability. These simulations, therefore, facilitate the development of adaptive strategies by farmers and agronomists associated with the selection of climate-resilient crop varieties, optimal scheduling of planting, and irrigation practice adjustment. GAN-based climate impact simulations help in making informed decisions that may reduce potential risks of climate variability on agriculture and ensure its long-term sustainability (Morales-García, Bueno-Crespo, Terroso-Sáenz, Arcas-Túnez, Martínez-España, & Cecilia, 2023).

6.4. Generative Models for Soil Moisture Estimation

It is useful in crop health, yield, irrigation management, and, at a larger scale, water resources. GANs are applied to the problem of estimating soil moisture by generating synthetic data featuring the interactions between physical properties of the soil, vegetation cover, and weather. These generative models can predict the soil moisture content more accurately than traditional methods even in areas where direct measurements are relatively scarce. This would be possible by fusing the data generated by GAN into satellite imagery and sensory readings for good monitoring of the level of soil moisture over large areas, thus optimizing irrigation schedules to avoid excess use of water resources and minimize the negative impacts of dry periods on crops (Jiang, Zhao, Ding, & Li, 2023).

6.5. Combining GANs with Deep Learning for Predictive Analytics

This integration of GANs with deep learning models is a strong trend in precision agriculture that essentially puts a premium on the accuracy and reliability of predictive analytics. In this respect, GANs can generate synthetic data to either fill gaps or complement existing datasets so deep learning models become completer and more varied during training. This will particularly help in predicting complicated agricultural outcomes, like crop yields or pest outbreaks and diseases. GANs improve the predictive capabilities of a model by generating more realistic simulations of possible future scenarios, aiding a farmer in proactive decision making. This integration will also support precision farming practices through more exact interventions, optimized use of resources, and reduction in the risks associated with crop failure (Fawakherji, Potena, Prevedello, Pretto, Bloisi, & Nardi, 2020).

7. EVALUATION

GANs have been well integrated into precision farming, showing huge improvements in all areas of agricultural applications. The results using the GAN-based methods were analysed with respect to the key areas, proving GANs to be highly effective in image quality enhancement, the accuracy of monitoring, and sustainability in agriculture.

7.1. Improvement in Image Quality and Resolution

In particular, the quality and resolution of satellite and drone imagery have been improved by GANs, more so the Super-Resolution GANs. GANs generate a high-resolution image from most low-resolution input images, hence allowing detailed visual field-level analysis of crop fields. This improvement in resolution enables the detection of fine-grained features such as early signs of pest infestations, subtle variations in soil moisture, and small-scale crop health issues that could not be detected earlier. Better image quality ensures that the monitoring will be real and, hence, the decision-making, for better management of the crops.

7.2. Enhanced Accuracy in Crop Health Monitoring

GANs have significantly improved crop health monitoring by providing the models with the capability to be trained on generated synthetic data and further advanced image analysis. For instance, GANs have been applied in generating NDVI images from standard RGB images, which are relevant to vegetation health status. This improves crop vigour monitoring, detection of stress factors, and prediction of possible yield losses. Since this gives better insight into crop health, timely interventions are possible with GANs, reducing the chance of crop failure and increasing overall productivity.

7.3. Accurate Yield Prediction Models

One of the most important applications of GANs in precision agriculture is in building yield prediction models with greater accuracy. GANs generate synthetic datasets simulating a wide range of crop growth scenarios that can be used to train machine-learning-based predictive models with a wide range of environmental and agronomic factors. These models have been shown to do a better job of predicting yields under a wide range of variable conditions, such as variable weather patterns and soil heterogeneity. Precise yield predictions can help guide resource planning, planning of the harvest, and maximizing the economic returns for a farmer.

7.4. Resource Optimization and Sustainable Practices

Finally, GANs have also been applied in precision farming to promote sustainable agriculture practices through resource optimization. GANs use crop health monitoring accuracy and yield prediction for the efficient use of resources like water, fertilizers, and pesticides. For example, GANs will use optimized irrigation schedules, which will significantly reduce water wastage with a precise soil moisture estimation. Further, GAN-generated insights enable the application of only the required amount of fertilizer and pesticides, thereby reducing their effects on the environment. Basically, application of GANs aids in enhancing sustainability in agriculture through a reduction in the use of resources as well as the promotion of eco-friendly farming.

Overall, with the application of GANs in precision farming, image quality, monitoring accuracy, predictive modelling, and resource optimization have grossly improved, hence supporting effective and sustainable agricultural management.

8. DISCUSSION

8.1. Benefits of Using GANs in Remote Sensing for Precision Farming

Necessities of using generative adversarial networks in remote sensing for precision farming include a variety of their significant benefits connected with the rise in quality of received data and possibilities of more accurate agricultural practices:

High-Resolution Images: Upscaling low-resolution satellite and drone images into high-resolution images using GANs or at least SRGANs can be done to obtain quality visuals for further identification of the crops and soil structure. This would identify problems at an incipient stage, such as pest infestations or nutrient deficiencies, which elude identification with lower resolution imagery (RFM-GAN, 2023).

Synthetic Data Generation: The application of GANs in generating synthetic remote sensing data is very useful in training machine learning models in cases where real-world data is limited or unavailable. This will increase the robustness and accuracy of predictive models that aid in precision farming (Feng, 2022).

Improved Vegetation Analysis: GANs like CycleGANs can convert an RGB image into an NDVI image, which becomes imperative in understanding plant health. This provides more accuracy within the crop conditions, and accordingly, better decision-making may be done in the agricultural management approach (Gurav, Karanjikar, Bhikle, & Deshpande, 2024).

Adaptive Climate Modelling: GANs allow for the simulating of different climate scenarios and their possible impacts on the crops. Making this facility available will, therefore, help farmers better equip and, therefore, mitigate against the varying impacts of climate, which ultimately would help boost agricultural resilience and sustainability (Jeyalakshmi, Kotra, Saiganesh, & Manohar, 2024).

Real-Time Processing and Decision-Making: When edge computing is incorporated into GANs, real-time data analysis is enabled for immediate responses to changing field conditions. This is particularly good in time-critical agriculture operations, such as pest control or irrigation adjustments (Lei, Zhu, Li, Tan, Wang, Zhou, & Zhang, 2024).

8.2. Challenges and Limitations

Although there are considerable advantages to the application of GANs in precision farming, there are associated challenges and limitations as well:

Data Requirements: GANs require a large amount of good quality data for training. In agricultural scenarios, it becomes difficult to collect enough labelled data, especially on particular crops or regions. Insufficient data will create less accurate models with unreliable predictions (Miletic & Sariyar, 2024).

Computational Complexity: GANs have generally become computationally very expensive and might require extensive processing powers while operating on high-resolution images or large datasets. This could prove to be a constraint on most farmers or organizations that lack access to high-performance computing resources (Kumar, Chaudhary, Karir, Díaz-Raviña, Choudhary, & Kumar, 2024).

Model Stability and Training: Training GANs could be quite hard due to problems like mode collapse, where the generator would produce very limited diversity at the output, with instabilities in the training process of the adversary. Deep learning expertise is required to master these challenges; therefore, GAN implementation remains rather inaccessible to non-experts (Pomar & Remus, 2023).

Generalization Across Different Conditions: A GAN model that has been trained with a certain dataset may not generalize across different environmental conditions. An example could be different types of soil, climate, and varieties of crops. In this case, GAN's applications in different agricultural conditions are limited (Tang, Asai, Kato, Fukano, & Guo, 2024).

Ethical and Environmental Issues: Although GANs really do optimize the expenditure of resources, their predictions inadvertently increase reliance on inputs, such as fertilizers or pesticides, which raises concerns in respect to environmental impacts and sustainability. In addition, synthetic data generated by GANs needs to be carefully validated in order to avoid biases or inaccuracies (Singh Klair, Agrawal, & Kumar, 2024).

In the end, even though GANs bear tremendous promise in improving agricultural practices, some practical challenges in terms of data requirement, computational demand, and model stability face it. However, overcoming those hurdles would significantly facilitate the use of GAN technology in any given agricultural setting (Huang, Chen, & Liu, 2023).

8.3. Comparison with Traditional Methods

Comparison to Traditional Methods GANs differ distinctly from the traditional methods used in precision farming, with a high edge in data processing and analysis, predictive accuracy, among others. They also pose some other unique challenges, such as the following:

Resolution in Images and Their Enhancement: Traditional methods most often use some pre-existing algorithms of image processing that might not substantially improve image resolution. GANs can generate high-resolution images from lower-resolution inputs, offering detailed insights into agricultural fields (Fawakherji, Potena, Prevedello, Pretto, Bloisi, & Nardi, 2020).

Data Augmentation: Traditional methods rely on the collection of real-world data, which can be constrained by time, cost, and environmental factors. GANs can generate synthetic data, thus allowing the creation of diversity in training datasets for better performance in machine learning models, especially if one is dealing with rare events or conditions (Sahoo, Dass, & Prakash, 2023).

Modelling Complexity: Most of the classical methods in remote sensing are simple calculations over the images, such as vegetation indices, for crop monitoring. GANs add higher complexity to the modelling of, for example, climate impacts or image conversion for various analysis goals, like RGB to NDVI. This added complexity comes at the cost of higher expertise and computational resources (Olaniyi, Chen, Lu, & Huang, 2022).

Real-Time Capabilities: Most of the traditional methods require post-processing of remote sensing data, which delays decision-making. GANs allow real-time analysis, especially when combined with edge computing, and immediate action in the field. This provides a large advantage for time-critical agricultural operations (Patibandla, Rao, & Malla, 2024).

Accuracy and Predictive Power: Traditional methods may lack the predictive power inherent in GANs, especially when representing complex, nonlinear relationships in the data. GANs can provide more accurate predictions related to crop health, yield estimation, and resource optimization. However, it requires large datasets and specialized training processes, which may become a limitation in themselves ("Machine Learning…", 2023).

In conclusion, though GANs do have transforming potential in image enhancement for resolutions higher than 1 and improving data augmentation abilities for real-time analytics in agriculture, they are pretty complex and resource-hungry. Balancing such advanced capabilities with practical implementation will be key to maximizing their impact in agricultural practices (Shankar Sahoo, Dass, & Prakash, 2023).

9. FUTURE PROSPECTS

A good number of enterprising habitats has been springing up concomitantly with GAN application in precision agriculture. In the same line, with time, several potentials have emerged that will significantly further a country's productivity and sustainability in agriculture. These future opportunities underline the revolution that GANs might have on precision agriculture, namely in developing simulations for weather conditions and real-time monitoring, besides integration with other cutting-edge technologies.

9.1. Potential for Advanced Weather Simulation

GANs can hugely be used to improve weather simulation models in order to provide more accurate and localized forecasts—very essential in precision farming. GANs generate synthetic weather data that masters both the complexity and variability of the climate conditions, hence helping simulate different weather scenarios and their impacts on agricultural systems. Advanced simulations will provide farmers with an idea of whether to adjust their schedules on planting, irrigation, and harvest time based on the weather forecast. In the near future, GANs will be useful for simulating the influence of extreme weather, such as drought or flooding, on agriculture and allowing risk management strategies (Tai, Wang, & Huang, 2023).

9.2. Real-Time Monitoring and Analysis

In this view, the future of precision farming will increasingly be oriented to real-time monitoring and analysis. GANs are going to be critical for this. Integrate GANs with edge computing and IoT devices, and continuous data analysis will then do all the work right in the field, providing instant insights on crop health, soil conditions, and resource needs to the farmer. All this real-time capability will now enable more dynamic decision-making and swift reaction to the emerging issues, be they pest attacks or nutrient deficiencies. The continuous monitoring and analysis of data in real time can reduce the time lag between data gathering and action taken, which basically does go on to result in efficient and effective farm management (Fan, Wang, Funk, et al., 2022).

9.3. Integration with Other Technologies (e.g., IoT, Automated Drones)

GANs, combined with other sophisticated technologies like the Internet of Things and automated drones, can be of great significance in the realm of precision farming. IoT devices are able to gather enormous amounts of data about soil moisture, temperature, and crop growth metrics directly from the field. For example, GANs and automated drones can be used to capture high-resolution images and to carry out aerial surveys. By integrating these data streams with GAN-based analysis, farmers can have a holistic view of the fields and make more accurate, data-driven decisions. For example, real-time processing using GANs from drone imagery might be used to identify areas of concern and guide automated drones for targeted interventions such as spraying or planting. These technologies integrated into farming operations will further automate and optimize production, reducing the need for manual labour in order to accomplish tasks, generally increasing efficiency (Khan, Abu Bakr, Aslam, & Saeed, n.d.).

10. CONCLUSION

The adoption of GANs in precision farming represents a significant advancement in agricultural AI applications. GANs have also been shown to increase remote sensing by refining image resolution, creating synthetic training data, and boosting predictive analytics for crop and yield monitoring. These advancements help to promote more efficient and sustainable agricultural practices by enabling the early diagnosis of crop diseases, detailed study of soil conditions, and accurate yield forecasts. Large-scale implementation will involve overcoming challenges like as computing complexity, data requirements, and model stability.

This forthcoming study will expand GAN application in precision agriculture. GAN-enhanced multi and hyperspectral imaging will reveal crop health and, consequently, soil conditions. IoT/UAV connection allows for real-time monitoring and in-field decision-making, resulting in more responsive, data-driven farming practice. GAN-based climate models will aid in the development of more accurate models for forecasting and anticipating the long-term effects of such climatic hazards on crops, potentially reducing climate-related agricultural risks. Another potential path is to use cloud and edge computing to analyze GANs, which addresses high computational demands without requiring substantial processing on-site and allows for real-time analysis. In large-scale agriculture, distributed computing can be used to efficiently use GANs for remote sensing applications. As technology advances, GANs will play an increasingly important role in data-driven farming, where decisions are optimised, resources are employed more efficiently, and resilience is built against environmental stresses. The next phase of precision agriculture will be characterized by more refinement and integration with future technology. GANs have the potential to convert farming into an intelligent, efficient, and climate-resilient business through ongoing research and innovation, ensuring long-term food production sustainability.

REFERENCES

Anantrasirichai, N., & Bull, D. (2022). Artificial intelligence in the creative industries: A review. *Artificial Intelligence Review*, *55*(1), 589–656. DOI: 10.1007/s10462-021-10039-7

Andrew, H. (2021). Generative Adversarial Network Synthesis of Hyperspectral Vegetation Data. *Remote Sensing*. Advance online publication. DOI: 10.3390/rs13122243

Anuradha, D., Kuchipudi, R., Ashreetha, B., Ramesh, J. V. N., & Rami, A. (2024) Enhancing Agricultural Yield Forecasting with Deep Convolutional Generative Adversarial Networks and Satellite Data. *International Journal of Advanced Computer Science and Applications*, *15*(2). Advance online publication. DOI: 10.14569/IJACSA.2024.0150269

Anuradha, D., Kuchipudi, R., Ashreetha, B., Ramesh, J. V. N., & Rami, A. (2024). Enhancing Agricultural Yield Forecasting with Deep Convolutional Generative Adversarial Networks and Satellite Data. *International Journal of Advanced Computer Science and Applications*, *15*(2). Advance online publication. DOI: 10.14569/IJACSA.2024.0150269

Awate, M. P. L., & Nagne, A. (2023). Recent trends in crop identification and analysis by using remote sensing and gis. *The Online Journal of Distance Education and e-Learning : TOJDEL*, *11*(1).

Bobadilla, J., & Gutiérrez, A. (2024). Wasserstein GAN-based architecture to generate collaborative filtering synthetic datasets. *Applied Intelligence*, *54*(3), 2472–2490.

Cheng, J., Zhu, Y., Zhao, Y., Li, T., Chen, M., Sun, Q., Gu, Q., & Zhang, X. (2024). Application of an improved U-Net with image-to-image translation and transfer learning in peach orchard segmentation. *International Journal of Applied Earth Observation and Geoinformation*, *130*, 103871. DOI: 10.1016/j.jag.2024.103871

DeVries, T., Romero, A., Pineda, L., Taylor, G. W., & Drozdzal, M. (2019). On the evaluation of conditional GANs. arXiv preprint arXiv:1907.08175.

Dou, X., Li, C., Shi, Q., & Liu, M. (2020). Super-resolution for hyperspectral remote sensing images based on the 3D attention-SRGAN network. *Remote Sensing (Basel)*, *12*(7), 1204. DOI: 10.3390/rs12071204

Fan, Y., Wang, X., Funk, T., Rashid, I., Herman, B., Bompoti, N., Mahmud, M. D. S., Chrysochoou, M., Yang, M., Vadas, T. M., Lei, Y., & Li, B. (2022). A critical review for real-time continuous soil monitoring: Advantages, challenges, and perspectives. *Environmental Science & Technology*, *56*(19), 13546–13564. DOI: 10.1021/acs.est.2c03562 PMID: 36121207

) Fawakherji, M., Potena, C., Prevedello, I., Pretto, A., Bloisi, D. D., & Nardi, D. (2020, August). Data augmentation using gans for crop/weed segmentation in precision farming. In 2020 IEEE conference on control technology and applications (CCTA) (pp. 279-284). IEEE.

Feng, J. (2022).. . *Generative Adversarial Networks for Remote Sensing.*, *108–112*, 108–112. Advance online publication. DOI: 10.1109/ICBAR58199.2022.00028

Ge, H., Xia, Y., Chen, X., Berry, R., & Wu, Y. (2018). Fictitious gan: Training gans with historical models. In *Proceedings of the European Conference on Computer Vision (ECCV)* (pp. 119-134). DOI: 10.1007/978-3-030-01246-5_8

Goodfellow, I., Pouget-Abadie, J., Mirza, M., Xu, B., Warde-Farley, D., Ozair, S., Courville, A., & Bengio, Y. (2020). Generative adversarial networks. *Communications of the ACM*, *63*(11), 139–144. DOI: 10.1145/3422622

) Gurav, A. S., Karanjikar, Y. M., Bhikle, O. A., & Deshpande, H. (2024). Comparing the Effectiveness of GAN-based Methods for Enhancing EuroSat Satellite Imagery Classification. 1611–1616. DOI: 10.1109/ICESC60852.2024.10689946

Han, X., Zhang, L., Zhou, K., & Wang, X. (2019). ProGAN: Protein solubility generative adversarial nets for data augmentation in DNN framework. *Computers & Chemical Engineering*, *131*, 106533. DOI: 10.1016/j.compchemeng.2019.106533

) Jeyalakshmi, J., Kotra, V., Saiganesh, B., & Manohar, E. (2024). AI Deployment Models in Precision Agriculture. 125–142. DOI: 10.1201/9781003504900-6

Jia, X., Wang, M., Khandelwal, A., Karpatne, A., & Kumar, V. (2019, August). Recurrent Generative Networks for Multi-Resolution Satellite Data: An Application in Cropland Monitoring. In IJCAI.

Jiang, C., Zhao, J., Ding, Y., & Li, G. (2023). Vis–NIR spectroscopy combined with GAN data augmentation for predicting soil nutrients in degraded Alpine Meadows on the Qinghai–Tibet Plateau. *Sensors (Basel)*, *23*(7), 3686. DOI: 10.3390/s23073686 PMID: 37050746

Jozdani, S., Chen, D., Pouliot, D., & Johnson, B. A. (2022). A review and meta-analysis of generative adversarial networks and their applications in remote sensing. *International Journal of Applied Earth Observation and Geoinformation, 108*, 102734. DOI: 10.1016/j.jag.2022.102734

Karras, T., Laine, S., Aittala, M., Hellsten, J., Lehtinen, J., & Aila, T. (2020). Analyzing and improving the image quality of stylegan. In *Proceedings of the IEEE/CVF conference on computer vision and pattern recognition* (pp. 8110-8119). DOI: 10.1109/CVPR42600.2020.00813

Kemp, T. (2024). Soil Analysis via Remote Sensing and Artificial Intelligence for Precision Regenerative Agriculture. Toronto Metropolitan University. Thesis. https://doi.org/DOI: 10.32920/26060842

Kerdegari, H., Razaak, M., Argyriou, V., & Remagnino, P. (2019). Semi-supervised gan for classification of multispectral imagery acquired by uavs. arXiv preprint arXiv:1905.10920.

Khan, J. A., Abu Bakr, M., Aslam, S., & Saeed, A. Recent Advancements and Challenges in Deep Learning and Internet of Things for Precision Agriculture. Sumera and Saeed, Adnan, Recent Advancements and Challenges in Deep Learning and Internet of Things for Precision Agriculture.

Klair, Y. S., Agrawal, K., & Kumar, A. (2024, March). Impact of generative ai in diagnosing diseases in agriculture. In 2024 2nd International Conference on Disruptive Technologies (ICDT) (pp. 870-875). IEEE.

Kumar, L., Chaudhary, S., Karir, S. R., Díaz-Raviña, M., Choudhary, A. K., & Kumar, S. G. (2024). *The prologue to precision livestock farming*., DOI: 10.58532/V3BCAG21P2CH11

Lakshmi, G., Brindha, S., Sathish, V., Kumar, V. M., & Santhoshkumar, P. (2024, April). Advancing Precision Agriculture: AI, IoT, and Machine Learning Solutions for Optimal Crop Yield. In 2024 International Conference on Communication, Computing and Internet of Things (IC3IoT) (pp. 1-5). IEEE.

Lei, D., Zhu, Q., Li, Y., Tan, J., Wang, S., Zhou, T., & Zhang, L. (2024). HPLTS-GAN: A High-Precision Remote Sensing Spatio-Temporal Fusion Method Based on Low Temporal Sensitivity. *IEEE Transactions on Geoscience and Remote Sensing, 1*, 1–16. Advance online publication. DOI: 10.1109/TGRS.2024.3416152

Liu, Y., Liu, Y., & Wang, J. (2023). RFM-GAN: Robust feature matching with GAN-based neighborhood representation for agricultural remote sensing image registration. *IEEE Geoscience and Remote Sensing Letters, 20*, 1–5.

Manikandababu, C. S., Preethi, V., Kanna, M. Y., Vedhathiri, K., & Kumar, S. S. (2024, May). Enhancing Crop Yield Prediction with IoT and Machine Learning in Precision Agriculture. In *2024 International Conference on Advances in Computing, Communication and Applied Informatics (ACCAI)* (pp. 1-6). IEEE.

Maqsood, M. H., Mumtaz, R., Haq, I. U., Shafi, U., Zaidi, S. M. H., & Hafeez, M. (2021). Super resolution generative adversarial network (Srgans) for wheat stripe rust classification. *Sensors (Basel), 21*(23), 7903. DOI: 10.3390/s21237903 PMID: 34883905

Miletic, M., & Sariyar, M. (2024). Challenges of Using Synthetic Data Generation Methods for Tabular Microdata. *Applied Sciences (Basel, Switzerland), 14*(14), 5975. DOI: 10.3390/app14145975

Morales-García, J., Bueno-Crespo, A., Terroso-Sáenz, F., Arcas-Túnez, F., Martínez-España, R., & Cecilia, J. M. (2023). Evaluation of synthetic data generation for intelligent climate control in greenhouses. *Applied Intelligence, 53*(21), 24765–24781. DOI: 10.1007/s10489-023-04783-2

) N., Desai., Dhara, O., Bhavsar., Himanshu, Pandya. (2023). A review: applications of remote sensing in agriculture. DOI: 10.56588/iabcd.v2i1.137

Olaniyi, E. O., Chen, D., Lu, Y., & Huang, Y.-Y. (2022). Generative Adversarial Networks for Image Augmentation in Agriculture: A Systematic Review. *Computers and Electronics in Agriculture, 200*, 107208. DOI: 10.1016/j.compag.2022.107208

Patibandla, R., Rao, B. T., & Malla, R. (2024). *Precision Agriculture and AI*. Informa., DOI: 10.1201/9781003451648-7

Pomar, C., & Remus, A. (2023). Review: Fundamentals, limitations and pitfalls on the development and application of precision nutrition techniques for precision livestock farming. *Animal, 100763*, 100763. Advance online publication. DOI: 10.1016/j.animal.2023.100763 PMID: 36966025

Qi, C., Gao, J., Chen, K., Shu, L., & Pearson, S. (2022). Tea chrysanthemum detection by leveraging generative adversarial networks and edge computing. *Frontiers in Plant Science, 13*, 850606. DOI: 10.3389/fpls.2022.850606 PMID: 35463441

Rodríguez-Suárez, B., Quesada-Barriuso, P., & Argüello, F. (2022). Design of CGAN models for multispectral reconstruction in remote sensing. *Remote Sensing (Basel), 14*(4), 816. DOI: 10.3390/rs14040816

Sahoo, G. S., Dass, S., & Prakash, F. (2023, November). The Synergy of Advanced Language Models, Sequential Skeleton Features, and GAN-Enhanced Data Augmentation for Unprecedented Precision. In *2023 International Conference on System, Computation, Automation and Networking (ICSCAN)* (pp. 1-6). IEEE.

) Sahoo, G. S., Dass, S., & Prakash, F. (2023). The Synergy of Advanced Language Models, Sequential Skeleton Features, and GAN-Enhanced Data Augmentation for Unprecedented Precision. 1–6. DOI: 10.1109/ICSCAN58655.2023.10395300

Salimans, T., Goodfellow, I., Zaremba, W., Cheung, V., Radford, A., & Chen, X. (2016). Improved techniques for training gans. *Advances in Neural Information Processing Systems*, •••, 29.

Sarmah, B., Baruah, V. J., & Sarmah, T. (2024). Remote sensing–aided disaster risk mitigation in agriculture. In *Remote Sensing in Precision Agriculture* (pp. 487–499). Academic Press.

Schnell, T., Bott, K., Puck, L., Buettner, T., Roennau, A., & Dillmann, R. (2022, October). RoBiGAN: A bidirectional Wasserstein GAN approach for online robot fault diagnosis via internal anomaly detection. In *2022 IEEE/RSJ International Conference on Intelligent Robots and Systems (IROS)* (pp. 4332-4337). IEEE. DOI: 10.1109/IROS47612.2022.9982240

Sharmilan, S., & Chaminda, H. T. (2017). Generate bioinformatics data using Generative Adversarial Network: a review. In 2nd international conference on information technology research. Colombo.

Sihan, R. (2024). *Applications and challenges of GAN in AI-powered artistry*. Applied and Computational Engineering., DOI: 10.54254/2755-2721/41/20230713

Sood, A. (2023). An overview of precision farming. *Int. J. Environ. Clim. Change*, *13*(12), 441–456.

Tai, C. Y., Wang, W. J., & Huang, Y. M. (2023). Using time-series generative adversarial networks to synthesize sensing data for pest incidence forecasting on sustainable agriculture. *Sustainability (Basel)*, *15*(10), 7834. DOI: 10.3390/su15107834

Tai, C. Y., Wang, W. J., & Huang, Y. M. (2023). Using time-series generative adversarial networks to synthesize sensing data for pest incidence forecasting on sustainable agriculture. *Sustainability (Basel)*, *15*(10), 7834. DOI: 10.3390/su15107834

Tang, L. (2024). Motoaki, Asai., Yoichiro, Kato., Yuya, Fukano., Wei, Guo. (2024). Channel Attention GAN-based Synthetic Weed Generation for Precise Weed Identification. *Plant Phenomics (Washington, D.C.)*, *6*, 0122. Advance online publication. DOI: 10.34133/plantphenomics.0122

Thirumagal, E., & Saruladha, K. (2021). GAN models in natural language processing and image translation. In *Generative Adversarial Networks for Image-to-Image Translation* (pp. 17–57). Academic Press. DOI: 10.1016/B978-0-12-823519-5.00001-4

Tran, N. T., Tran, V. H., Nguyen, N. B., Nguyen, T. K., & Cheung, N. M. (2021). On data augmentation for GAN training. *IEEE Transactions on Image Processing*, *30*, 1882–1897. DOI: 10.1109/TIP.2021.3049346 PMID: 33428571

Vuletić, M., Prenzel, F., & Cucuringu, M. (2024). Fin-gan: Forecasting and classifying financial time series via generative adversarial networks. *Quantitative Finance*, *24*(2), 175–199. DOI: 10.1080/14697688.2023.2299466

Wang, Z., She, Q., & Ward, T. E. (2021). Generative adversarial networks in computer vision: A survey and taxonomy. *ACM Computing Surveys*, *54*(2), 1–38.

Yican, H. (2023, December). Zhengguang, Chen., Jinming, Liu. (2023). Limited agricultural spectral dataset expansion based on generative adversarial networks. *Computers and Electronics in Agriculture*, *215*, 108385. Advance online publication. DOI: 10.1016/j.compag.2023.108385

Zualkernan, I., Abuhani, D. A., Hussain, M. H., Khan, J., & ElMohandes, M. (2023). Machine Learning for Precision Agriculture Using Imagery from Unmanned Aerial Vehicles (UAVs): A Survey. *Drones (Basel)*, *2023*(7), 382.

Chapter 12
Harnessing the Power of Generative Adversarial Networks in Remote Sensing: Challenges, Opportunities, and Future Directions

C. V. Suresh Babu
https://orcid.org/0000-0002-8474-2882
Hindustan Institute of Technology and Science, India

P. Yukesh Kumar
Hindustan Institute of Technology and Science, India

K. Arjun
Hindustan Institute of Technology and Science, India

ABSTRACT

This chapter aims to explore the potential of Generative Adversarial Networks (GANs) in enhancing remote sensing applications, addressing key challenges such as data scarcity and low image resolution. The study employs a comprehensive review of existing literature and case studies to analyze the effectiveness of GANs in image generation, super-resolution, and anomaly detection. Key findings indicate that GANs significantly improve the quality of remote sensing imagery, enabling better data synthesis and interpretation. The results underscore the importance of high-quality datasets and the need for robust evaluation metrics to assess GAN performance. The implications of this research highlight the transformative role of

DOI: 10.4018/979-8-3693-6900-5.ch012

GANs in remote sensing, paving the way for future advancements in environmental monitoring and disaster management.

1.0. INTRODUCTION

1.1 Overview of Generative Adversarial Networks (GANs)

Generative Adversarial Networks (GANs) have emerged as a ground-breaking approach in the field of machine learning, particularly for their ability to generate high-dimensional data that closely resembles real-world distributions. Introduced by Goodfellow et al. (2014) and Suresh Babu, C. V., (2003), GANs consist of two neural networks—the generator and the discriminator—that engage in a competitive process. The generator creates synthetic data, while the discriminator evaluates its authenticity, leading to iterative improvements in both models. This unique architecture allows GANs to be applied across various domains, including image synthesis, data augmentation, and anomaly detection (Shorten & Khoshgoftaar, 2019).

1.2 Importance of Remote Sensing

Remote sensing plays a crucial role in monitoring and analysing environmental changes, urban development, and disaster management. However, challenges such as data scarcity, low-resolution imagery, and the complexity of multi-sensor data fusion hinder the effectiveness of traditional remote sensing techniques (Mash & Weavers, 2011). The integration of advanced machine learning methods, particularly GANs, offers promising solutions to these challenges by enhancing data quality and enabling more accurate analyses. Generative Adversarial Networks (GANs) are a generative AI approach that can mimic the patterns of generating new, realistic data from existing datasets. Most of the time, this comes in handy in the domains of generating images, videos, or even medical records for research, which are difficult or expensive to collect in terms of real data. The usability of GANs stems from the quality of output and diversity, irrespective of labelled data

1.3 Objectives of the Study

The primary objective of this study is to investigate the application of GANs in remote sensing, focusing on their potential to improve image quality, facilitate data synthesis, and enhance anomaly detection. By reviewing existing literature and case studies, this research aims to provide insights into the effectiveness of GANs in addressing the limitations of current remote sensing methodologies and to identify future research directions in this evolving field. Following are the objectives of the chapter:

1. To investigate how Generative Adversarial Networks (GANs) can be utilized to enhance various aspects of remote sensing, including image quality and data synthesis.
2. To outline the primary challenges faced in remote sensing, such as data scarcity, low-resolution imagery, and the complexities of multi-sensor data integration.
3. To review existing literature and case studies that demonstrate the successful application of GANs in remote sensing tasks, including image enhancement, anomaly detection, and data augmentation.
4. To assess the performance of GANs in improving remote sensing outcomes, focusing on metrics such as image resolution, classification accuracy, and data reliability.
5. To identify potential areas for future research and development in the application of GANs within remote sensing, emphasizing the need for advancements in model architectures and interpretability.
6. To discuss the implications of GANs for real-world applications in environmental monitoring, disaster management, and urban planning, showcasing their transformative potential in these fields.

1.4. Need and Importance of the Study

The integration of Generative Adversarial Networks (GANs) in remote sensing is increasingly vital due to the growing demand for high-quality data in various applications, including environmental monitoring, urban planning, and disaster management. Traditional remote sensing methods often face significant challenges, such as limited data availability, low image resolution, and difficulties in data fusion from multiple sensors (Zhang et al., 2018). These challenges can hinder accurate analysis and decision-making processes, making it essential to explore innovative solutions that can enhance data quality and usability.

GANs offer a promising approach to address these challenges by generating realistic synthetic data and improving the resolution of existing images (Frid-Adar et al., 2018). Their ability to perform data augmentation can significantly enhance the training datasets available for machine learning models, leading to better performance in tasks such as land cover classification and anomaly detection (Shorten & Khoshgoftaar, 2019). Furthermore, GANs can facilitate cloud removal and improve domain adaptation, which are crucial for ensuring the robustness of remote sensing applications across different geographic regions and imaging conditions (Li et al., 2020).

The importance of this study lies in its potential to provide a comprehensive understanding of how GANs can transform remote sensing practices. By identifying effective applications and evaluating their impact, this research can contribute to the development of more reliable and efficient remote sensing technologies, ultimately benefiting various sectors that rely on accurate environmental data (Zhang et al., 2018).

1.5 Structure of the chapter

This chapter is organized into several key sections. Following this introduction, a literature review will summarize existing research on GANs in remote sensing. The methodology section will outline the approach taken in this study, followed by a discussion of the findings and their implications. Finally, the paper will conclude with recommendations for future research and practical applications of GANs in remote sensing.

2.0. THEORETICAL BACKGROUND OF THE STUDY

The theoretical foundation of this study is rooted in the principles of Generative Adversarial Networks (GANs) and their application in remote sensing. GANs, introduced by Goodfellow et al. (2014), consist of two neural networks—the generator and the discriminator—that engage in a zero-sum game. The generator's role is to create synthetic data that mimics real data, while the discriminator's task is to distinguish between real and generated data. This adversarial process drives both networks to improve iteratively, resulting in the generation of high-quality synthetic data that can be used in various applications, including remote sensing. The nature theory concerned with tapping the power of generative AI within adversarial networks is heavily-dependent on game theory and machine learning principles. A generative adversarial network (GAN) is a type of model introduced by Goodfellow et al. in 2014 .The operates on an adversarial model in which two neural networks-the generator and the discriminator-are trained simultaneously.

Machine learning and game theory bases the theoretical bedrock on which harnessing the power of generative AI into the adversarial networks was built. Generative adversarial networks (GANs), introduced by Goodfellow et al. in 2014, are mostly based on the competitive model, where two neural networks are trained on a competitive basis: the generator and the discriminator.

2.1 Generative Adversarial Networks (GANs)

GANs operate on the premise of competition between the generator and the discriminator. The generator learns to produce data that is indistinguishable from real data, while the discriminator learns to identify the authenticity of the data it receives. This dynamic leads to a continuous improvement cycle, where the generator enhances its output based on feedback from the discriminator (Goodfellow et al., 2014). The mathematical formulation of GANs involves optimizing a minimax game, where the generator aims to minimize the probability of the discriminator making a correct classification, while the discriminator seeks to maximize its accuracy (Goodfellow et al., 2014).

2.2 Applications of GANs in Remote Sensing

In the context of remote sensing, GANs have been applied to various tasks, including image enhancement, data augmentation, and anomaly detection. For instance, GANs can be utilized to improve the resolution of low-quality satellite images through a process known as super-resolution (Zhang et al., 2018). This application is particularly important in remote sensing, where high-resolution imagery is crucial for accurate analysis and decision-making.

Moreover, GANs can generate synthetic data to augment limited datasets, which is a common challenge in remote sensing due to the scarcity of labeled data (Frid-Adar et al., 2018). By creating realistic synthetic samples, GANs can enhance the training of machine learning models, leading to improved performance in tasks such as land cover classification and change detection (Shorten & Khoshgoftaar, 2019).

2.3 Challenges and Considerations

Despite their potential, the application of GANs in remote sensing is not without challenges. Issues such as mode collapse, where the generator produces limited variations of data, and the need for extensive computational resources can hinder the effectiveness of GANs (Karras et al., 2019). Additionally, ensuring the interpretability and explainability of GAN-generated data is crucial for gaining trust among users and stakeholders in remote sensing applications (Lipton, 2016).

2.4 Future Directions

The theoretical exploration of GANs in remote sensing highlights the need for ongoing research to address these challenges. Future studies should focus on enhancing the robustness and generalization of GAN models, developing hybrid approaches that integrate GANs with CNNs and transformers for enhancing the extraction of features and representations from the data. Therefore, there is a need to establish new objectives of loss functions for the specific remote sensing applications with a view of enhancing spatial and spectral information retention. It is also essential to investigate the state-of-the-art approaches for domain adaptation mechanisms with the purpose of improving the generalization performance in the cross-sensor and the cross-region.

3.0. LITERATURE REVIEW

3.1 Overview of Generative Adversarial Networks (GANs) in Remote Sensing

The application of Generative Adversarial Networks (GANs) in remote sensing has garnered significant attention in recent years due to their ability to handle complex data transformations and generate realistic synthetic data. GANs have been recognized for their innovative approaches to addressing inherent challenges in remote sensing, such as data scarcity, image quality enhancement, and anomaly detection (Shorten & Khoshgoftaar, 2019). The foundational work by Goodfellow et al. (2014) established the theoretical framework for GANs, which has since been adapted for various remote sensing applications. The application of GANs for super-resolution reconstruction stands as a primary advancement since this method enhances spatial quality in remote sensing images. The technique provides valuable results with scanning data at reduced resolution. The application of GANs in remote sensing allows investigators to complete image inpainting tasks that reconstruct damaged or missing areas of imagery to improve overall data integrity.

3.2 Applications of GANs in Data Augmentation

One of the primary applications of GANs in remote sensing is data augmentation, which is crucial for balancing datasets and generating synthetic data. Previous studies have demonstrated the effectiveness of GANs in creating synthetic images to enhance the training of machine learning models. For instance, Frid-Adar et al. (2018) utilized GANs to generate synthetic medical images, which improved classification performance. Similarly, Mariana et al. (2018) explored the use of GANs for augmenting remote sensing datasets, highlighting their potential to address the limitations posed by the scarcity of labelled data. They have general application to various types of data and have found a common application to aid data scarcity using Generative Adversarial Networks (GAN) for data augmentation, given their results to improve the hyperparameters of a model. They create synthetic data that looks like real data and helps with better training. In agriculture, for example, GANs generate high-resolution crop images to train models for disease monitoring. This enhances model accuracy and robustness in the real-world applications.

3.3 Image Translation and Anomaly Detection

In addition to data augmentation, GANs have been successfully applied in image translation and anomaly detection within remote sensing contexts. For example, GAN-based models have been employed for hyperspectral image classification and super-resolution tasks, significantly improving the quality and resolution of remote sensing images (Zhang et al., 2018). Furthermore, the ability of GANs to generate realistic synthetic data has proven beneficial for anomaly detection, allowing for the identification of unusual patterns in remote sensing imagery (Karras et al., 2019).

Generative Adversarial Network (GANs) have become the general workhorse for image translation and anomaly detection in the last time as can be seen from recent studies. By generating realistic images, they learn the perturbations that indicate deviations from normal, allowing them to do things like transform images between domains, or identify abnormalities in medical imaging, industrial defect detection, or surveillance.

- GANs are leveraged to transform low-quality MRI scans into high-res images.
- Comparing the patterns in normal tissue against what was generated, they assist in revealing cancerous regions. Ganokratanaa(2020)

3.4 Challenges and Limitations

Despite the promising applications of GANs in remote sensing, several challenges remain. Mode collapse, where the generator produces limited variations of data, poses a significant hurdle in ensuring the diversity of generated samples (Karras et al., 2019). Additionally, the computational resources required for training GANs can be substantial, which may limit their accessibility for some researchers and practitioners. Addressing these challenges is essential for the broader adoption of GANs in remote sensing applications.

3.5. Limitations Encountered in the Literature Review

1. **Scope of Research**: The literature review may be limited by the scope of existing research on GANs in remote sensing. While there is a growing body of work, many studies focus on specific applications or datasets, which may not provide a comprehensive understanding of the broader implications and potential of GANs across various remote sensing tasks.
2. **Variability in Methodologies**: The methodologies employed in different studies can vary significantly, making it challenging to compare results and draw generalized conclusions. Differences in GAN architectures, training protocols, and evaluation metrics can lead to inconsistencies in findings, complicating the synthesis of knowledge in the field.
3. **Data Scarcity**: Many studies highlight the issue of data scarcity in remote sensing, which can limit the robustness of GAN models. The reliance on limited datasets for training can lead to overfitting and reduced generalization capabilities, impacting the effectiveness of GANs in real-world applications.
4. **Computational Complexity**: The computational resources required for training GANs can be substantial, which may restrict access for researchers with limited resources. This limitation can hinder the exploration of more complex GAN architectures and their potential applications in remote sensing.
5. **Mode Collapse**: A common challenge in GANs is mode collapse, where the generator produces a limited variety of outputs. This issue can affect the diversity and quality of synthetic data generated, which is critical for applications such as data augmentation and anomaly detection.
6. **Interpretability and Trust**: The black-box nature of GANs poses challenges in interpretability, making it difficult for practitioners to understand how models arrive at specific outputs. This lack of transparency can hinder trust and acceptance among users, particularly in critical applications like environmental monitoring and disaster response.
7. **Ethical Considerations**: The use of synthetic data generated by GANs raises ethical concerns, particularly regarding the potential for misuse or misrepresentation of generated imagery. Addressing these ethical considerations is essential for ensuring responsible use of GAN technology in remote sensing.
8. **Limited Longitudinal Studies**: Many existing studies focus on short-term applications of GANs without exploring their long-term impacts or effectiveness over time. Longitudinal studies are necessary to assess the sustainability and reliability of GAN applications in remote sensing.

9. **Emerging Trends and Technologies**: The rapid evolution of GAN architectures and techniques may outpace the literature, leading to gaps in knowledge regarding the latest advancements and their implications for remote sensing. Continuous updates and reviews are necessary to keep pace with these developments.
10. **Interdisciplinary Integration**: The literature may lack comprehensive discussions on the integration of GANs with other emerging technologies, such as deep learning and machine learning frameworks. Exploring interdisciplinary approaches could enhance the effectiveness of GANs in remote sensing applications.

By acknowledging these limitations, future research can aim to address these gaps, leading to a more robust understanding of GANs in remote sensing and their potential to transform the field.

3.6 Future Research Directions

Future research should focus on enhancing the robustness and generalization of GAN models, as well as developing hybrid approaches that integrate GANs with other machine learning techniques. Moreover, improving the interpretability of GAN-generated data is crucial for building trust among users and stakeholders in remote sensing applications (Lipton, 2016). By advancing the theoretical understanding of GANs and their applications, researchers can contribute to the development of more effective remote sensing technologies.

4.0. RESEARCH METHODOLOGY

The research methodology employed in the study incorporates a mixed-methods approach, which combines both quantitative and qualitative research methods to provide a comprehensive evaluation of GAN applications in remote sensing.

4.1. Theoretical Framework

The foundational theoretical framework for the research is based on the principles of Generative Adversarial Networks (GANs), which consist of two competing neural networks: a generator and a discriminator. This framework operates in a zero-sum game, where the generator aims to produce synthetic data that is indistinguishable from real data, while the discriminator seeks to differentiate between real and generated data (Goodfellow et al., 2014). This iterative process enhances the quality of the generated outputs, making GANs a powerful tool for various remote sensing tasks, such as image enhancement and data synthesis (Ma et al., 2020).

4.2. Mixed-Methods Approach

The study employs a mixed-methods approach, integrating quantitative metrics with qualitative insights. Quantitative methods involve the use of performance metrics, such as accuracy, Structural Similarity Index (SSIM), and Peak Signal-to-Noise Ratio (PSNR), to assess the efficacy of GANs in generating realistic remote sensing imagery (Creswell & Plano Clark, 2018). These metrics provide objective measures of model performance and scalability.

In contrast, qualitative methods include expert interviews and case studies to gather insights on the practicality, usability, and ethical considerations of deploying GANs in real-world scenarios. This qualitative data helps address challenges related to user interpretation and operational feasibility, ensuring that the technological innovations align with domain-specific needs in remote sensing (Johnson & Onwuegbuzie, 2004).

4.3. Case Study Selection

The research methodology emphasizes the selection of case studies that highlight tasks where GANs can offer significant improvements, such as image generation, enhancement, and super-resolution. The selected case studies are designed to represent diverse geographical and environmental contexts, ensuring the robustness of GAN models across various terrains and atmospheric conditions (Goodfellow et al., 2014).

4.4. Addressing Ethical Considerations

The methodology also incorporates a focus on ethical considerations surrounding the use of synthetic data generated by GANs. This aspect is crucial for ensuring responsible use of the technology, particularly in applications that may impact environmental monitoring and resource management (Denton et al., 2015).

4.5. Limitations and Future Directions

The methodology acknowledges the limitations of current research, such as data scarcity and the need for improved interpretability of GAN-generated data. It also outlines future research directions aimed at advancing model architectures, enhancing efficiency, and addressing ethical concerns to facilitate broader adoption of GANs in remote sensing applications (Mash & Weavers, 2011).

In summary, the research methodology employed in this study is characterized by a mixed-methods approach that integrates theoretical frameworks, quantitative assessments, qualitative insights, and ethical considerations, providing a holistic evaluation of GAN applications in remote sensing.

5.0. CASE STUDIES OVERVIEW

The research includes several case studies that illustrate the practical applications of GANs in enhancing remote sensing data. These case studies are critical for understanding the effectiveness of GANs in real-world scenarios and provide valuable insights into their potential benefits and limitations.

5.1. Image Generation and Enhancement

One of the primary case studies focuses on the use of GANs for generating high-resolution satellite imagery from low-resolution inputs. This application is particularly relevant in remote sensing, where high-quality imagery is essential for accurate analysis. The study demonstrates that GANs can effectively upscale low-resolution images, improving their clarity and detail, which is crucial for applications such as urban planning and environmental monitoring (Goodfellow et al., 2014).

5.2. Data Augmentation for Disaster Management

Another significant case study involves the use of GANs to generate synthetic disaster imagery, which enriches datasets for disaster prediction and management. This application addresses the challenge of data scarcity in emergency situations, where real disaster data may be limited. By generating realistic disaster scenarios, GANs can help improve the training of predictive models, leading to better preparedness and response strategies (Rui et al., 2021).

5.3. Scene Classification

A third case study explores the application of semi-supervised GAN frameworks for scene classification in remote sensing. This approach leverages both labelled and unlabelled data to enhance the classification accuracy of high-resolution remote sensing images. The findings indicate that GANs can effectively bridge the gap caused by limited labelled datasets, making them a valuable tool for improving the efficiency of remote sensing applications (Han et al., 2018).

5.4. Key Takeaways

Improved Image Quality: The case studies highlight that GANs significantly enhance the quality of remote sensing imagery, making low-resolution images more usable for analysis. This improvement is vital for various applications, including environmental monitoring and urban planning (Goodfellow et al., 2014).

Addressing Data Scarcity: The ability of GANs to generate synthetic data is particularly beneficial in scenarios where real data is scarce, such as during disasters. This capability allows for better training of machine learning models, ultimately leading to improved disaster response and management (Rui et al., 2021).

Enhanced Classification Accuracy: The use of semi-supervised GAN frameworks demonstrates that GANs can effectively utilize both labelled and unlabelled data, improving the accuracy of scene classification in remote sensing. This approach is crucial for overcoming the challenges posed by limited labelled datasets (Han et al., 2018).

Ethical Considerations: The case studies also underscore the importance of ethical considerations when deploying GANs in remote sensing. The potential for misuse of synthetic data, particularly in surveillance or misleading stakeholders, necessitates careful oversight and responsible deployment practices (Denton et al., 2015).

6.0. ANALYSIS OVERVIEW

The analysis conducted in the study focuses on evaluating the effectiveness of GANs in various applications within the field of remote sensing. This analysis not only highlights the capabilities of GANs but also discusses their implications for future research and practical applications. It is well established that GANs can be useful in various fields of remote sensing especially image synthesis, augmentation, and domain transfer. The work by Goodfellow et al., 2014 was the first to introduce GANs and since then has been adopted in addressing the remote sensing challenges. GANs have played the significant role in the improvement of satellite images, data interpolation and for the enhancement quality of remote sensing images. However, despite the given benefits, there are certain limitations and drawbacks Some of them include difficulties concerning the mode collapse, increased computational complexity of DNs, and the requirement of large training datasets. However, the GAN-generated outputs remain difficult to interpret relying solely on the results, especially in applications of significant importance such as natural calamity surveillance and climate research. Therefore, future research should investigate more about the mounting of GAN with over composite models and employ self-supervision techniques for more stable outcomes in the field of remote sensing. Improving the model's generalization will be helpful in applying GANs in actual remote sensing applications (Goodfellow et al., 2014).

6.1. Performance Metrics

The methodologies chosen in the investigation include measures like accuracy, Structural Similarity Index and Peak Signal-to-Noise Ratio to correctly measure the effectiveness of the GANs in providing synthetic raw remote sensing images. Even though analysing these metrics enables the quantification of the image quality, M0: The research would be helpful to consider other criteria for evaluating such models including FID and IS which are more precise for generative models. In addition, the comparison with other deep learning methods of image enhancement could be beneficial to understand the efficiency and drawbacks of GANs. Considering the results for each case and the depicted figures of the generated images would enhance the credibility of the evaluation. Such directions would also help to strengthen the computational complexity and the balance between high resolution and restricted resources that can be helpful for improving the study's realism (Creswell & Plano Clark, 2018).

6.2. Qualitative Insights

Apart from the quantitative measurements, qualitative analysis gives more detailed analysis of GAN-generated remote sensing images. The case study discusses specific operational consequences and issues of perceiving GANs based on interviews with experts, and the main issues concern the interpretability of the produced images, as well as the possible existence of biases in the training sample. Honestly, while GANs enhances the image synthesis quality, researchers pointed out that by applying GANs in practice, there are several issues, such as computational cost and the need for domain-specific tuning. In user case scenarios, it was found that users failed to rely on synthetic images, which required a way to implement both the means of automation as well as human checking. Moreover, the issues of ethical nature including biased representation of the material and the General public's trust in AI-created images are still contentious. Implementing the knowledge of these challenges, it is possible to enhance using GANs in practice with the help of integrating experts' point of view and comparison with classical RS methods. There are several qualitative benefits of using GANs, and understanding how they work in practice allows to simultaneously maintain both their efficiency and exclude meaningless biases and usability issues (Johnson & Onwuegbuzie, 2004).

6.3. Scalability and Transferability

The analysis also examines the scalability of GAN models across different remote sensing tasks and their ability to transfer knowledge between various applications. This aspect is particularly important as it determines the versatility of GANs in handling diverse datasets and tasks, such as image enhancement, data augmentation, and scene classification (Ma et al., 2020).

6.4. Implications of the Analysis

Enhanced Decision-Making: The findings from the analysis suggest that GANs can significantly improve the quality of remote sensing data, which can lead to better decision-making in critical areas such as environmental monitoring, disaster management, and urban planning. By providing high-quality imagery and synthetic data, GANs enable more accurate assessments and predictions (Mash & Weavers, 2011).

Research Directions: The analysis identifies key areas for future research, including the need for energy-efficient GAN designs, the development of explainable models, and the exploration of ethical considerations in the deployment of GANs. These directions are essential for ensuring the responsible use of GANs in remote sensing applications (Denton et al., 2015).

Collaboration between Academia and Industry: The study emphasizes the importance of collaboration between researchers and industry professionals. By aligning research efforts with practical needs, both sectors can benefit from shared expertise, leading to innovations that enhance the efficiency and effectiveness of remote sensing technologies (Santoro & Chakrabarti, 2002).

Ethical Frameworks: The analysis highlights the necessity of establishing ethical frameworks for the use of GANs in remote sensing. As GANs can generate highly realistic synthetic data, there is a risk of misuse, particularly in surveillance or misinformation. Addressing these ethical concerns is crucial for fostering trust and ensuring the responsible application of GAN technologies (Creswell & Plano Clark, 2018).

7.0. FINDINGS AND RESULTS

The paper focuses on discussing how GANs have been used in remote sensing and the benefits of them such as improvement of image quality, data augmentation as well as to detect anomalies and improve model defences. The findings were aimed at presenting the effectiveness of the works regarding GANs aimed at enhancing interpretability and the use of hybrid models as well as addressing the relevant ethical issues.

1. **Enhanced Resolution:** GAN use has been reported to enhance the quality of the remotely sensed image hence enhancing its applications such as the land cover classification and urban planning (Ma et al., 2020).
2. **Data Augmentation:** The study also establishes GAN's efficiency in producing composite data to enhance the effectiveness of a machine learning model on remote sensing (Shorten & Khoshgoftaar, 2019).
3. **Anomaly Detection:** The GANs use helps in the detection of anomalies in the remote sensing data for purposes of monitoring the environment as well as disaster management (Frid-Adar et al., 2018).
4. **Time Series Generation:** GANs help in the generation of time series data that exists in vegetation growth and urban planning (Suresh Babu, C. V., Koti Reddy, et al., 2024).
5. **mModel robustness:** Adversarial training methods improve the security of GANs so that they produce quality outcomes (Suresh Babu, Cosing &V. Abhinaba Pal 2024).
6. **Interpretability and Explainability:**This paper underscores the need to develop better methods of data presentation and different interfaces to enhance comprehension of GAN-generated information (Suresh Babu, C. V., Sowmi Saltonya, et al., 2024).
7. **Hybrid Approaches:** GAN simultaneous combination with CNN and reinforcement learning enhances the classification and object detection features in remote sensing (Suresh Babu, C. V., Mathew, S., et al., 2024).
8. **Ethical issue:** As the study, it calls for better governance of ethical issues and regulation of data bias and misuse with particular regard to sound ethical principles (Suresh Babu et al., 2023).

Therefore, the study has shown that GANs play a major role in contributing to remote sensing due to issues like image enhancement, data synthesis and anomaly detection. Also, generating Temporal Data and combined modeling makes them applicable for dynamic environmental monitoring. This paper is based on data collected from different resources therefore it is credible. Therefore, based on the advancements in data augmentation, robustness, and explanations, the authenticity of GAN applications is evident. Thus, the ethical issues have to be discussed to avoid such risks that may involve bias or deception.

8.0. SUGGESTIONS AND RECOMMENDATIONS

Based on the findings of the study the following suggestions and recommendations are proposed to enhance the application and effectiveness of GANs in this field:

1. **Enhance Model Architectures**: Future research should focus on advancing GAN architectures to improve their efficiency and effectiveness in generating high-quality remote sensing data. This includes exploring novel designs that can better capture the complexities of remote sensing imagery (Suresh Babu, C. V., Koti Reddy, et. al. 2024).
2. **Integrate Multispectral and Hyperspectral Data**: Researchers should investigate the integration of GANs with multispectral and hyperspectral imagery to enhance the generation of detailed remote sensing data. This integration can significantly improve applications in agriculture, forestry, and land cover classification (Suresh Babu, C. V., Koti Reddy, et. al. 2024).
3. **Develop Robustness Techniques**: Implementing adversarial training and other robustness techniques is essential to enhance the resilience of GANs against potential attacks. Future studies should prioritize developing methods that maintain the quality of generated imagery while ensuring model security (Suresh Babu, C. V. & Yadavamuthiah, K. 2023)..
4. **Focus on Interpretability**: There is a critical need for developing advanced visualization tools and user-friendly interfaces that enhance the interpretability of GAN-generated data. This will facilitate broader adoption and understanding among non-experts in remote sensing applications (Suresh Babu, C. V. & Abhinaba Pal. 2024)..
5. **Explore Hybrid Models**: Future research should explore hybrid models that combine GANs with other machine learning techniques, such as Convolutional Neural Networks (CNNs) and reinforcement learning. This approach can enhance the capabilities of remote sensing applications, particularly in classification and object detection tasks (Suresh Babu, C. V., Suruthi, G., et. al. 2024b)..
6. **Address Ethical Considerations**: Establishing ethical frameworks for the deployment of GANs in remote sensing is crucial. Researchers and practitioners should work together to address issues related to data bias, privacy, and potential misuse of generated data (Suresh Babu, C. V. & Adhithya, S. 2023).

7. **Conduct Longitudinal Studies**: Longitudinal studies should be conducted to assess the long-term impacts of using GANs in remote sensing applications. This will provide insights into the sustainability and effectiveness of GAN technologies over time (Suresh Babu, C. V., Koti Reddy, et. al. 2024).
8. **Foster Collaboration between Academia and Industry**: Encouraging collaboration between academic researchers and industry professionals can drive innovation and ensure that research findings are effectively translated into practical applications. This partnership can enhance the relevance and applicability of GAN technologies in real-world scenarios (Sureh Babu, C. V., Sowmi Saltonya, et. al. 2024).

9.0. FUTURE RESEARCH SCOPE

The study opens several avenues for future research. The following points outline the potential research scope that can be explored to further enhance the application of GANs in remote sensing:

1. **Advancement of GAN Architectures**: Future research can focus on developing more sophisticated GAN architectures that improve the quality and efficiency of generated images. This includes exploring variations such as Conditional GANs (cGANs) and CycleGANs, which can be tailored for specific remote sensing tasks like image translation and super-resolution (Suresh Babu, C. V., Koti Reddy, et. al. 2024).
2. **Integration with Other Data Types**: There is significant potential for integrating GANs with multispectral and hyperspectral data to enhance the richness of remote sensing applications. Research can investigate how GANs can leverage additional spectral information to improve classification accuracy and environmental monitoring (Suresh Babu, C. V., Koti Reddy, et. al. 2024).
3. **Temporal Data Generation**: Exploring the capability of GANs to generate time-series data for remote sensing applications can facilitate better monitoring of dynamic environmental changes. This research could focus on developing models that can simulate temporal variations in land use, vegetation growth, and climate impacts (Suresh Babu, C. V. & Yadavamuthiah, K. 2023)..
4. **Robustness and Generalization**: Future studies should address the robustness of GANs against adversarial attacks and their ability to generalize across different geographic and environmental conditions. Research can focus on techniques that enhance model resilience while maintaining high-quality output (Suresh Babu, C. V. & Abhinaba Pal. 2024)..

5. **Interpretability and Explainability**: As GANs become more prevalent in remote sensing, there is a growing need for research focused on interpretability. Developing visualization techniques and user-friendly interfaces can help users understand how GANs generate data and the implications of their outputs (Suresh Babu, C. V., Subhash, S., et. al. 2024).
6. **Ethical and Societal Implications**: Investigating the ethical considerations surrounding the use of GANs in remote sensing is crucial. Future research can explore issues related to data privacy, bias in generated data, and the potential societal impacts of deploying GAN technologies in environmental monitoring and disaster management (Suresh Babu, C. V. & Adhithya, S. 2023).
7. **Hybrid Approaches**: Research can explore hybrid models that combine GANs with other machine learning techniques, such as Convolutional Neural Networks (CNNs) and reinforcement learning. This can enhance the capabilities of remote sensing applications, particularly in complex tasks like object detection and scene classification (Suresh Babu, C. V., Koti Reddy, et. al. 2024).
8. **Real-time Processing and Applications**: There is a need for research focused on the real-time application of GANs in remote sensing. Developing efficient algorithms that can process data in real-time will be essential for applications in disaster response, environmental monitoring, and urban planning (Suresh Babu, C. V., Sharon Manoj et. al 2023).
9. **Longitudinal Studies**: Conducting longitudinal studies to assess the long-term effectiveness and sustainability of GAN applications in remote sensing can provide valuable insights. This research can help in understanding the evolving role of GANs in addressing emerging challenges in the field (Suresh Babu, C. V., Koti Reddy, et. al. 2024).
10. **Collaboration between Disciplines**: Encouraging interdisciplinary collaboration between remote sensing experts, data scientists, and domain specialists can lead to innovative applications of GANs. Future research can focus on how these collaborations can enhance the development and implementation of GAN technologies in various sectors (Sureh Babu, C. V., Sowmi Saltonya, et. al. 2024).

The future research scope for GANs in remote sensing is vast and multifaceted. By addressing the outlined areas, researchers can significantly contribute to the advancement of remote sensing technologies, ultimately leading to improved environmental monitoring, disaster management, and urban planning.

10. CONCLUSION

This paper aimed at investigating the effects of GANs on remote sensing with special focus on issues such as improvement of image quality, data augmentation, anomaly detection, and temporal data generation. They also find that GANs enhance image resolution in the area of remote sensing (Ma et al., 2020), help increase the data set in the use of models such as in machine learning (Shorten & Khoshgoftaar, 2019) and assist in anomaly detection (Frid-Adar et al., 2018). Also, the nature of contact they have with temporal data would also facilitate an easy monitoring of changes within the environment (Suresh Babu C.V and Koti reddy et al., 2024). The first study is aware of the effect of adversarial training on the model robustness (Suresh Babu, C. V. & Abhinaba Pal, 2024) while the second one is on the need to better understand the GAN-generated data (Suresh Babu, C. V., Sowmi Saltonya, et al., 2024). Hence, by using GANs together with other deep learning approaches, like CNNs, the capability is advanced to classify and detect objects and features (Suresh Babu, C. V., Mathew, S. et al., 2024). Concerning the ethical issues arising from data bias and data misuse, there is a need for appropriate regulation (Suresh Babu, C. V. & Adhithya, S., 2023). These findings highlight the possibility of achieving great breakthroughs with GANs in remote sensing applications and indicate the need for future work on the integration of other methods, the rights and wrongs of utilizing GANs, and the need for better model interpretability.

REFERENCES

Babu, C. S., Reddy, K., Chandu, G., Kowshik, T., & Rao, L. N. (2024). Trustworthy AI for Optimizing Agriculture: Leveraging IoT Sensory Data. In Developments Towards Next Generation Intelligent Systems for Sustainable Development (pp. 124-147). IGI Global.

Babu, C. S., Yadavamuthiah, K., Sathiyanarayana, S., & Mathew, S. (2024). Harnessing Agricultural Data: Advancing Sustainability Through the Application of Find S Algorithm. In Advanced Computational Methods for Agri-Business Sustainability (pp. 190-202). IGI Global.

Creswell, A., & Plano Clark, V. L. (2018). *Generative Adversarial Networks for Image Synthesis: Principles and Applications*. Springer.

Creswell, J. W., & Plano Clark, V. L. (2018). *Designing and conducting mixed methods research* (3rd ed.). SAGE Publications.

Denton, E., Chintala, S., Szlam, A., & Olah, C. (2015). *Deep generative image models using a laplacian pyramid of adversarial networks*. In Advances in Neural Information Processing Systems (pp. 1486-1494).

Frid-Adar, M., Ganin, A., Kahn, S., & Goldberger, J. (2018). GAN-based synthetic medical image augmentation for increased CNN performance in liver lesion classification. *Neurocomputing*, *321*, 321–331. DOI: 10.1016/j.neucom.2018.09.013

Goodfellow, I., Pouget-Abadie, J., Mirza, M., Xu, B., Warde-Farley, D., Ozair, S., . . . Courville, A. (2014). *Generative adversarial nets*. In Advances in Neural Information Processing Systems (pp. 27-34).

Goodfellow, I., Pouget-Abadie, J., Mirza, M., Xu, B., Warde-Farley, D., Ozair, S., Courville, A., & Bengio, Y. (2014). Generative adversarial nets. *Advances in Neural Information Processing Systems*, *27*, 2672–2680.

Han, X., Zhang, Y., & Wang, Y. (2018). Semi-supervised generative adversarial networks for remote sensing image classification. *Remote Sensing*, *10*(5), 1–15.

Johnson, R. B., & Onwuegbuzie, A. J. (2004). Mixed methods research: A research paradigm whose time has come. *Educational Researcher*, *33*(7), 14–26. DOI: 10.3102/0013189X033007014

Karras, T., Aila, T., Laine, S., & Lehtinen, J. (2019). A style-based generator architecture for generative adversarial networks. *Proceedings of the IEEE/CVF Conference on Computer Vision and Pattern Recognition*, 4401-4410. DOI: 10.1109/CVPR.2019.00453

Li, Y., Zhang, Y., & Wang, Y. (2020). A GAN-based approach for cloud removal in remote sensing images. *Remote Sensing*, *12*(3), 456.

Lipton, Z. C. (2016). The mythos of model interpretability. *Communications of the ACM*, *59*(10), 36–43. DOI: 10.1145/3233231

Ma, Y., Zhang, Y., & Wang, Y. (2020). Generative adversarial networks for remote sensing applications: A review. *Remote Sensing*, *12*(1), 1–20.

Mash, H. E., & Weavers, L. K. (2011). Water resources. *Water Research*, 45(1), 1-10. Shorten, C., & Khoshgoftaar, T. M. (2019). A survey on image data augmentation for deep learning. *Journal of Big Data*, *6*(1), 1–48.

Rui, Y., Zhang, Y., & Wang, Y. (2021). DisasterGAN: A generative adversarial network for generating synthetic disaster images. *Remote Sensing*, *13*(2), 1–15. DOI: 10.3390/rs13214284

Santoro, M. A., & Chakrabarti, A. (2002). The role of research in the development of new technologies: A case study of the telecommunications industry. *Technological Forecasting and Social Change*, *69*(5), 487–507.

Shorten, C., & Khoshgoftaar, T. M. (2019). A survey on image data augmentation for deep learning. *Journal of Big Data*, *6*(1), 1–48. DOI: 10.1186/s40537-019-0197-0

Suresh Babu, C. V. (2003). *Artificial Intelligence and Expert Systems*. Anniyappa Publicaitons.

Suresh Babu, C. V., Abhinaba Pal, A., Vinith, A., Muralirajan, V., & Gunasekaran, S. (2024). Enhancing Cloud and IoT Security: Leveraging IoT Technology for Multi-Factor User Authentication. In Ahmed Nacer, A., & Abdmeziem, M. (Eds.), *Emerging Technologies for Securing the Cloud and IoT* (pp. 258–282). IGI Global., DOI: 10.4018/979-8-3693-0766-3.ch011

. Suresh Babu, C. V., Abirami, S., & Sharon Manoj (2023). AI-Based Carthage Administration Towards Smart City. In C. Chowdhary, B. Swain, & V. Kumar (Eds.), Investigations in Pattern Recognition and Computer Vision for Industry 4.0 (pp. 1-17). IGI Global. DOI: 10.4018/978-1-6684-8602-3.ch001

Suresh Babu, C. V., & Adhithya, S. (2023). ChatGPT and Other Generative AI Tools in Education: Transformative Potential and Ethical Considerations. In Keengwe, J. (Ed.), *Creative AI Tools and Ethical Implications in Teaching and Learning* (pp. 135–152). IGI Global., DOI: 10.4018/979-8-3693-0205-7.ch007

Suresh Babu, C. V., Sowmi Saltonya, M., Ganapathi, S., & Gunasekar, A. (2024). AIoT Revolution: Transforming Networking Productivity for the Digital Age. In Rezaei, S., & Ansary, A. (Eds.), *Artificial Intelligence of Things (AIoT) for Productivity and Organizational Transition* (pp. 108–143). IGI Global., DOI: 10.4018/979-8-3693-0993-3.ch005

Suresh Babu, C. V., Suruthi, G., Indhumathi, C., & Sushruth, S. (2024b). Integrated approach for network intrusion detection: Combining KNN algorithm, Karhunen-Loève transform, and genetic algorithm optimization. In Metaheuristic and machine learning optimization strategies for complex systems (p. 21). DOI: 10.4018/979-8-3693-7842-7.ch002

Suresh Babu, C. V., & Yadavamuthiah, K. (2023). Precision Agriculture and Farming Using Cyber-Physical Systems: A Systematic Study. In Karthick, G. (Ed.), *Contemporary Developments in Agricultural Cyber-Physical Systems* (pp. 184–203). IGI Global., DOI: 10.4018/978-1-6684-7879-0.ch010

Zhang, Y., Wang, Y., & Li, X. (2018). A review of deep learning in remote sensing: Current applications and future directions. *ISPRS Journal of Photogrammetry and Remote Sensing*, *145*, 1–16.

Chapter 13
Privacy-Security Considerations and Adversarial Attacks on Remote Sensing With GANs

Ajay Devidas Nagne
https://orcid.org/0000-0002-9652-5631
JSPM University, India

Pradnya L. Awate
MGM University, India

Shriram P. Kathar
MGM University, India

ABSTRACT

Remote sensing has revolutionized Earth observation, improving environmental monitoring, agriculture, urban planning, and disaster management. However, its increasing use raises significant privacy and security concerns. This chapter examines these issues, focusing on adversarial attacks using Generative Adversarial Networks (GANs). GANs, comprising generator and discriminator networks, enhance remote sensing by improving image resolution, anomaly detection, and data augmentation. Yet, it also enable security threats, such as evasion and poisoning attacks, which compromise data integrity, privacy, and trust. Defending against these attacks involves detection, mitigation, and resilience strategies. Future research should explore robust refences, ethical frameworks, and regulations to secure remote sensing systems.

DOI: 10.4018/979-8-3693-6900-5.ch013

1. INTRODUCTION

Remote sensing means the acquisition and analysis of data without physical contact related to the Earth's surface, mainly through sensors mounted on satellites, aircraft, or drones. Such sensors capture multispectral, hyperspectral, or radar imagery to monitor the environment and anthropogenic processes in different places. This technique has developed as a key tool of environmental monitoring, precision agriculture, disaster management, and urban planning by revealing significant insights to spatial and temporal dynamics. Advanced data processing techniques like machine learning and deep learning in the interpretation of remote sensing data have improved the utility of remote sensing, providing detailed accuracy for applications involving complexity (Awate, 2023).

1.2 Importance of Privacy and Security in Remote Sensing

This increased dependence on remote sensing for key decision-making has only strengthened the need for robust privacy and security measures. Figure 1, represents a graphic description of it. These data hold sensitive information related to strategic locations, natural resources, and critical infrastructure and may therefore become an attractive target for malicious players. Access or theft of remote sensing data can lead to serious negative impacts, including economic and national security losses, which could be followed by mistrust of the technical systems. The advancement in technology such as cloud computing and AI applied to the processes of remote sensing introduced a risk because of cyber-attacks, so there is an increased need for comprehensive security frameworks put in place with respect to data integrity, confidentiality, and availability (Li et al., 2016).

Figure 1. Privacy and Security in Remote Sensing

1.3 Role of Generative Adversarial Networks (GANs) in Remote Sensing

Generative Adversarial Networks are a type of deep learning model that have really proven useful in dealing with a wide range of remote sensing challenges. GANs are used for data augmentation, generate synthesis images, and super-resolution to enhance low resolution images which helps to improve analysis and training the model for applications where we have limited or noisy data. GANs can also useful in creating high-resolution imagery from coarser inputs images, which improves the accuracy for applications like land-use classification and crop monitoring. However, it allows the adversarial attacks by possibly generating fake satellite imagery or altering datasets in remote sensing applications raising the questions of security and trust of the data. It is therefore important that there must be a balanced approach that utilizes the strength of GANs while simultaneously reducing the potential of misuse through antagonistic (Jozdani, Chen, Pouliot, & Johnson, 2022).

2. PRIVACY IN REMOTE SENSING

Remote sensing data privacy refers to the protecting the sensitive and personal information which can be derived from high resolution satellite imagery, aerial platforms, or unmanned aerial vehicles (UAVs) which can be potentially used for identifying individuals and their activities. The remote sensing system captures high-resolution imagery data over large geographical regions that might contain private properties, human activities, and other confidential or sensitive details of person, organization, community. With the rapid development of sensor technologies, the capacity to capture subtle spatial and temporal information has also increased considerably; this also raises a great concern related to unauthorized data exposure, surveillance, and breach of privacy. The intersection of remote sensing with privacy requires responsible data collection, data processing, and distribution that maintains confidentiality rights of individuals and organizations in society (Zhang, Ren, Shafiq, & Gu, 2022).

Privacy issues in remote sensing are based on the potential misapplication of high-resolution imagery and the monitoring of activities without explicit consent. Many applications, such as defence, critical infrastructure monitoring, or urban planning, involve data whose release can compromise individual privacy, national security, or commercial interests. Moreover, satellite data from space, when combined with AI models like GANs, can be manipulated or enhanced to reveal details beyond the capability of the original sensor, thereby increasing risks of privacy. Open-access satellite data availability makes it even more complicated for the regulation of use as it becomes challenging to monitor what the user intends and what happens as a consequence of its application (Zhang, Shafiq, Wang, Srivastava, & Yin, 2023).

All such concerns need to be adequately framed in comprehensive frameworks so as to protect privacy in full-scale while allowing the potentialities of remote sensing to grow. Strategies for an acceptable reduction in privacy risk through anonymization of private information, access controls to sensitive data, and ethical guidelines related to data usage would require further emphasis. Legal framework with regulations like the GDPR helps to provide a structural base towards balancing technological progressions along with individual rights. As remote sensing continues to advance, the need to balance innovation and preserve privacy remains a priority for policymakers, researchers, and stakeholders (Zhang, Ren, Shafiq, & Gu, 2022; Zhang, Shafiq, Wang, Srivastava, & Yin, 2023).

2.1. Definition of Privacy in the Context of Remote Sensing

Privacy in remote sensing involves keeping sensitive information captured using the remote sensing technologies out of unauthorized hands, especially if this information pertains to specific individuals, organizations, or strategic places. This refers to safeguarding spatial, temporal, and contextual information derived from the satellite images, aerial photography, and drone sensors against unauthorized access, disclosure, and misuse. In the context of geospatial data, privacy would extend beyond traditional personal data protection, ensuring that data about particular areas, activities, or patterns of interest remains confidential, especially where military, commercial, or individual interests are involved ("P3S", 2023).

2.2 Privacy Risks and Concerns

The widespread deployment of remote sensing systems comes with several privacy risks, such as the unintentional collection of sensitive information and exploitation by malicious parties. High-resolution imagery combined with advances in data fusion techniques enables the interpretation of activities of individuals or property details with respect to critical infrastructure, which then elevates surveillance and misuse of data concerns. Additional to all the above, risk also arises while merging the data with others from remote sensing, which could include social media records or commercial data and so forth. That can hint at the personal or strategic information available. Further making it hard is that a jurisdiction may run into others' territories in case of a problem related to cross-border violation (Huang & Cao, 2023).

2.3 Case Studies of Privacy Breaches

There have been various instances where the vulnerabilities in remote sensing systems regarding privacy were highlighted. For instance, the unapproved use of high-resolution satellite imagery for the surveillance of private properties brought up great ethical and legal concerns (Crews, 2013). In another case, drones that used imaging sensors were employed for the unauthorized gathering of information related to sensitive facilities that later resulted in security breaches (Lee, Lee, Park, & Kim, 2016). Such incidents speak to the dual-edged nature of remote sensing technologies wherein the same tools used for legitimate purposes can be exploited for surveillance or data manipulation. These case studies demonstrate why strong privacy safeguards are urgently required to reduce potential hazards (McAmis, Sim, Bennett, & Kohno, 2024).

2.4 Mitigation Strategies

To solve privacy concerns in remote sensing, a multifaceted solution combining technological, legal, and ethical approaches is required. Data anonymization, encryption, and safe access methods must be included in remote sensing workflows. Techniques such as differential privacy can also be applied so that aggregated data insights do not reveal sensitive details. The comprehensive policy and international agreement will provide the frame on data collection, sharing, and usage so that both sovereignty and the principles of privacy are ensured. Deploying the systems of remote sensing has an ethical foundation and, where necessary, will demand transparency and informed consent of the population to whom they provide service. There should be collaboration between governments, the academia, and private entities involved in establishing a holistic approach to privacy protection within remote sensing applications (Khan, Kumar, Rath, & Zaib, 2024; Huang & Cao, 2023).

3. SECURITY CONSIDERATIONS IN REMOTE SENSING

Remote sensing security includes protection against unauthorized access, alteration, and adversarial attacks of geographic data, processing systems, and communication networks. Since remote sensing technologies employ satellite platforms, aerial vehicles, and advanced sensors to acquire and transmit high-resolution data, they are highly susceptible to cyberattacks, data breaches, and system intrusions. This calls for the necessity of ensuring that geospatial information is not compromised and maintained for integrity, confidentiality, and availability as such applications have been at the heart of their uses in military, disaster management, precision agriculture, and urban planning (Avtar et al., 2021).

The key security threats to remote sensing involve data interception, tampering, and adversarial attacks targeting machine learning algorithms. For example, a cyberattack that involves jamming a signal or spoofing during the acquisition or transmission process can lead to inaccuracies or unreliability of collected information. Moreover, evasive or poisoning-based attacks on the machine learning aspect can affect the reliability of the classification or predictions of the models in remote sensing, causing flaws in decision-making processes. In cases of sensitive infrastructures or defense-related imagery, the security breaches will have a far-reaching impact beyond just data loss; they would also have national security implications.

Tackling these issues demands highly secure frameworks incorporating encryption protocols, secure methods of transmitting data, and resilience approaches for both hardware and software. Advanced cryptographic techniques, quantum encryption, and blockchain-based approaches are some promising solutions toward securing remote sensing data against unauthorized access or tampering. Secure data pipelines, multi-factor authentication, and anomaly detection mechanisms add strength to the system against malicious intrusion. With increasing reliance on remote sensing, the development of such secure and resilient geospatial information systems will require considerable interdisciplinary collaboration among stakeholders, namely policymakers, scientists, and cybersecurity experts (Tasneem & Islam, 2024).

3.1. Definition of Security in Remote Sensing

In remote sensing, security entails securing geographic data, sensing systems, and infrastructures from unwanted access, disruption, or manipulation. It refers to all steps for protecting the integrity, confidentiality, and availability of remote sensing data and systems. It would safeguard the entire data lifecycle (collection, processing, storage, and distribution) from both physical and cyber dangers. Remote sensing data in military, disaster management, and resource monitoring sectors is strategically important, thus the systems must be trusted and operated efficiently (Chun & Atluri, 2000).

3.2. Types of Security Threats

Remote sensing systems encounter a variety of security vulnerabilities, which might jeopardize their functioning and dependability. Hacking, ransomware, and denial-of-service assaults all damage the hardware and software components of remote sensing systems. Physical dangers, like as sabotage or theft of ground stations and data storage facilities, represent serious risks. Furthermore, spoofing and jamming techniques can interrupt satellite communication and data delivery. The expanding use of adversarial machine learning techniques, such as Generative Adversarial Networks (GANs), raises concerns about attempts to inject misleading or manipulative modifications into remote sensing pictures, compromising decision processes (Li, Ma, & Zhang, 2021).

3.3. Impact of Security Breaches

Remote sensing security breaches will have a wide-ranging impact. Compromises in geographic integrity may result in inaccurate conclusions when evaluating data. Critical applications include disaster response and recovery, urban planning, and agricultural monitoring. In defence and national security, hacks can expose sensitive information about military actions or strategic assets, jeopardizing national interests. In fact, service disruption would have a significant impact on several economic sectors that rely on reliable geospatial information in real time, such as logistics, construction, and environmental monitoring. Furthermore, a loss of confidence in these remote sensing techniques erodes community trust, reducing overall acceptance and creativity in their use (Kasula, Yadulla, Konda, & Yenugula, 2024).

3.4. Best Practices for Enhancing Security

To reduce security concerns, a holistic approach must be implemented that incorporates technological, procedural, and policy-driven solutions. Advanced encryption methods, secure communication routes, and multi-factor authentication all contribute to stronger system defences. Continuous vulnerability assessments and penetration testing are required to discover and remedy potential problems. Blockchain technology can improve traceability and integrity of remote sensing data throughout its lifecycle. To provide operational resilience, procedures such as rigorous access controls, employee training, and incident response processes are used. At the international level, coordination is essential to establish standard security frameworks that would stimulate information exchange among stakeholders, all with the purpose of providing uniform defence against these new dangers in the remote sensing area (Cheng, Liu, & Yao, 2017).

4. GENERATIVE ADVERSARIAL NETWORKS (GANS)

Generative Adversarial Networks are a particular sort of deep learning model. These models generate synthetic data that closely resembles actual data. The invention of GANs in 2014 featured two neural networks: a discriminator and a generator. It operates on a competitive platform. The generator creates synthetic samples, such as photographs, while the discriminator attempts to authenticate the samples by distinguishing between actual and produced data. The notion is expressed pictorially in figure 2. This adversarial training process is repeated iteratively until the generator provides outputs that are indistinguishable from the original data, making GANs particularly well-suited for applications that need realistic data augmentation, enhancement, or reconstruction (Jozdani, Chen, Pouliot, & Johnson, 2022).

Figure 2. GANs discriminator and generator

GANs have shown tremendous abilities in remote sensing, mainly applied in image synthesis, super-resolution, and data augmentation. They are used to obtain satellite images of high resolution using the low-resolution inputs, so more accurate and visual applications become possible, such as the classification of land-use or crop health monitoring, among others. GANs fill gaps in incomplete datasets due to malfunctions of sensors or obstruction by the atmosphere, which in turn improves data continuity and reliability. Conditional GANs have also been very promising in domain adaptation tasks. This includes translating multispectral data into other forms of geospatial imagery, which may aid in cross-sensor interoperability (Zhang, Song, Han, & Zhang, 2020).

While there are numerous benefits, GANs are not without pitfalls. The training of a GAN is very computationally costly and often leads to mode collapse, wherein the generator only produces minor variations on one of the possible outputs; also, generated data carries along issues of data integrity by potential malicious individuals, creating adverse conditions for malicious intent creation by producing faked geospatial images. Addressing these challenges requires the development of robust training strategies, interpretability techniques, and ethical guidelines that ensure responsible implementation of GANs in remote sensing. In the future, as advancement continues, GANs have great potential to be used to enhance accuracy and efficiency in scalable applications for remote sensing while maintaining fidelity with data (Jozdani, Chen, Pouliot, & Johnson, 2022; Zhang, Song, Han, & Zhang, 2020).

4.1. Introduction to GANs

In 2014, Ian Goodfellow et al. introduced a class of deep learning models called generative adversarial networks (GAN). They are the first models in this class to use adversarial training. GANs are described as two neural networks: one, called the generator, tries to create synthetic data that would look like real data, and another, the discriminator, tries to validate the authenticity of the created data. This dynamic interplay enhances the quality of generators in producing high-quality output. As pointed out in Figure 3, GANs became popular because they could develop realistic synthetic data, fill out data shortages, and produce different machine learning applications (Goodfellow et al., 2020).

Figure 3. GAN for high quality imagery

4.2. Applications of GANs in Remote Sensing

GANs have found massive utility in remote sensing as they have revolutionized the processing and analysis of data. They are widely utilized for super-resolution enhancement, a critical application in land-use classification and urban planning, because they can generate high-resolution images from low-resolution inputs. Moreover, GANs help with data augmentation since they generate synthetic samples which improve the training of a machine learning model in case of a limited number of labelled datasets. GANs further enhance change detection in the identification of subtle variations within multi-temporal satellite imagery. Secondly, GANs prove very useful in noise elimination and data reconstruction processes in remote sensing imagery; hence ensuring that analytical results obtained are sound. These applications underscore how GANs can prove to be transformational solutions for long-standing challenges facing remote sensing research (Dash, Ye, & Wang, 2023).

4.3. Benefits and Limitations of Using GANs

The deployment of GANs in remote sensing gives quite a number of advantages including better quality data, higher ability to train the models and filling data gaps. Reduced field data collection allows reducing GAN dependency and increasing remote sensing workflows efficiency as well as cost-effectiveness. What is more, generating very realistic imagery provides adequate facilities for detailed analysis and also assists in high-accuracy prediction. However, GANs are not without some of its limitations. Training GANs is quite computationally intensive and is not easy to balance, requiring a lot of skills to prevent mode collapse. A more significant concern is the potential misuse of GANs for adversarial attacks or the generation of deceptive imagery; this raises ethical and security concerns. Overcoming these weaknesses calls for continuous research effort to develop robust, interpretable, and secure models of GANs suited specifically to the needs of the remote sensing application (Jozdani, Chen, Pouliot, & Johnson, 2022).

5. ADVERSARIAL ATTACKS ON REMOTE SENSING

Adversarial attacks in remote sensing refer to an intentional action designed to cause confusion for the machine learning model or destroy the precision and integrity of remote sensing systems. It targets vulnerabilities of deep learning model, by simply adding a minor perturbation to the input data and which cause significant misclassifications or system failures, remote sensing depends more on AI and ML models for tasks like environmental monitoring, land cover classification, and object recognition, their susceptibility for adversarial attacks is a serious concern; Evasion, poisoning, and model inversion are some examples of adversarial attacks on remote sensing data. Evasion attacks involve introducing invisible changes to the input images for deceiving trained algorithms at the time of inference. For instance, altering pixel values in a satellite image may lead to misclassification of land cover as pixel values are used for training and accuracy assessment, which may have direct impact on decision-making processes of an deep learning model. Poisoning attacks occur at the training phase, where adversaries introduce malicious data into the training set in order to disrupt the model's learning process, leading to poor performance on further inputs. In contrast, model inversion attacks try to reconstruct sensitive input data or pictures using the outputs of the model and create major privacy issues in the remote sensing application (Xu & Ghamisi, 2022).

The adversarial attacks on remote sensing systems results in compromising the reliability of data, performance of the model, and decision-making process during operation. Applications in defense surveillance, disaster management, and precision agriculture can lead to inaccurate assessment, delayed response, or resources being wasted due to misallocation. Additionally, the robustness of machine learning models is tested, and AI-driven remote sensing technologies lose confidence in their integrity, especially in domains that rely heavily on accuracy and integrity. It requires a holistic understanding of how attacks are conducted and defense in advance. Techniques to counter adversarial vulnerabilities include adversarial training, input sanitization, and anomaly detection. This also calls for robust architecture and uncertainty quantification for improved model robustness against adversarial manipulations. As remote sensing integrates more advanced AI techniques, the area of research in adversarial attacks to ensure integrity, reliability, and trustworthiness of data in geospatial systems is more than ever critical (Xu, Du, & Zhang, 2020).

5.1. Definition and Types of Adversarial Attacks

Adversarial attack threats in remote sensing are intentional attempts to tamper or modify geospatial data and mislead the machine learning models used for remote sensing applications; these attacks on the weakness of machine learning algorithms or data pipelines to misleading results, interrupt analysis, or compromise the decision-making processes of an model. Adversarial attacks can be classified into different categories; Evasion attacks distort model predictions by making tiny alterations to input data that are often undetectable, Poisoning attacks are the manipulation of training datasets to reduce model performance or introduce biases, inversion attacks try to create image using output of an model. In the context of remote sensing, these attacks can target satellite images, classification algorithms used, or data transmission channels, posing major issues to systems reliability and security (Kazmi et al., 2023).

5.2. Techniques Used in Adversarial Attacks

Adversarial attack use advanced and sophisticated methods, typically using complex and intricate computational methodologies to exploit the algorithm flaws and loopholes. For instance, one strategy for creating adversarial instances is to optimize perturbations with gradient-based algorithms like the Fast Gradient Sign strategy and Projected Gradient Descent. In contrast, the black-box attacks does not require access to model parameters; instead it uses a query-based strategy to repeatedly refine perturbations in model parameter. GANs are increasingly being utilized to create adversarial instances by training generators to provide inputs that confuse discriminators. In remote sensing, these techniques can be used to create or edit satellite imagery, which cause the change in classification outputs, or introduce flaws in derived results such as vegetation indices loke NDVI, SAVI or land-cover maps for LULC classification (Sattar & Oymak, 2020; Naqvi, Shabaz, Khan, & Hassan, 2023).

5.3. Impact of Adversarial Attacks on Data Integrity and Privacy

The adversarial attacks compromise the integrity, reliability and privacy of geospatial data, attacks could jeopardize important applications that depends on fidelity in the manipulation of satellite images or derived analytics from remote sensing data for applications, such as disaster response, urban planning, and environmental monitoring. Evasion, for example, could result in a mistaken assessment of land cover class for another class. Furthermore, adversarial attacks jeopardize privacy by revealing or modifying critical geographical information, potentially leading to unauthorized surveillance or exploitation. The consequences of compromised data may include economic loss, loss of stakeholder trust, and security vulnerabilities in defence and national infrastructure. Effective adversarial defensive measures and proactive system monitoring will be required to mitigate the effects of developing threats (Hathaliya, Tanwar, & Sharma, 2022).

6. DEFENDING AGAINST ADVERSARIAL ATTACKS

In order to protect remote sensing from adversarial attacks, there is a need for the development and deployment of sophisticated techniques to make sure that machine learning models and geospatial data systems are robust, resilient, and secure. Adversarial attacks take advantage of the vulnerabilities of deep learning algorithms in data integrity, classification accuracy, and operational reliability. These challenges have to be addressed by defense mechanisms both in the detection of adversarial manipulations and mitigation of their impacts for ensuring the trustworthiness of remote sensing applications (Xu, Du, & Zhang, 2020; Xue et al., 2022).

6.1. Detection and Mitigation Techniques

To counter adversarial assaults in remote sensing, advanced detection and mitigation mechanisms are required to protect the integrity of data and algorithms. Detection methods detect anomalies or adversarial patterns in datasets or model outputs. Statistical methods for outlier detection and distribution analysis were utilized to find anomalous perturbations. Machine learning-based techniques, which include adversarial detectors trained on labelled datasets, perform significantly better at differentiating between adversarial and real inputs. Mitigation techniques often include pre-processing steps such as sanitizing inputs to remove noise or disturbance. Adversarial training improves model robustness to adversarial examples. Feature space regularization and gradient masking reduce an adversary's ability to exploit a model (Shi et al., 2022).

6.2. Robustness and Resilience Strategies

Building robust and resilient systems is crucial for combating adversarial attacks in remote sensing applications. Robustness improves the system's resistance to attacks, which are often algorithmic. The most prominent strategies are ensemble learning, which combines several models to reduce weaknesses, and defensive distillation, which reduces model susceptibility to input perturbations. Resilience techniques primarily focus on the system's capability to recover from attacks, while still maintaining operational functionality of system; in this sense, redundancy in data processing workflows and fail-safe measures are critical to maintain the continuous service delivery; additionally, blockchain technology can be used for data traceability, reducing the risk of adversarial change made in the data and maintain its integrity (Mei et al., 2024).

6.3. Case Studies of Successful Defence Mechanisms

Different case studies demonstrated applicability of the effectiveness of protecting systems against adversarial attacks in remote sensing; in one case study, adversarial training was used to land cover classification models, which resulted in excellent accuracy under adversarial settings of training model, another work was used in GAN-based defences to identify and neutralize hostile examples in high-resolution satellite imagery, ensuring data dependability for urban planning applications. Additionally, the use of blockchain technology for secure data transfer in catastrophe monitoring systems showed the potential for distributed ledger technology to improve system resilience against adversarial attacks. These case study examples demonstrate the importance of proactive defensive measures which are tailored to specific vulnerabilities of different algorithm and models and application scenarios in remote sensing applications (Kumaran, Thangam, Prabhakar, Selvaganesan, & Vishwas, 2023).

7. ETHICAL AND LEGAL CONSIDERATIONS

Ethical and legal considerations in remote sensing include critical concerns about balancing technological advances with social responsibility, data privacy, and compliance with regulatory requirements. As technologies of remote sensing, particularly machine learning and GANs, are further developed, ethical issues concerning data usage, ownership, and transparency have become important to prevent risks associated with surveillance, unauthorized exploitation of data, and unintended consequences of applying advanced analytical tools on geospatial data (Hossain, Islam, & Martin, 2024).

7.1. Ethical Implications of Using GANs in Remote Sensing

Ethically, this use of GANs in remote sensing raises problems. The legitimacy of data is called into question, as are the societal repercussions. GANs, on the other hand, bring value to data through synthetic data generation and super-resolution imaging; but, they also have the potential for misuse, such as the ability to create misleading imagery or manipulate geospatial data to inflict harm. The ethical quandary arises because such technologies have the ability to blur the distinction between what is genuine and what is synthetic, lowering trust in remote sensing results. Furthermore, using GANs in surveillance and monitoring applications would violate people's rights to privacy and autonomy, therefore there was a strong demand for transparency and accountability in these applications (Paolanti et al., 2024).

7.2. Legal Frameworks and Regulations

The legal framework governing GANs and remote sensing technology will evolve to meet issues of data governance, intellectual property, and national security. International accords, such as the Outer Space Treaty, provide broad principles for the peaceful use of space technology, including satellite-based remote sensing. However, when synthetic data generation involves personal or sensitive information, region-specific legislation, such as GDPR in Europe, impose stricter data privacy and permission requirements. A complete lack of legal frameworks addressing adversarial applications of GANs, such as the generation of faked imagery, indicates that proactive policy development will be required to avoid any legal ambiguity and assure responsible usage (Abdulmuttalib et al., 2024).

7.3. Balancing Innovation with Privacy and Security

The use of GANs in remote sensing requires balance between innovation promotion and privacy and security concerns, while adopting GANs has enormous capabilities such as improved data analysis and model performance, but they must be done in consideration with the framework of ethical norms and robust safeguards ensuring integrity and privacy data. Achieving this type of balance we must include privacy-preserving safeguards into GAN process workflows, maintain transparency about the synthetic data which we used, and following the legal and ethical guideline. collaboration between governments, academics, and business is crucial to integrate scientific innovation with societal values, making sure that the benefits of GANs in remote sensing are realized without compromising individual or organization privacy rights or global security (Vyawahare, Rathod, Dhande, Bajaj, & Khandelwal, 2024).

8. FUTURE DIRECTIONS AND RESEARCH OPPORTUNITIES

The field of remote sensing, especially with the integration of advanced machine learning models such as Generative Adversarial Networks (GANs), opens up more opportunities for future research and innovation prospects. Future directions will be towards improving the robustness, scalability, and security of remote sensing systems while dealing with emerging challenges such as adversarial vulnerabilities, ethical concerns, and data privacy.

8.1. Emerging Trends in Remote Sensing Security

The evolving environment of remote sensing security is fuelled by technology breakthroughs and the increasing complexity of cyber-attacks. Emerging themes include using artificial intelligence and machine learning to improve real-time anomaly detection and threat mitigation. Blockchain technology is gaining popularity as a way to secure data transmission and maintain the traceability of geospatial datasets, hence lowering vulnerabilities to hostile attacks. Furthermore, quantum cryptography has the potential to contribute to previously unattainable levels of security in data encryption and communication. The emergence of such technologies, together with increased international cooperation in developing standard security protocols, highlights the critical need for remote sensing security frameworks to be adaptive and innovative (Ratnaparkhi, Nagne, & Gawali, 2014; Zhang & Zhang, 2022).

8.2. Potential Developments in GANs

Generative adversarial networks (GANs) will transform remote sensing research in the near future, considering ongoing research on their robustness, scalability, and interpretability. Advances in the future may include hybrid architectures for GANs that integrate the strengths of standard GANs and other deep learning models for better synthetic data quality and computational efficiency. Better adversarial training techniques will make GANs more resilient to adversarial attacks, hence limiting their potential misuse. Other area the novel conditional GAN might have in producing domain specific artificial data that would be really pertinent in special applications of remote sensing like accurate vegetation map models or infrastructure urban construction and modelling. In any way improvement here might augment the application to actually take back for their present inadequacy (Wang, Sun, Chehri, & Song, 2023).

8.3. Open Research Questions

Despite tremendous progress, there are numerous unanswered concerns at the interface of remote sensing and GANs that deserve further investigation. Examples include reconciling the trade-offs between GAN performance and computing efficiency, as seen in real-time remote sensing applications. Furthermore, the ethical implications of synthetic data generation, particularly in sensitive sectors such as medicine, must be thoroughly investigated in order to set clear criteria for responsible use. Further tough topics include fully investigating GAN vulnerabilities to adversarial attacks and developing effective security methods. The integration of GANs with emerging technologies such as edge computing and quantum processing may offer up new possibilities, but it will also create additional levels of complexity that require substantial research (Oubara, Wu, Amamra, & Yang, 2022).

9. CONCLUSION

Conclusion Remote sensing technologies, with the enhanced capabilities of Generative Adversarial Networks (GANs), have tremendous potential in advancing geospatial analysis and applications. However, there is a significant challenge regarding privacy, security, and adversarial attacks. It is this discussion that puts together the delicate balance in utilizing the benefits of GANs and mitigating their vulnerabilities. Main things that have become integral measures are robust strategies of detection and mitigation; adherence to new norms of ethics and evolving laws. The ongoing progressions in the remote sensing discipline would involve innovated defences, along with interdisciplinary collaborative work, and will mark a path so that all kinds of technologies, deployed through it, would remain effective along with security as well as ethics. Proactive consideration of these challenges will better position the remote sensing community to advance in ways that are sustainable and responsible in terms of relevance to societal and scientific progress. Through the Generative Adversarial Networks, this discussion has highlighted the pivotal relationship between privacy, security, and adversarial attacks in remote sensing. Remote sensing technologies play a critical role in many applications but are prone to privacy and security violations. GANs which offer unprecedented functionality for data augmentation and image enhancement also introduces new avenues for adversaries to manipulate the integrity & authenticity of data. The defence methods against such threats are improved detection systems, strong adversarial training, and ethical frameworks of transparency and accountability. Further, the integration of major security trends as well as legal regulations illustrates the importance of multifaceted approach for security of remote sensing workflows.

9.2. Final Thoughts on Privacy, Security, and Adversarial Attacks

The rapid advancement in remote sensing and GAN technologies presents an opportunity for a proactive approach to the balancing act of innovation with responsible and secure practices. Technological sophistication of adversarial threats necessitates an integrated framework of legal definition and ethical oversight. Therefore, more research has to be focused on the dual-use potential of GANs in order to gain capabilities with less risk. The future of a safer and more ethical remote sensing technology will be determined by the balance of power between academia, industry, and policymakers. In an ever more dependent society where geography will remain a keystone, the technical question of privacy and security quickly matures into a societal question of the appropriate use of these transformative technologies.

REFERENCES

Abdulmuttalib, H., Al Doori, M., Barsi, Á., Blaschke, T., Gao, Y., Kugler, Z., Lang, S., Szabo, G., & Tiede, D. (2024). Remote Sensing Data Quality in the Era of AI. *ISPRS Annals of the Photogrammetry, Remote Sensing and Spatial Information Sciences*, *10*, 1–11. DOI: 10.5194/isprs-annals-X-3-2024-1-2024

Algarni, F., Khan, M. A., Alawad, W., & Halima, N. B. (2023). P3S: Pertinent Privacy Preserving Scheme for Remotely Sensed Environmental Data in Smart Cities. *IEEE Journal of Selected Topics in Applied Earth Observations and Remote Sensing*, *16*, 1–13. DOI: 10.1109/JSTARS.2023.3288743

Avtar, R., Kouser, A., Kumar, A., Singh, D., Misra, P., Gupta, A., Yunus, A. P., Kumar, P., Johnson, B. A., Dasgupta, R., Sahu, N., & Besse Rimba, A. (2021). Remote sensing for international peace and security: Its role and implications. *Remote Sensing (Basel)*, *13*(3), 439. DOI: 10.3390/rs13030439

Awate, M. P. L. (2023). Recent Trends In Crop Identification And Analysis By Using Remote Sensing and GIS. *The Online Journal of Distance Education and e-Learning : TOJDEL*, *11*(1).

Bosung, L. (2016). *Joongyeup, Lee., Yujin, Park., Beomsoo, Kim.* Privacy Protection from Unmanned Aerial Vehicle., DOI: 10.13089/JKIISC.2016.26.4.1057

Cheng, L., Liu, F., & Yao, D. (2017). Enterprise data breach: Causes, challenges, prevention, and future directions. *Wiley Interdisciplinary Reviews. Data Mining and Knowledge Discovery*, *7*(5), e1211. DOI: 10.1002/widm.1211

Crews, K. A. (2013). *Public/Private: The Ethics.* Legalities, and Practice of Airborne and Satellite Surveillance Technologies., DOI: 10.1016/B978-0-12-409548-9.10422-1

Dash, A., Ye, J., & Wang, G. (2023). A review of generative adversarial networks (GANs) and its applications in a wide variety of disciplines: From medical to remote sensing. *IEEE Access : Practical Innovations, Open Solutions.*

Goodfellow, I., Pouget-Abadie, J., Mirza, M., Xu, B., Warde-Farley, D., Ozair, S., Courville, A., & Bengio, Y. (2020). Generative adversarial networks. *Communications of the ACM*, *63*(11), 139–144. DOI: 10.1145/3422622

Hathaliya, J. J., Tanwar, S., & Sharma, P. (2022). Adversarial learning techniques for security and privacy preservation: A comprehensive review. *Security and Privacy*, *5*(3), e209. DOI: 10.1002/spy2.209

Hossain, I., Islam, M. M., & Martin, M. H. H. (2024). Potential Applications and Limitations of Artificial Intelligence in Remote Sensing Data Interpretation: A Case Study. *Control Systems and Optimization Letters*, *2*(3), 295–302. DOI: 10.59247/csol.v2i3.128

Jozdani, S., Chen, D., Pouliot, D., & Johnson, B. A. (2022). A review and meta-analysis of generative adversarial networks and their applications in remote sensing. *International Journal of Applied Earth Observation and Geoinformation*, *108*, 102734. DOI: 10.1016/j.jag.2022.102734

Jozdani, S., Chen, D., Pouliot, D., & Johnson, B. A. (2022). A review and meta-analysis of generative adversarial networks and their applications in remote sensing. *International Journal of Applied Earth Observation and Geoinformation*, *108*, 102734. DOI: 10.1016/j.jag.2022.102734

Jozdani, S., Chen, D., Pouliot, D., & Johnson, B. A. (2022). A review and meta-analysis of generative adversarial networks and their applications in remote sensing. *International Journal of Applied Earth Observation and Geoinformation*, *108*, 102734. DOI: 10.1016/j.jag.2022.102734

Kasula, V. K., Yadulla, A. R., Konda, B., & Yenugula, M. (2024). Fortifying cloud environments against data breaches: A novel AI-driven security framework. *World Journal of Advanced Research and Reviews*, *24*(01), 1613–1626. DOI: 10.30574/wjarr.2024.24.1.3194

Kazmi, S. M. K. A., Aafaq, N., Khan, M. A., Khalil, M., & Saleem, A. (2023). From Pixel to Peril: Investigating Adversarial Attacks on Aerial Imagery Through Comprehensive Review and Prospective Trajectories. *IEEE Access : Practical Innovations, Open Solutions*, *11*, 81256–81278. DOI: 10.1109/ACCESS.2023.3299878

Khan, M. J., Rath, S., & Zaib, M. H. (2024, April). Privacy-Enhanced Image Restoration in Remote Sensing via Federated Learning. In 2024 12th International Symposium on Digital Forensics and Security (ISDFS) (pp. 1-6). IEEE.

Kumaran, U., Thangam, S., Prabhakar, T. N., Selvaganesan, J., & Vishwas, H. N. (2023). Adversarial Defense: A GAN-IF Based Cyber-security Model for Intrusion Detection in Software Piracy. *Journal of Wireless Mobile Networks, Ubiquitous Computing and Dependable Applications*, *14*(4), 96–114. DOI: 10.58346/JOWUA.2023.I4.008

Li, S., Dragicevic, S., Castro, F. A., Sester, M., Winter, S., Coltekin, A., Pettit, C., Jiang, B., Haworth, J., Stein, A., & Cheng, T. (2016). Geospatial big data handling theory and methods: A review and research challenges. *ISPRS Journal of Photogrammetry and Remote Sensing*, *115*, 119–133. DOI: 10.1016/j.isprsjprs.2015.10.012

Li, Y., Ma, J., & Zhang, Y. (2021). Image retrieval from remote sensing big data: A survey. *Information Fusion*, *67*, 94–115. DOI: 10.1016/j.inffus.2020.10.008

McAmis, R., Sim, M., Bennett, M., & Kohno, T. (2024). Over fences and into yards: Privacy threats and concerns of commercial satellites. Proceedings on Privacy Enhancing Technologies.

) Mei, S., Lian, J., Wang, X., Su, Y., Ma, M., & Chau, L. P. (2024). A Comprehensive Study on the Robustness of Deep Learning-Based Image Classification and Object Detection in Remote Sensing: Surveying and Benchmarking. *Journal of Remote Sensing, 4*, 0219.

Naqvi, S. M. A., Shabaz, M., Khan, M. A., & Hassan, S. I. (2023). Adversarial attacks on visual objects using the fast gradient sign method. *Journal of Grid Computing*, *21*(4), 52. DOI: 10.1007/s10723-023-09684-9

Oubara, A., Wu, F., Amamra, A., & Yang, G. (2022, May). Survey on remote sensing data augmentation: Advances, challenges, and future perspectives. In *International Conference on Computing Systems and Applications* (pp. 95-104). Cham: Springer International Publishing. DOI: 10.1007/978-3-031-12097-8_9

Paolanti, M., Tiribelli, S., Giovanola, B., Mancini, A., Frontoni, E., & Pierdicca, R. (2024). Ethical Framework to Assess and Quantify the Trustworthiness of Artificial Intelligence Techniques: Application Case in Remote Sensing. *Remote Sensing (Basel)*, *16*(23), 4529. DOI: 10.3390/rs16234529

Ratnaparkhi, N. S., Nagne, A. D., & Gawali, B. (2014). A land use land cover classification system using remote sensing data. changes, 7(8).

Sattar, Y., & Oymak, S. (2020). Quickly finding the best linear model in high dimensions via projected gradient descent. *IEEE Transactions on Signal Processing*, *68*, 818–829. DOI: 10.1109/TSP.2020.2964216

Shi, C., Dang, Y., Fang, L., Zhao, M., Lv, Z., Miao, Q., & Pun, C. M. (2022). Multifeature collaborative adversarial attack in multimodal remote sensing image classification. *IEEE Transactions on Geoscience and Remote Sensing*, *60*, 1–15. DOI: 10.1109/TGRS.2022.3208337

Soon, Ae. (2000). *Chun., Vijayalakshmi, Atluri*. Protecting Privacy from Continuous High-resolution Satellite Surveillance., DOI: 10.1007/0-306-47008-X_21

Sumaiya, T. (2024, August 30). Kazi, Aminul, Islam. (2024). Improve Adversarial Robustness of AI Models in Remote Sensing via Data-Augmentation and Explainable-AI Methods. *Remote Sensing (Basel)*, *16*(17), 3210. Advance online publication. DOI: 10.3390/rs16173210

Vyawahare, H. R., Rathod, S. B., Dhande, S., Bajaj, K. A., & Khandelwal, S. (2024). Exploring Generative Adversarial Networks (GANs) in the Context of Public Space Protection. In *Enhancing Security in Public Spaces Through Generative Adversarial Networks (GANs)* (pp. 231-245). IGI Global.

Wang, X., Sun, L., Chehri, A., & Song, Y. (2023). A review of GAN-based super-resolution reconstruction for optical remote sensing images. *Remote Sensing (Basel), 15*(20), 5062. DOI: 10.3390/rs15205062

Xu, Y., Du, B., & Zhang, L. (2020). Assessing the threat of adversarial examples on deep neural networks for remote sensing scene classification: Attacks and defenses. *IEEE Transactions on Geoscience and Remote Sensing, 59*(2), 1604–1617. DOI: 10.1109/TGRS.2020.2999962

Xu, Y., & Ghamisi, P. (2022). Universal adversarial examples in remote sensing: Methodology and benchmark. *IEEE Transactions on Geoscience and Remote Sensing, 60*, 1–15. DOI: 10.1109/TGRS.2022.3156392

Xue, W., Chen, Z., Tian, W., Wu, Y., & Hua, B. (2022). A cascade defense method for multidomain adversarial attacks under remote sensing detection. *Remote Sensing (Basel), 14*(15), 3559. DOI: 10.3390/rs14153559

Yujian, H. (2023, September 15). Lei, Cao. (2023). Privacy-Preserving Remote Sensing Image Generation and Classification with Differentially Private GANs. *IEEE Sensors Journal, 23*(18), 20805–20816. Advance online publication. DOI: 10.1109/JSEN.2023.3267001

Zhang, D., Ren, L., Shafiq, M., & Gu, Z. (2022). A lightweight privacy-preserving system for the security of remote sensing images on iot. *Remote Sensing (Basel), 14*(24), 6371. DOI: 10.3390/rs14246371

Zhang, D., Shafiq, M., Wang, L., Srivastava, G., & Yin, S. (2023). Privacy-preserving remote sensing images recognition based on limited visual cryptography. *CAAI Transactions on Intelligence Technology, 8*(4), 1166–1177. DOI: 10.1049/cit2.12164

Zhang, H., Song, Y., Han, C., & Zhang, L. (2020). Remote sensing image spatiotemporal fusion using a generative adversarial network. *IEEE Transactions on Geoscience and Remote Sensing, 59*(5), 4273–4286. DOI: 10.1109/TGRS.2020.3010530

Zhang, L., & Zhang, L. (2022). Artificial intelligence for remote sensing data analysis: A review of challenges and opportunities. *IEEE Geoscience and Remote Sensing Magazine, 10*(2), 270–294. DOI: 10.1109/MGRS.2022.3145854

Compilation of References

) Fawakherji, M., Potena, C., Prevedello, I., Pretto, A., Bloisi, D. D., & Nardi, D. (2020, August). Data augmentation using gans for crop/weed segmentation in precision farming. In 2020 IEEE conference on control technology and applications (CCTA) (pp. 279-284). IEEE.

) Gurav, A. S., Karanjikar, Y. M., Bhikle, O. A., & Deshpande, H. (2024). Comparing the Effectiveness of GAN-based Methods for Enhancing EuroSat Satellite Imagery Classification. 1611–1616. DOI: 10.1109/ICESC60852.2024.10689946

) Jeyalakshmi, J., Kotra, V., Saiganesh, B., & Manohar, E. (2024). AI Deployment Models in Precision Agriculture. 125–142. DOI: 10.1201/9781003504900-6

) Mei, S., Lian, J., Wang, X., Su, Y., Ma, M., & Chau, L. P. (2024). A Comprehensive Study on the Robustness of Deep Learning-Based Image Classification and Object Detection in Remote Sensing: Surveying and Benchmarking. *Journal of Remote Sensing, 4*, 0219.

) N., Desai., Dhara, O., Bhavsar., Himanshu, Pandya. (2023). A review: applications of remote sensing in agriculture. DOI: 10.56588/iabcd.v2i1.137

) Sahoo, G. S., Dass, S., & Prakash, F. (2023). The Synergy of Advanced Language Models, Sequential Skeleton Features, and GAN-Enhanced Data Augmentation for Unprecedented Precision. 1–6. DOI: 10.1109/ICSCAN58655.2023.10395300

. Ben Ayed, R., Hanana, M. 2021. Artificial intelligence to improve the food and agriculture sector. Journal of Food Quality. Article ID 5584754, 7 pages. https://doi.org/DOI: 10.1155/2021/5584754

. Kumar, S., Meena, R.S., Sheoran, S., Jangir, C.K., Jhariya, M.K., Banerjee, A., Raj, A. 2022. Remote sensing for agriculture and resource management. Natural Resources Conservation and Advances for Sustainability. 91-135. https://doi.org/DOI: 10.1016/B978-0-12-822976-7.00012-0

. Suresh Babu, C. V., Abirami, S., & Sharon Manoj (2023). AI-Based Carthage Administration Towards Smart City. In C. Chowdhary, B. Swain, & V. Kumar (Eds.), Investigations in Pattern Recognition and Computer Vision for Industry 4.0 (pp. 1-17). IGI Global. DOI: 10.4018/978-1-6684-8602-3.ch001

Abbasi, R., Martinez, P., & Ahmad, R. (2022). The digitization of agricultural industry–a systematic literature review on agriculture 4.0. *Smart Agricultural Technology*, *2*, 100042. DOI: 10.1016/j.atech.2022.100042

Abd El-Ghany, N. M., Abd El-Aziz, S. E., & Marei, S. S. (2020). A review: Application of remote sensing as a promising strategy for insect pests and diseases management. *Environmental Science and Pollution Research International*, *27*(27), 33503–33515. DOI: 10.1007/s11356-020-09517-2 PMID: 32564316

Abdal, R., Qin, Y., & Wonka, P. (2019). Image2StyleGAN: How to embed images into the StyleGAN latent space? In *Proceedings of the IEEE/CVF International Conference on Computer Vision* (pp. 4432-4441). DOI: 10.1109/ICCV.2019.00453

Abdal, R., Zhu, P., Mitra, N. J., & Wonka, P. (2021). Styleflow: Attribute conditioned exploration of stylegan-generated images using conditional continuous normalizing flows. [ToG]. *ACM Transactions on Graphics*, *40*(3), 1–21. DOI: 10.1145/3447648

Abdelhalim, I. S. A., Mohamed, M. F., & Mahdy, Y. B. (2021). Data augmentation for skin lesion using self-attention based progressive generative adversarial network. *Expert Systems with Applications*, *165*, 113922. DOI: 10.1016/j.eswa.2020.113922

Abdulmuttalib, H., Al Doori, M., Barsi, Á., Blaschke, T., Gao, Y., Kugler, Z., Lang, S., Szabo, G., & Tiede, D. (2024). Remote Sensing Data Quality in the Era of AI. *ISPRS Annals of the Photogrammetry, Remote Sensing and Spatial Information Sciences*, *10*, 1–11. DOI: 10.5194/isprs-annals-X-3-2024-1-2024

Adler, J., & Lunz, S. (2018). Banach Wasserstein GAN. *Advances in Neural Information Processing Systems*, 31. https://doi.org/10.48550/arXiv.1806.06621

Afaq, Y., & Manocha, A. (2021). Analysis on change detection techniques for remote sensing applications: A review. *Ecological Informatics*, *63*, 101310. DOI: 10.1016/j.ecoinf.2021.101310

Aggarwal, A., Mittal, M., & Battineni, G. (2021). Generative adversarial network: An overview of theory and applications. *International Journal of Information Management Data Insights*, *1*(1), 100004. DOI: 10.1016/j.jjimei.2020.100004

Ahmad, Z., Jaffri, Z. U. A., Chen, M., & Bao, S. (2024). Understanding GANs: Fundamentals, variants, training challenges, applications, and open problems. *Multimedia Tools and Applications*, •••, 1–77. DOI: 10.1007/s11042-024-19361-y

Ahsan, M., Ali, M., & Siddique, Z. (2022). Imbalanced Class Data Performance Evaluation and Improvement using Novel Generative Adversarial Network-based Approach: SSG and GBO. *ArXiv*, abs/2210.12870. https://doi.org//arXiv.2210.12870.DOI: 10.48550

Ajayi, O. G., & Ojima, A. (2022). Performance evaluation of selected cloud occlusion removal algorithms on remote sensing imagery. *Remote Sensing Applications: Society and Environment*, *25*, 100700. DOI: 10.1016/j.rsase.2022.100700

Akintuyi, O. B. (2024). Adaptive AI in precision agriculture: a review: investigating the use of self-learning algorithms in optimizing farm operations based on real-time data. *Research Journal of Multidisciplinary Studies.*, *7*(2), 16–30. DOI: 10.53022/oarjms.2024.7.2.0023

Alajaji, S. A., Khoury, Z. H., Elgharib, M., Saeed, M., Ahmed, A. R., Khan, M. B., Tavares, T., Jessri, M., Puche, A. C., Hoorfar, H., Stojanov, I., Sciubba, J. J., & Sultan, A. S. (2024). Generative adversarial networks in digital histopathology: Current applications, limitations, ethical considerations, and future directions. *Modern Pathology*, *37*(1), 100369. DOI: 10.1016/j.modpat.2023.100369 PMID: 37890670

Algarni, F., Khan, M. A., Alawad, W., & Halima, N. B. (2023). P3S: Pertinent Privacy Preserving Scheme for Remotely Sensed Environmental Data in Smart Cities. *IEEE Journal of Selected Topics in Applied Earth Observations and Remote Sensing*, *16*, 1–13. DOI: 10.1109/JSTARS.2023.3288743

Ali-Gombe, A., & Elyan, E. (2019). MFC-GAN: Class-imbalanced dataset classification using Multiple Fake Class Generative Adversarial Network. *Neurocomputing*, *361*, 212–221. DOI: 10.1016/j.neucom.2019.06.043

Ali, S., DiPaola, D., & Breazeal, C. (2021, May). What are GANs?: Introducing generative adversarial networks to middle school students. *Proceedings of the AAAI Conference on Artificial Intelligence*, *35*(17), 15472–15479. DOI: 10.1609/aaai.v35i17.17821

Allen, D. E., Mushunje, L., & Peiris, S. (2024). GANs and synthetic financial data: Calculating VaR. *Applied Economics*, •••, 1–16. DOI: 10.1080/00036846.2024.2365456

Alqahtani, H., Kavakli-Thorne, M., & Kumar, G. (2021). Applications of generative adversarial networks (gans): An updated review. *Archives of Computational Methods in Engineering*, *28*(2), 525–552. DOI: 10.1007/s11831-019-09388-y

Anantrasirichai, N., & Bull, D. (2022). Artificial intelligence in the creative industries: A review. *Artificial Intelligence Review*, *55*(1), 589–656. DOI: 10.1007/s10462-021-10039-7

An, D., Zhang, L., Liu, Z., Liu, J., & Wei, Y. (2023). Advances in infrared spectroscopy and hyperspectral imaging combined with artificial intelligence for the detection of cereals quality. *Critical Reviews in Food Science and Nutrition, 63*(29), 9766–9796. DOI: 10.1080/10408398.2022.2066062 PMID: 35442834

Andrew, H. (2021). Generative Adversarial Network Synthesis of Hyperspectral Vegetation Data. *Remote Sensing*. Advance online publication. DOI: 10.3390/rs13122243

Ankan, D., Junyi, Y., & Guiling, W. (2024). *A Review of Generative Adversarial Networks (GANs) and Its Applications in a Wide Variety of Disciplines: From Medical to Remote Sensing*. IEEE., DOI: 10.1109/ACCESS.2023.3346273

Anuradha, D., Kuchipudi, R., Ashreetha, B., Ramesh, J. V. N., & Rami, A. (2024) Enhancing Agricultural Yield Forecasting with Deep Convolutional Generative Adversarial Networks and Satellite Data. *International Journal of Advanced Computer Science and Applications, 15*(2). Advance online publication. DOI: 10.14569/IJACSA.2024.0150269

Araújo, S. O., Peres, R. S., Barata, J., Lidon, F., & Ramalho, J. C. (2021). Characterising the agriculture 4.0 landscape—Emerging trends, challenges and opportunities. *Agronomy (Basel), 11*(4), 667. DOI: 10.3390/agronomy11040667

Arjovsky, M., Chintala, S., & Bottou, L. (2017). Wasserstein generative adversarial networks. In *International conference on machine learning* (pp. 214–223). PMLR.

Arjovsky, M., Chintala, S., & Bottou, L. (2017, July). Wasserstein generative adversarial networks. In *International conference on machine learning* (pp. 214-223). PMLR.

Arora, S., Ge, R., Liang, Y., Ma, T., & Zhang, Y. (2017, July). Generalization and equilibrium in generative adversarial nets (gans). In *International conference on machine learning* (pp. 224-232). PMLR.

Avtar, R., Kouser, A., Kumar, A., Singh, D., Misra, P., Gupta, A., Yunus, A. P., Kumar, P., Johnson, B. A., Dasgupta, R., Sahu, N., & Besse Rimba, A. (2021). Remote sensing for international peace and security: Its role and implications. *Remote Sensing (Basel), 13*(3), 439. DOI: 10.3390/rs13030439

Awate, M. P. L. (2023). Recent Trends In Crop Identification And Analysis By Using Remote Sensing and GIS. *The Online Journal of Distance Education and e-Learning : TOJDEL, 11*(1).

Awate, M. P. L., & Nagne, A. (2023). Recent trends in crop identification and analysis by using remote sensing and gis. *The Online Journal of Distance Education and e-Learning : TOJDEL, 11*(1).

Babu, C. S., Reddy, K., Chandu, G., Kowshik, T., & Rao, L. N. (2024). Trustworthy AI for Optimizing Agriculture: Leveraging IoT Sensory Data. In Developments Towards Next Generation Intelligent Systems for Sustainable Development (pp. 124-147). IGI Global.

Babu, C. S., Yadavamuthiah, K., Sathiyanarayana, S., & Mathew, S. (2024). Harnessing Agricultural Data: Advancing Sustainability Through the Application of Find S Algorithm. In Advanced Computational Methods for Agri-Business Sustainability (pp. 190-202). IGI Global.

Babu, T., Gupta, D., Singh, T., Hameed, S., Zakariah, M., & Alotaibi, Y. A. (2021). Robust magnification independent colon biopsy grading system over multiple data sources. *Computers, Materials & Continua, 69*(1), 99–128. DOI: 10.32604/cmc.2021.016341

Babu, T., Singh, T., & Gupta, D. (2020). Colon cancer prediction using 2dr e ca segmentation and hybrid features on histopathology images. *IET Image Processing, 14*(16), 4144–4157. DOI: 10.1049/iet-ipr.2019.1717

Backman, J., Koistinen, M., & Ronkainen, A. (2023). Agricultural process data as a source for knowledge: Perspective on artificial intelligence. *Smart Agricultural Technology, 5*, 100254. DOI: 10.1016/j.atech.2023.100254

Balaji Prabhu, B. V., Shashank, R., Shreyas, B., & Narsipura, O. S. J. (2024). ARIA: Augmented Reality and Artificial Intelligence enabled mobile application for Yield and grade prediction of tomato crops. *Procedia Computer Science, 235*, 2693–2702. DOI: 10.1016/j.procs.2024.04.254

Barratt, S., & Sharma, R. (2018). A note on the inception score. *arXiv preprint arXiv:1801.01973.*

Baskar, D., Parambalath, N. K., & Krishnanunni, S. O. (2024). Variational mode decomposition-enhanced thin cloud removal using UNet vision transformer cycle-consistent generative adversarial network. *Journal of Applied Remote Sensing, 18*(2), 026504–026504.

Bendaoud, N. M. M., Farah, N., & Ahmed, S. B. (2021). Comparing generative adversarial networks architectures for electricity demand forecasting. *Energy and Building, 247*, 111152. DOI: 10.1016/j.enbuild.2021.111152

Bengesi, S., El-Sayed, H., Sarker, M. K., Houkpati, Y., Irungu, J., & Oladunni, T. (2024). Advancements in Generative AI: A Comprehensive Review of GANs, GPT, Autoencoders, Diffusion Model, and Transformers. *IEEE Access*, 69812 – 69837, https://doi.org/DOI: 10.1109/ACCESS.2024.3397775

Bernard, N., & Leprévost, F. (2019). Evolutionary algorithms for convolutional neural network visualisation. In *High Performance Computing:5th Latin American Conference, CARLA 2018,Bucaramanga, Colombia,September 26–28, 2018, Revised Selected Papers 5* (pp. 18-32). Springer International Publishing DOI: 10.1007/978-3-030-16205-4_2

Bharti, V., Biswas, B., & Shukla, K. K. (2020, January). Recent trends in nature inspired computation with applications to deep learning. In *2020 10th International Conference on Cloud Computing, Data Science & Engineering (Confluence)* (pp. 294-299). IEEE. DOI: 10.1109/Confluence47617.2020.9057841

Bhat, S. A., & Huang, N. F. (2021). Big data and ai revolution in precision agriculture: Survey and challenges. *IEEE Access : Practical Innovations, Open Solutions, 9*, 110209–110222. DOI: 10.1109/ACCESS.2021.3102227

Bhat, S. A., Qadri, S. A. A., Dubbey, V., Sofi, I. B., & Huang, N. F. (2024). Impact of crop management practices on maize yield: Insights from farming in tropical regions and predictive modeling using machine learning. *Journal of Agriculture and Food Research, 18*, 101392. DOI: 10.1016/j.jafr.2024.101392

Bi, Y., Abrol, A., Sui, J., & Calhoun, V. (2024, April). Cross-Modal Synthesis of Structural MRI and Functional Connectivity Networks via Conditional ViT-GANs. In *ICASSP 2024-2024 IEEE International Conference on Acoustics, Speech and Signal Processing (ICASSP)*, Seoul, Korea, (pp. 1756-1760). IEEE. https://doi.org/ DOI: 10.1109/ICASSP48485.2024.10446450

Bihlo, A. (2021). A generative adversarial network approach to (ensemble) weather prediction. *Neural Networks, 139*, 1–16. DOI: 10.1016/j.neunet.2021.02.003 PMID: 33662648

Bindu, N. M., & Satyanarayana, S. (2025). Designing of GAN for a real-time image processing in neuromorphic system. In *Primer to Neuromorphic Computing* (pp. 21–44). Academic Press. DOI: 10.1016/B978-0-443-21480-6.00007-9

Bishshash, P., Nirob, A. S., Shikder, H., Sarower, A. H., Bhuiyan, T., & Noori, S. R. H. (2024). A comprehensive cotton leaf disease dataset for enhanced detection and classification. *Data in Brief, 57*, 110913. DOI: 10.1016/j.dib.2024.110913 PMID: 39328970

Bobadilla, J., & Gutiérrez, A. (2024). Wasserstein GAN-based architecture to generate collaborative filtering synthetic datasets. *Applied Intelligence*, *54*(3), 2472–2490.

Bodrud-Doza, M., Yang, W., de Queiroga Miranda, R., Martin, A., DeVries, B., & Fraser, E. D. G. (2023). Towards implementing precision conservation practices in agricultural watersheds: A review of the use and prospects of spatial decision support systems and tools. *The Science of the Total Environment*, *905*, 167118. DOI: 10.1016/j.scitotenv.2023.167118 PMID: 37717782

Boroujeni, S. P. H., & Razi, A. (2024). Ic-gan: An improved conditional generative adversarial network for rgb-to-ir image translation with applications to forest fire monitoring. *Expert Systems with Applications*, *238*, 121962. DOI: 10.1016/j.eswa.2023.121962

Bosung, L. (2016). *Joongyeup, Lee., Yujin, Park., Beomsoo, Kim.* Privacy Protection from Unmanned Aerial Vehicle., DOI: 10.13089/JKIISC.2016.26.4.1057

Bouteska, A., Seranto, M. L., Hajek, P., & Abedin, M. Z. (2023). Data-driven decadal climate forecasting using Wasserstein time-series generative adversarial networks. *Annals of Operations Research*. Advance online publication. DOI: 10.1007/s10479-023-05722-7

Brock, A., Donahue, J., & Simonyan, K. (2018). Large scale GAN training for high fidelity natural image synthesis. arXiv preprint arXiv:1809.11096.

Brophy, E., Wang, Z., She, Q., & Ward, T. (2023). Generative adversarial networks in time series: A systematic literature review. *ACM Computing Surveys*, *55*(10), 1–31.

Buduma, N. (2019), Fundamentals of Deep Learning, O'Reilly publications, November 2019, ISBN:978-93-5213-560-8.

Cai, B., Xu, X., Jia, K., Qing, C., & Tao, D. (2016). DehazeNet: An End-to-End System for Single Image Haze Removal. *IEEE Transactions on Image Processing*, *25*(11), 5187–5198. DOI: 10.1109/TIP.2016.2598681 PMID: 28873058

Campbell, J. B., & Wynne, R. H. (2011). *Introduction to remote sensing*. Guilford press.

Cao, J., Luo, M., Yu, J., Yang, M., & He, R. (2022). ScoreMix: A Scalable Augmentation Strategy for Training GANs With Limited Data. *IEEE Transactions on Pattern Analysis and Machine Intelligence*, *45*, 8920–8935. DOI: 10.1109/TPAMI.2022.3231649 PMID: 37015400

Caparros-Santiago, J. A., Rodriguez-Galiano, V., & Dash, J. (2021). Land surface phenology as indicator of global terrestrial ecosystem dynamics: A systematic review. *ISPRS Journal of Photogrammetry and Remote Sensing*, *171*, 330–347. DOI: 10.1016/j.isprsjprs.2020.11.019

Carlini, N., & Wagner, D. (2018), Audio Adversarial Examples: Targeted Attacks on Speech-to-Text, IEEE Deep Learning and Security Workshop, http://bit.ly/21FXTIW

Carneiro, G. (2020, November). Why are generative adversarial networks so fascinating and annoying? In 2020 33rd SIBGRAPI Conference on Graphics, Patterns and Images (SIBGRAPI) (pp. 1-8). IEEE.

Cate, M. Fundamentals of Deep Generative Models.

Cesco, S., Sambo, P., Borin, M., Basso, B., Orzes, G., & Mazzetto, F. (2023). Smart agriculture and digital twins: Applications and challenges in a vision of sustainability. *European Journal of Agronomy*, *146*, 126809. DOI: 10.1016/j.eja.2023.126809

Chahal, A., & Gulia, P. (2019). Machine learning and deep learning. *International Journal of Innovative Technology and Exploring Engineering*, *8*(12), 4910–4914.

Chakraborty, T., Reddy K S, U., Naik, S. M., Panja, M., & Manvitha, B. (2024). Ten years of generative adversarial nets (GANs): A survey of the state-of-the-art. *Machine Learning: Science and Technology*, *5*(1), 011001. DOI: 10.1088/2632-2153/ad1f77

Cheng, J., Yang, Y., Tang, X., Xiong, N., Zhang, Y., & Lei, F. (2020), Generative Adversarial Networks: A Literature Review, Ksii Transactions On Internet And Information Systems Vol. 14, No, , ISSN: 1976-7277.DOI: 10.3837/tiis.2020.12.001

Cheng, J., Zhu, Y., Zhao, Y., Li, T., Chen, M., Sun, Q., Gu, Q., & Zhang, X. (2024). Application of an improved U-Net with image-to-image translation and transfer learning in peach orchard segmentation. *International Journal of Applied Earth Observation and Geoinformation*, *130*, 103871. DOI: 10.1016/j.jag.2024.103871

Cheng, L., Liu, F., & Yao, D. (2017). Enterprise data breach: Causes, challenges, prevention, and future directions. *Wiley Interdisciplinary Reviews. Data Mining and Knowledge Discovery*, *7*(5), e1211. DOI: 10.1002/widm.1211

Cheng, W., Li, Y., & Ma, T. (2024). S-KDGAN: Series-Knowledge Distillation with GANs for Anomaly Detection of Sensor Time-Series Data in Smart IoT. *IEEE Sensors Journal*, *24*(15), 24344–24354. DOI: 10.1109/JSEN.2024.3415390

Chen, H. Y., Sharma, K., Sharma, C., & Sharma, S. (2023a). Integrating explainable artificial intelligence and blockchain to smart agriculture: Research prospects for decision making and improved security. *Smart Agricultural Technology*, *6*, 100350. DOI: 10.1016/j.atech.2023.100350

Chen, J., Dai, X., Guo, Y., Zhu, J., Mei, X., Deng, M., & Sun, G. (2023). Urban built environment assessment based on scene understanding of high-resolution remote sensing imagery. *Remote Sensing (Basel)*, *15*(5), 1436. DOI: 10.3390/rs15051436

Chen, L., Chen, Z., Zhang, Y., Liu, Y., Osman, A. I., Farghali, M., Hua, J., Al-Fatesh, A., Ihara, I., Rooney, D. W., & Yap, P. (2023b). Artificial intelligence-based solutions for climate change: A review. *Environmental Chemistry Letters*, *21*(5), 2525–2557. DOI: 10.1007/s10311-023-01617-y

Chen, S., Chen, X., Chen, J., Jia, P., Cao, X., & Liu, C. (2016). An Iterative Haze Optimized Transformation for Automatic Cloud/Haze Detection of Landsat Imagery. *IEEE Transactions on Geoscience and Remote Sensing*, *54*(5), 2682–2694. DOI: 10.1109/TGRS.2015.2504369

Chen, S., Tong, B., Russell, L. M., Wei, J., Guo, J., Mao, F., Liu, D., Huang, Z., Xie, Y., Qi, B., Zhang, H., Sun, Y., Zhang, B., Xu, C., Wu, L., & Liu, D. (2022a). Lidar-based daytime boundary layer height variation and impact on the regional satellite-based pm2. 5 estimate. *Remote Sensing of Environment*, *281*, 113224. DOI: 10.1016/j.rse.2022.113224

Chen, X., Li, Y., Yao, L., Adeli, E., Zhang, Y., & Wang, X. (2022b). Generative adversarial u-net for domain-free few-shot medical diagnosis. *Pattern Recognition Letters*, *157*, 112–118. DOI: 10.1016/j.patrec.2022.03.022

Choi, R. Y., Coyner, A. S., Kalpathy-Cramer, J., Chiang, M. F., & Peter Campbell, J. (2024), Introduction to Machine Learning, Neural Networks, and Deep Learning, TVST | Special Issue | Vol. 9 | No. 2 |, Article 14.

Choi, H., Jung, D., Kim, S., & Yoon, S. (2021). Imbalanced Data Classification via Cooperative Interaction Between Classifier and Generator. *IEEE Transactions on Neural Networks and Learning Systems*, *33*(8), 3343–3356. DOI: 10.1109/TNNLS.2021.3052243 PMID: 33531305

Choi, Y., Choi, M., Kim, M., Ha, J. W., Kim, S., & Choo, J. (2018). Stargan: Unified generative adversarial networks for multi-domain image-to-image translation. In *Proceedings of the IEEE conference on computer vision and pattern recognition* (pp. 8789-8797). DOI: 10.1109/CVPR.2018.00916

Choi, Y., Choi, M., Kim, M., Ha, J.-W., Kim, S., & Choo, J. (2018). Star gan: Unified generative adversarial networks for multi-domain image-to-image translation. In *Proceedings of the IEEE conference on computer vision and pattern recognition*, pages 8789–8797.

Congalton, R. G., & Green, K. (2019). *Assessing the accuracy of remotely sensed data: principles and practices*. CRC press. DOI: 10.1201/9780429052729

Craswell, E. (2021). Fertilizers and nitrate pollution of surface and ground water: An increasingly pervasive global problem. *SN Applied Sciences*, *3*(4), 518. DOI: 10.1007/s42452-021-04521-8

Creswell, A., & Plano Clark, V. L. (2018). *Generative Adversarial Networks for Image Synthesis: Principles and Applications*. Springer.

Creswell, A., White, T., Dumoulin, V., Arulkumaran, K., Sengupta, B., & Bharath, A. A. (2018). Generative adversarial networks: An overview. *IEEE Signal Processing Magazine*, *35*(1), 53–65. DOI: 10.1109/MSP.2017.2765202

Creswell, J. W., & Plano Clark, V. L. (2018). *Designing and conducting mixed methods research* (3rd ed.). SAGE Publications.

Crews, K. A. (2013). *Public/Private: The Ethics*. Legalities, and Practice of Airborne and Satellite Surveillance Technologies., DOI: 10.1016/B978-0-12-409548-9.10422-1

Crivellari, A., Wei, H., Wei, C., & Shi, Y. (2023). Super-resolution GANs for upscaling unplanned urban settlements from remote sensing satellite imagery–the case of Chinese urban village detection. *International Journal of Digital Earth*, *16*(1), 2623–2643. DOI: 10.1080/17538947.2023.2230956

Cui, X., Han, W., Zhang, H., Dong, Y., Ma, W., Zhai, X., Zhang, L., & Li, G. (2023). Estimating and mapping the dynamics of soil salinity under different crop types using Sentinel-2 satellite imagery. *Geoderma*, *440*, 116738. DOI: 10.1016/j.geoderma.2023.116738

Dash, A., Ye, J., & Wang, G. (2023). A review of generative adversarial networks (GANs) and its applications in a wide variety of disciplines: From medical to remote sensing. *IEEE Access : Practical Innovations, Open Solutions*.

De Natale, F., Alilla, R., Parisse, B., & Nardi, P. (2023). A bibliometric analysis on drought and heat indices in agriculture. *Agricultural and Forest Meteorology*, *341*, 109626. DOI: 10.1016/j.agrformet.2023.109626

Demiray, B. Z., Sit, M., & Demir, I. (2021). D-srgan: Dem super-resolution with generative adversarial networks. *SN Computer Science*, *2*(1), 1–11. DOI: 10.1007/s42979-020-00442-2

Deng, Q., Huang, Z., Tsai, C.-C., & Lin, C.-W. (2020). HardGAN: A Haze-Aware Representation Distillation GAN for Single Image Dehazing. In Vedaldi, A., Bischof, H., Brox, T., & Frahm, J.-M. (Eds.), *Computer Vision – ECCV 2020* (pp. 722–738). Springer International Publishing. DOI: 10.1007/978-3-030-58539-6_43

Denton, E., Chintala, S., Szlam, A., & Olah, C. (2015). *Deep generative image models using a laplacian pyramid of adversarial networks*. In Advances in Neural Information Processing Systems (pp. 1486-1494).

DeVries, T., Romero, A., Pineda, L., Taylor, G. W., & Drozdzal, M. (2019). On the evaluation of conditional GANs. *arXiv preprint arXiv:1907.08175*.

Dewitte, S., Cornelis, J. P., Müller, R., & Munteanu, A. (2021). Artificial intelligence revolutionises weather forecast, climate monitoring and decadal prediction. *Remote Sensing (Basel)*, *13*(16), 3209. DOI: 10.3390/rs13163209

Dhanush, G., Khatri, N., Kumar, S., & Shukla, P. K. (2023). A comprehensive review of machine vision systems and artificial intelligence algorithms for the detection and harvesting of agricultural produce. *Scientific African*, *21*, e01798. DOI: 10.1016/j.sciaf.2023.e01798

Donahue, J., Krähenbühl, P., & Darrell, T. (2017). Adversarial feature learning. In 5th International Conference on Learning Representations, ICLR 2017, Toulon, France, April 24-26, 2017, Conference Track Proceedings. OpenReview.net.

Dou, X., Li, C., Shi, Q., & Liu, M. (2020). Super-resolution for hyperspectral remote sensing images based on the 3D attention-SRGAN network. *Remote Sensing (Basel)*, *12*(7), 1204. DOI: 10.3390/rs12071204

Dubovik, O., Schuster, G. L., Xu, F., Hu, Y., Bösch, H., Landgraf, J., & Li, Z. (2021). Grand challenges in satellite remote sensing. *Frontiers in Remote Sensing*, *2*, 619818. DOI: 10.3389/frsen.2021.619818

Du, W.-L., Zhou, Y., Zhao, J., Tian, X., Yang, Z., & Bian, F. (2021). Exploring the potential of unsupervised image synthesis for sar-optical image matching. *IEEE Access : Practical Innovations, Open Solutions*, *9*, 71022–71033. DOI: 10.1109/ACCESS.2021.3079327

Du, Z., Li, X., Miao, J., Huang, Y., Shen, H., & Zhang, L. (2023). Concatenated deep learning framework for multi-task change detection of optical and sar images. *IEEE Journal of Selected Topics in Applied Earth Observations and Remote Sensing*.

Ebrahimzadeh, G., Mahabadi, N. Y., & Bayat, H. (2023). Estimating pre-compression stress in agricultural Soils: Integrating spectral indices and soil properties through machine learning. *Computers and Electronics in Agriculture*, *215*, 108393. DOI: 10.1016/j.compag.2023.108393

Eiben, A. E., & Smith, J. (2015). From evolutionary computation to the evolution of things. *Nature*, *521*(7553), 476–482. DOI: 10.1038/nature14544 PMID: 26017447

Elgammal, A., Liu, B., Elhoseiny, M., & Mazzone, M. (2017). CAN: Creative adversarial networks, generating "art" by learning about styles and deviating from style norms. arXiv preprint arXiv:1706.07068.

Eltehewy, R., Abouelfarag, A., & Saleh, S. N. (2023). Efficient classification of imbalanced natural disasters data using generative adversarial networks for data augmentation. *ISPRS International Journal of Geo-Information*, *12*(6), 245. DOI: 10.3390/ijgi12060245

Engin, D., Genc, A., & Ekenel, H. K. (2018). Cycle-Dehaze: Enhanced CycleGAN for Single Image Dehazing. *2018 IEEE/CVF Conference on Computer Vision and Pattern Recognition Workshops (CVPRW)*, (pp. 938-9388). DOI: 10.1109/CVPRW.2018.00127

Fang, B., Chen, G., Pan, L., Kou, R., & Wang, L. (2020). Gan-based siamese framework for landslide inventory mapping using bi-temporal optical remote sensing images. *IEEE Geoscience and Remote Sensing Letters*, *18*(3), 391–395. DOI: 10.1109/LGRS.2020.2979693

Fan, Y., Wang, X., Funk, T., Rashid, I., Herman, B., Bompoti, N., Mahmud, M. D. S., Chrysochoou, M., Yang, M., Vadas, T. M., Lei, Y., & Li, B. (2022). A critical review for real-time continuous soil monitoring: Advantages, challenges, and perspectives. *Environmental Science & Technology*, *56*(19), 13546–13564. DOI: 10.1021/acs.est.2c03562 PMID: 36121207

Feng, J. (2022)... *Generative Adversarial Networks for Remote Sensing.*, *108–112*, 108–112. Advance online publication. DOI: 10.1109/ICBAR58199.2022.00028

Feuerriegel, S., Hartmann, J., Janiesch, C., & Zschech, P. (2024). Generative ai. *Business & Information Systems Engineering*, *66*(1), 111–126. DOI: 10.1007/s12599-023-00834-7

Fogel, D. B. (1995, December). Phenotypes, genotypes, and operators in evolutionary computation. In *Proc. 1995 IEEE Int. Conf. Evolutionary Computation (ICEC'95)* (pp. 193-198). DOI: 10.1109/ICEC.1995.489143

Foroumandi, E., Gavahi, K., & Moradkhani, H. (2024). Generative adversarial network for real-time flash drought monitoring: A deep learning study. *Water Resources Research, 60*(5). Advance online publication. DOI: 10.1029/2023WR035600

Fowlis, N., Barnett, N., Banks, S., & Jubraj, B. (2024). A qualitative evaluation of weekly reflective practice sessions for the intensive care unit pharmacy team during the COVID-19 pandemic. *European Journal of Hospital Pharmacy. Science and Practice, 31*(1), 57–62. DOI: 10.1136/ejhpharm-2021-003164 PMID: 35428696

Frid-Adar, M., Ganin, A., Kahn, S., & Goldberger, J. (2018). GAN-based synthetic medical image augmentation for increased CNN performance in liver lesion classification. *Neurocomputing, 321*, 321–331. DOI: 10.1016/j.neucom.2018.09.013

Fuentes-Peñailillo, F., Gutter, K., Vega, R., & Silva, G. C. (2024). Transformative technologies in digital agriculture: Leveraging Internet of Things, remote sensing, and artificial intelligence for smart crop management. *Journal of Sensor and Actuator Networks., 13*(4), 39. DOI: 10.3390/jsan13040039

Gafurov, A., Mukharamova, S., Saveliev, A., & Yermolaev, O. (2023). Advancing Agricultural Crop Recognition: The Application of LSTM Networks and Spatial Generalization in Satellite Data Analysis. *Agriculture, 13*(9), 1672. DOI: 10.3390/agriculture13091672

Gamage, A., Gangahagedara, R., Subasinghe, S., Gamage, J., Guruge, C., Senaratne, S., Randika, T., Rathnayake, C., Hameed, Z., Madhujith, T., & Merah, O. (2024). Advancing sustainability: The impact of emerging technologies in agriculture. *Current Plant Biology, 40*, 100420. DOI: 10.1016/j.cpb.2024.100420

Gao, Y., Wang, Y., Lv, H., & Qian, J. (2019). A Revised RTM-Based Algorithm to Remove Thin Clouds within Visible Band Data of Sentinel-2A. IGARSS 2019 - 2019 IEEE International Geoscience and Remote Sensing Symposium, (pp. 1418-1421). DOI: 10.1109/IGARSS.2019.8898499

Gao, G., Lai, H., & Jia, Z. (2023). Unsupervised image dedusting via a cycle-consistent generative adversarial network. *Remote Sensing (Basel), 15*(5), 1311. DOI: 10.3390/rs15051311

Gao, L., Wang, X., Johnson, B. A., Tian, Q., Wang, Y., Verrelst, J., Mu, X., & Gu, X. (2020). Remote sensing algorithms for estimation of fractional vegetation cover using pure vegetation index values: A review. *ISPRS Journal of Photogrammetry and Remote Sensing, 159*, 364–377. DOI: 10.1016/j.isprsjprs.2019.11.018 PMID: 36082112

Gao, W., Liu, F., Liu, J., Xiao, L., & Tang, X. (2023). Spatial–spectral adaptive learning with pixelwise filtering for hyperspectral image classification. *IEEE Transactions on Geoscience and Remote Sensing*, *61*, 1–15. DOI: 10.1109/TGRS.2023.3284074

Gao, Y., Hu, Z., Shi, S., Chen, W. A., & Liu, M. (2024). Adversarial discriminative domain adaptation for solar radiation prediction: A cross-regional study for zero-label transfer learning in Japan. *Applied Energy*, *359*, 122685. DOI: 10.1016/j.apenergy.2024.122685

Garg, S., Rumjit, N. P., & Roy, S. (2024). Smart agriculture and nanotechnology: Technology, challenges, and new perspective. *Advanced Agrochem.*, *3*(2), 115–125. DOI: 10.1016/j.aac.2023.11.001

Gatys, L. A., Ecker, A. S., & Bethge, M. (2016). Image style transfer using convolutional neural networks. In *Proceedings of the IEEE conference oncomputer vision and pattern recognition*, pages 2414–2423. DOI: 10.1109/CVPR.2016.265

Ge, H., Xia, Y., Chen, X., Berry, R., & Wu, Y. (2018). Fictitious gan: Training gans with historical models. In *Proceedings of the European Conference on Computer Vision (ECCV)* (pp. 119-134). DOI: 10.1007/978-3-030-01246-5_8

Giovos, R., Tassopoulos, D., Kalivas, D., Lougkos, N., & Priovolou, A. (2021). Remote Sensing Vegetation Indices in Viticulture: A Critical. *Agriculture*, *11*(5), 457. DOI: 10.3390/agriculture11050457

Gireesh Babu, C., Guru Dutt, A., Pushpa, S., & Manjunath, T. (2023). Conditional generative adversarial networks for image transformation. In Advances in Data-driven Computing and Intelligent Systems: Selected Papers from AD CIS 2022, Volume 2, pages 547–560. Springer. DOI: 10.1007/978-981-99-0981-0_42

Gong, Z., Ge, W., Guo, J., & Liu, J. (2024). Satellite remote sensing of vegetation phenology: Progress, challenges, and opportunities. *ISPRS Journal of Photogrammetry and Remote Sensing*, *217*, 149–164. DOI: 10.1016/j.isprsjprs.2024.08.011

Gonog, L., & Zhou, Y. (2019, June). A review: generative adversarial networks. In *2019 14th IEEE conference on industrial electronics and applications (ICIEA)* (pp. 505-510). IEEE. DOI: 10.1109/ICIEA.2019.8833686

Goodfellow, I., Pouget-Abadie, J., Mirza, M., Xu, B., Warde-Farley, D., Ozair, S., . . . Bengio, Y. (2014). Generative Adversarial Nets. In Z. Ghahramani, M. Welling, C. Cortes, N. Lawrence, & K. Q. Weinberger (Ed.), Advances in Neural Information Processing Systems. 27. Curran Associates, Inc. Retrieved from https://proceedings.neurips.cc/paper_files/paper/2014/file/5ca3e9b122f61f8f06494c97b1afccf3-Paper.pdf

Goodfellow, I., Pouget-Abadie, J., Mirza, M., Xu, B., Warde-Farley, D., Ozair, S., . . . Courville, A. (2014). *Generative adversarial nets*. In Advances in Neural Information Processing Systems (pp. 27-34).

Goodfellow, I., Bengio, Y., Courville, A., & Bengio, Y. (2016). Deep learning: Vol. 1. *No. 2*. MIT press.

Goodfellow, I., Pouget-Abadie, J., Mirza, M., Xu, B., Warde-Farley, D., Ozair, S., & Bengio, Y. (2014). Generative adversarial nets. *Advances in Neural Information Processing Systems*, •••, 27.

Goodfellow, I., Pouget-Abadie, J., Mirza, M., Xu, B., Warde-Farley, D., Ozair, S., Courville, A., & Bengio, Y. (2014). Generative adversarial nets. In *Advances in neural information processing systems* (Vol. 27). Curran Associates, Inc.

Goodfellow, I., Pouget-Abadie, J., Mirza, M., Xu, B., Warde-Farley, D., Ozair, S., Courville, A., & Bengio, Y. (2020). Generative adversarial networks. *Communications of the ACM*, *63*(11), 139–144. DOI: 10.1145/3422622

Goswami, A., Sharma, D., Mathuku, H., Gangadharan, S. M. P., Yadav, C. S., Sahu, S. K., Pradhan, M. K., Singh, J., & Imran, H. (2022). Change detection in remote sensing image data comparing algebraic and machine learning methods. *Electronics (Basel)*, *11*(3), 431. DOI: 10.3390/electronics11030431

Greza, M., Bhattacharya, I., Hoegner, L., & Jutzi, B. (2024). GAN-Based dual image super resolution for satellite imagery decreasing radiometric uncertainty. *ISPRS Annals of the Photogrammetry, Remote Sensing and Spatial Information Sciences*, *X-3–2024*, 155–162. https://doi.org/DOI: 10.5194/isprs-annals-X-3-2024-155-2024

Gui, J., Sun, Z., Wen, Y., Tao, D., & Ye, J. (2021). A review on generative adversarial networks: Algorithms, theory, and applications. *IEEE Transactions on Knowledge and Data Engineering*, *35*(4), 3313–3332. DOI: 10.1109/TKDE.2021.3130191

Gulrajani, I., Ahmed, F., Arjovsky, M., Dumoulin, V., & Courville, A. C. (2017). Improved training of wasserstein gans. *Advances in Neural Information Processing Systems*, •••, 30.

Guo, H., Jia, X., Dong, Y., Ye, J., Chen, D., Zhang, R., & Zheng, Y. (2020). Applications of AlGaN/GaN high electron mobility transistor-based sensors in water quality monitoring. *Semiconductor Science and Technology*, *35*(12), 123001. DOI: 10.1088/1361-6641/abb8fb

Guo, X., & Mounjid, O. (2024). GANs training: A game and stochastic control approach. *Mathematical Finance*, *34*(2), 522–556. DOI: 10.1111/mafi.12427

Guo, Y., He, W., Xia, Y., & Zhang, H. (2023). Blind single-image-based thin cloud removal using a cloud perception integrated fast Fourier convolutional network. *ISPRS Journal of Photogrammetry and Remote Sensing*, *206*, 63–86. DOI: 10.1016/j.isprsjprs.2023.10.014

Hachana, R., & Rasheed, B. (2025). Probe-Assisted Fine-Grained Control for Non-Differentiable Features in Symbolic Music Generation. *IEEE Access : Practical Innovations, Open Solutions*, *13*, 28059–28070. DOI: 10.1109/ACCESS.2025.3540543

Hadid, A., Chakraborty, T., & Busby, D. (2024). When geoscience meets generative AI and large language models: Foundations, trends, and future challenges. *Expert Systems: International Journal of Knowledge Engineering and Neural Networks*, *41*(10), e13654.

Han, J., Zhou, J., Wang, L., Wang, Y., & Ding, Z. (2023). Fe-gan: Fast and efficient underwater image enhancement model based on conditional gan. *Electronics (Basel)*, *12*(5), 1227. DOI: 10.3390/electronics12051227

Han, X., Zhang, L., Zhou, K., & Wang, X. (2019). ProGAN: Protein solubility generative adversarial nets for data augmentation in DNN framework. *Computers & Chemical Engineering*, *131*, 106533. DOI: 10.1016/j.compchemeng.2019.106533

Han, X., Zhang, Y., & Wang, Y. (2018). Semi-supervised generative adversarial networks for remote sensing image classification. *Remote Sensing*, *10*(5), 1–15.

Hassan, A., Ali, M., Trigui, A., Savaria, Y., & Sawan, M. (2019). A GaN-based wireless monitoring system for high-temperature applications. *Sensors (Basel)*, *19*(8), 1785. DOI: 10.3390/s19081785 PMID: 31013978

Hathaliya, J. J., Tanwar, S., & Sharma, P. (2022). Adversarial learning techniques for security and privacy preservation: A comprehensive review. *Security and Privacy*, *5*(3), e209. DOI: 10.1002/spy2.209

He, C., Huang, S., Cheng, R., Tan, K. C., & Jin, Y. (2020). Evolutionary multi-objective optimization driven by generative adversarial networks (GANs). *IEEE Transactions on Cybernetics*, *51*(6), 3129–3142. DOI: 10.1109/TCYB.2020.2985081 PMID: 32365041

Henry, J., Natalie, T., & Madsen, D. (2021). *Pix2pix gan for image-to-image translation*. Research Gate Publication.

He, X., Ji, W., & Xie, J. (2022). Unsupervised Haze Removal for Aerial Imagery Based on Asymmetric Contrastive CycleGAN. *IEEE Access : Practical Innovations, Open Solutions*, *10*, 67316–67328. DOI: 10.1109/ACCESS.2022.3186004

Hong, Y., Hwang, U., Yoo, J., & Yoon, S. (2019). How generative adversarial networks and their variants work: An overview. *ACM Computing Surveys*, *52*(1), 1–43. DOI: 10.1145/3301282

Hossain, I., Islam, M. M., & Martin, M. H. H. (2024). Potential Applications and Limitations of Artificial Intelligence in Remote Sensing Data Interpretation: A Case Study. *Control Systems and Optimization Letters*, *2*(3), 295–302. DOI: 10.59247/csol.v2i3.128

https://www.coursera.org/in/articles/ai-vs-deep-learning-vs-machine-learning-beginners-guide on 15/1/2025 at 10.06 pm

Huang, S., Tang, L., Hupy, J. P., Wang, Y., & Shao, G. (2021). A commentary review on the use of normalized difference vegetation index (NDVI) in the era of popular remote sensing. *Journal of Forestry Research*, *32*(1), 1–6. DOI: 10.1007/s11676-020-01155-1

Huang, X., & Belongie, S. (2017). Arbitrary style transfer in real-time with adaptive instance normalization. In *Proceedings of the IEEE International Conference on Computer Vision* (pp. 1501-1510). DOI: 10.1109/ICCV.2017.167

Huang, Y., & Cao, L. (2023). Privacy-Preserving Remote Sensing Image Generation and Classification With Differentially Private GANs. *IEEE Sensors Journal*, *23*(18), 20805–20816. DOI: 10.1109/JSEN.2023.3267001

Huang, Z., Zhang, X., Tang, Z., Xu, F., Datcu, M., & Han, J. (2024). Gen erative artificial intelligence meets synthetic aperture radar: A survey. *IEEE Geoscience and Remote Sensing Magazine*.

Hughes, R. T., Zhu, L., & Bednarz, T. (2021). Generative Adversarial Networks–Enabled Human–Artificial Intelligence Collaborative Applications for Creative and Design Industries: A Systematic Review of Current Approaches and Trends. *Frontiers in Artificial Intelligence*, *4*, 604234. DOI: 10.3389/frai.2021.604234 PMID: 33997773

Huo, D., Malik, A. W., Ravana, S. D., Rahman, A. U., & Ahmedy, I. (2024). Mapping smart farming: Addressing agricultural challenges in data-driven era. *Renewable & Sustainable Energy Reviews*, *189*, 113858. DOI: 10.1016/j.rser.2023.113858

Iglesias, G., Talavera, E., & Díaz-Álvarez, A. (2023). A survey on GANs for computer vision: Recent research, analysis and taxonomy. *Computer Science Review*, *48*, 100553. DOI: 10.1016/j.cosrev.2023.100553

Imaging, E. S. (2024). What is sar imagery? Isola, P., Zhu, J.-Y., Zhou, T., and Efros, A. A. (2017). Image-to-image translation with conditional adversarial networks. In Proceedings of the IEEE conference on computer vision and pattern recognition, pages 1125–1134.

Isola, P., Zhu, J. Y., Zhou, T., & Efros, A. A. (2017). Image-to-image translation with conditional adversarial networks. In *Proceedings of the IEEE conference on computer vision and pattern recognition* (pp. 1125-1134).

Isola, P., Zhu, J.-Y., Zhou, T., & Efros, A. A. (2018). Image-to-Image Translation with Conditional Adversarial Networks. Image-to-Image Translation with Conditional Adversarial Networks. Retrieved from https://arxiv.org/abs/1611.07004

Jafarigol, E., & Trafalis, T. B. (2024). A distributed approach to meteorological predictions: Addressing data imbalance in precipitation prediction models through federated learning and GANs. *Computational Management Science*, *21*(1), 22. DOI: 10.1007/s10287-024-00504-3

Janga, B., Asamani, G. P., Sun, Z., & Cristea, N. (2023). A Review of Practical AI for Remote Sensing in Earth Sciences. *Remote Sensing (Basel)*, *2023*(15), 4112. DOI: 10.3390/rs15164112

Javaid, M., Haleem, A., Khan, I. H., & Suman, R. (2023). Understanding the potential applications of Artificial Intelligence in Agriculture Sector. *Advanced Agrochem.*, *2*(1), 15–30. DOI: 10.1016/j.aac.2022.10.001

Javaid, M., Haleem, A., Singh, R. P., & Suman, R. (2022). Enhancing smart farming through the applications of Agriculture 4.0 technologies. *International Journal of Intelligent Networks.*, *3*, 150–164. DOI: 10.1016/j.ijin.2022.09.004

Jebara, T. (2012). *Machine learning: discriminative and generative* (Vol. 755). Springer Science & Business Media.

Jeong, J., & Shin, J. (2021). Training gans with stronger augmentations via contrastive discriminator. *arXiv preprint arXiv:2103.09742.*

Jia, X., Wang, M., Khandelwal, A., Karpatne, A., & Kumar, V. (2019, August). Recurrent Generative Networks for Multi-Resolution Satellite Data: An Application in Cropland Monitoring. In IJCAI.

Jiang, B., Chong, H., Tan, Z., An, H., Yin, H., Chen, S., Yin, Y., & Chen, X. (2024). FDT-Net: Deep-Learning Network for Thin-Cloud Removal in Remote Sensing Image Using Frequency-Domain Training Strategy. *IEEE Geoscience and Remote Sensing Letters*, *21*, 1–5. DOI: 10.1109/LGRS.2024.3389995

Jiang, C., Zhao, J., Ding, Y., & Li, G. (2023). Vis–NIR spectroscopy combined with GAN data augmentation for predicting soil nutrients in degraded Alpine Meadows on the Qinghai–Tibet Plateau. *Sensors (Basel), 23*(7), 3686. DOI: 10.3390/s23073686 PMID: 37050746

Jinqi, H. (2024). Thin Cloud Removal Generative Adversarial Network Based on Sparse Transformer in Remote Sensing Images. *Remote Sensing (Basel), 16*(19), 3658–3658. DOI: 10.3390/rs16193658

Johnson, B. A. and Ma, L. (2020). Image segmentation and object-based image analysis for environmental monitoring: Recent areas of interest, researchers' views on the future priorities.

Johnson, R. B., & Onwuegbuzie, A. J. (2004). Mixed methods research: A research paradigm whose time has come. *Educational Researcher, 33*(7), 14–26. DOI: 10.3102/0013189X033007014

Jordan, M. I., & Mitchell, T. M. (2015). Machine learning: Trends, perspectives, and prospects. *Science, 349*(6245), 255–260. DOI: 10.1126/science.aaa8415 PMID: 26185243

Jozdani, S., Chen, D., Pouliot, D., & Johnson, B. A. (2022). A review and meta-analysis of generative adversarial networks and their applications in remote sensing. *International Journal of Applied Earth Observation and Geoinformation, 108*, 102734. DOI: 10.1016/j.jag.2022.102734

Ju, M., Ding, C., Ren, W., Yang, Y., Zhang, D., & Guo, Y. J. (2021). Ide: Image dehazing and exposure using an enhanced atmospheric scattering model. *IEEE Transactions on Image Processing, 30*, 2180–2192. DOI: 10.1109/TIP.2021.3050643 PMID: 33476267

Kaewunruen, S., AbdelHadi, M., Kongpuang, M., Pansuk, W., & Remennikov, A. M. (2022). Digital twins for managing railway bridge maintenance, resilience, and climate change adaptation. *Sensors (Basel), 23*(1), 252. DOI: 10.3390/s23010252 PMID: 36616851

Karamouz, M., Alipour, R. S., Roohinia, M., & Fereshtehpour, M. (2022). A remote sensing driven soil moisture estimator: Uncertain downscaling with geostatistically based use of ancillary data. *Water Resources Research, 58*(10), e2022WR031946. DOI: 10.1029/2022WR031946

Karras, T., Aila, T., Laine, S., & Lehtinen, J. (2017). Progressive growing of gans for improved quality, stability, and variation. *arXiv preprint arXiv:1710.10196.*

Karras, T., Aila, T., Laine, S., & Lehtinen, J. (2017). Progressive growing of GANs for improved quality, stability, and variation. arXiv preprint arXiv:1710.10196.

Karras, T., Aittala, M., Hellsten, J., Laine, S., Lehtinen, J., & Aila, T. (2020). Training generative adversarial networks with limited data. *Advances in Neural Information Processing Systems*, *33*, 12104–12114.

Karras, T., Aittala, M., Laine, S., Härkönen, E., Hellsten, J., Lehtinen, J., & Aila, T. (2021). Alias-free generative adversarial networks. *Advances in Neural Information Processing Systems*, *34*, 852–863.

Karras, T., Laine, S., & Aila, T. (2019), A style-based generator architecture for generative adversarial networks. In *Proceedings of the IEEE/CVF conference on computer vision and pattern recognition* (pp. 4401-4410). DOI: 10.1109/CVPR.2019.00453

Karras, T., Laine, S., Aittala, M., Hellsten, J., Lehtinen, J., & Aila, T. (2020). Analyzing and improving the image quality of stylegan. In *Proceedings of the IEEE/CVF conference on computer vision and pattern recognition* (pp. 8110-8119). DOI: 10.1109/CVPR42600.2020.00813

Karunathilake, E. M. B. M., Le, A. T., Heo, S., Chung, Y. S., & Mansoor, S. (2023). Chung, Y.S.; Mansoor, S. The Path to Smart Farming: Innovations and Opportunities in Precision Agriculture. *Agriculture*, *13*(8), 1593. DOI: 10.3390/agriculture13081593

Kasula, V. K., Yadulla, A. R., Konda, B., & Yenugula, M. (2024). Fortifying cloud environments against data breaches: A novel AI-driven security framework. *World Journal of Advanced Research and Reviews*, *24*(01), 1613–1626. DOI: 10.30574/wjarr.2024.24.1.3194

Kazeminia, S., Baur, C., Kuijper, A., Ginneken, B., Navab, N., Albarqouni, S., & Mukhopadhyay, A. (2018). GANs for Medical Image Analysis. *Artificial Intelligence in Medicine*, *109*, 101938. DOI: 10.1016/j.artmed.2020.101938 PMID: 34756215

Kazmi, S. M. K. A., Aafaq, N., Khan, M. A., Khalil, M., & Saleem, A. (2023). From Pixel to Peril: Investigating Adversarial Attacks on Aerial Imagery Through Comprehensive Review and Prospective Trajectories. *IEEE Access : Practical Innovations, Open Solutions*, *11*, 81256–81278. DOI: 10.1109/ACCESS.2023.3299878

Kemp, T. (2024). Soil Analysis via Remote Sensing and Artificial Intelligence for Precision Regenerative Agriculture. Toronto Metropolitan University. Thesis. https://doi.org/DOI: 10.32920/26060842

Kerdegari, H., Razaak, M., Argyriou, V., & Remagnino, P. (2019). Semi-supervised gan for classification of multispectral imagery acquired by uavs. arXiv preprint arXiv:1905.10920.

Khan, J. A., Abu Bakr, M., Aslam, S., & Saeed, A. Recent Advancements and Challenges in Deep Learning and Internet of Things for Precision Agriculture. Sumera and Saeed, Adnan, Recent Advancements and Challenges in Deep Learning and Internet of Things for Precision Agriculture.

Khan, M. J., Rath, S., & Zaib, M. H. (2024, April). Privacy-Enhanced Image Restoration in Remote Sensing via Federated Learning. In 2024 12th International Symposium on Digital Forensics and Security (ISDFS) (pp. 1-6). IEEE.

Khan, A., Patil, C. H., Vibhute, A. D., & Mali, S. (2024). A U-Net Based Approach for High-Accuracy Land Use Land Cover Classification in Hyperspectral Remote Sensing. In Patel, K. K., Santosh, K., Patel, A., & Ghosh, A. (Eds.), *Soft Computing and Its Engineering Applications. icSoftComp 2023. Communications in Computer and Information Science* (Vol. 2031). Springer., DOI: 10.1007/978-3-031-53728-8_8

Khan, N., Ray, R. L., Sargani, G. R., Ihtisham, M., Khayyam, M., & Ismail, S. (2021). Current progress and future prospects of agriculture technology: Gateway to sustainable agriculture. *Sustainability (Basel)*, *13*(9), 4883. DOI: 10.3390/su13094883

Khetkeeree, S., Petchthaweetham, B., Liangrocapart, S., & Srisuk, S. (2020). Sentinel-2 Image Dehazing using Correlation between Visible and Infrared Bands. 2020 8th International Electrical Engineering Congress (iEECON), (pp. 1-4). DOI: 10.1109/iEECON48109.2020.229585

Kim, E., Cho, H.-H., Kwon, J., Oh, Y.-T., Ko, E. S., & Park, H. (2022). Tumor attentive segmentation-guided gan for synthesizing breast contrast-enhanced mri without contrast agents. *IEEE Journal of Translational Engineering in Health and Medicine*, *11*, 32–43. DOI: 10.1109/JTEHM.2022.3221918 PMID: 36478773

Kim, S. E., & Seo, I. W. (2015). Artificial neural network ensemble modeling with exploratory factor analysis for streamflow forecasting. *Journal of Hydroinformatics*, *17*(4), 614–639. DOI: 10.2166/hydro.2015.033

Kim, T., Cha, M., Kim, H., Lee, J. K., & Kim, J. (2017). Learning to discover cross-domain relations with generative adversarial networks. In *International conference on machine learning* (pp. 1857-1865). PMLR.

Kirova, V. D., Ku, C. S., Laracy, J. R., & Marlowe, T. J. (2023). The ethics of artificial intelligence in the era of generative AI. *Journal of Systemics, Cybernetics and Informatics*, *21*(4), 42–50. DOI: 10.54808/JSCI.21.04.42

Klair, Y. S., Agrawal, K., & Kumar, A. (2024, March). Impact of generative ai in diagnosing diseases in agriculture. In 2024 2nd International Conference on Disruptive Technologies (ICDT) (pp. 870-875). IEEE.

Klopries, H., & Schwung, A. (2024). ITF-GAN: Synthetic time series dataset generation and manipulation by interpretable features. *Knowledge-Based Systems, 283*, 111131. DOI: 10.1016/j.knosys.2023.111131

Kong, J., Hu, G., & Liang, D. (2010). Thin Cloud Removing Approach of Color Remote Sensing Image Based on Support Vector Machine. *2010 Asia-Pacific Conference on Wearable Computing Systems*, (pp. 131-135). DOI: 10.1109/APWCS.2010.39

Koochali, A., Bakhshipour, A. E., Bakhshizadeh, M., Habermehl, R., Dilly, T. C., Dittmer, U., Ahmed, S., Haghighi, A., & Dengel, A. (2024). Water depth prediction in combined sewer networks, application of generative adversarial networks. *Discover Applied Sciences*, *6*(3), 123. DOI: 10.1007/s42452-024-05787-4

Kossale, Y., Airaj, M., & Darouichi, A. (2022). Mode collapse in generative adversarial networks: An overview. In *2022 8th International Conference on Optimization and Applications (ICOA)* (pp. 1-6). IEEE.

Kouadio, L., Deo, R. C., Byrareddy, V., Adamowski, J. F., Mushtaq, S., & Phuong Nguyen, V. (2018). Artificial intelligence approach for the prediction of Robusta coffee yield using soil fertility properties. *Computers and Electronics in Agriculture*, *155*, 324–338. DOI: 10.1016/j.compag.2018.10.014

Kulkarni, P., Pathak, P., Pillai, S., & Tigga, V. (2025). Role of Generative AI for Fraud Detection and Prevention. *Generative Artificial Intelligence in Finance: Large Language Models, Interfaces, and Industry Use Cases to Transform Accounting and Finance Processes*, 175-198.

Kumaran, U., Thangam, S., Prabhakar, T. N., Selvaganesan, J., & Vishwas, H. N. (2023). Adversarial Defense: A GAN-IF Based Cyber-security Model for Intrusion Detection in Software Piracy. *Journal of Wireless Mobile Networks, Ubiquitous Computing and Dependable Applications*, *14*(4), 96–114. DOI: 10.58346/JOWUA.2023.I4.008

Kumar, L., Chaudhary, S., Karir, S. R., Díaz-Raviña, M., Choudhary, A. K., & Kumar, S. G. (2024). *The prologue to precision livestock farming*., DOI: 10.58532/V3BCAG21P2CH11

Kumar, P., Bhatnagar, R., Gaur, K., & Bhatnagar, A. (2021). Classification of imbalanced data: Review of methods and applications. [IOP Publishing.]. *IOP Conference Series. Materials Science and Engineering*, *1099*(1), 012077. DOI: 10.1088/1757-899X/1099/1/012077

Kunkel, L., & Trabs, M. (2024). A Wasserstein perspective of Vanilla GANs. *arXiv preprint arXiv:2403.15312*.

Kuntalp, M., & Düzyel, O. (2024). A new method for GAN-based data augmentation for classes with distinct clusters. *Expert Systems with Applications*, *235*, 121199. DOI: 10.1016/j.eswa.2023.121199

Kurach, K., Lucic, M., Zhai, X., Michalski, M., & Gelly, S. (2018). The gan landscape: Losses, architectures, regularization, and normalization.

Kwon, G., Han, C., & Kim, D. S. (2019, October). Generation of 3D brain MRI using auto-encoding generative adversarial networks. In *International Conference on Medical Image Computing and Computer-Assisted Intervention* (pp. 118-126). Cham: Springer International Publishing. DOI: 10.1007/978-3-030-32248-9_14

Lakshmi, G., Brindha, S., Sathish, V., Kumar, V. M., & Santhoshkumar, P. (2024, April). Advancing Precision Agriculture: AI, IoT, and Machine Learning Solutions for Optimal Crop Yield. In 2024 International Conference on Communication, Computing and Internet of Things (IC3IoT) (pp. 1-5). IEEE.

Lanucara, S., Praticò, S., Pioggia, G., Di Fazio, S., & Modica, G. (2024). Web-based spatial decision support system for precision agriculture: A tool for delineating dynamic management unit zones (MUZs). *Smart Agricultural Technology*, *8*, 100444. DOI: 10.1016/j.atech.2024.100444

Ledig, C., Theis, L., Huszár, F., Caballero, J., Cunningham, A., Acosta, A., & Shi, W. (2017). Photo-realistic single image super-resolution using a generative adversarial network. In *Proceedings of the IEEE conference on computer vision and pattern recognition* (pp. 4681-4690). DOI: 10.1109/CVPR.2017.19

Lei, D., Zhu, Q., Li, Y., Tan, J., Wang, S., Zhou, T., & Zhang, L. (2024). HPLTS-GAN: A High-Precision Remote Sensing Spatio-Temporal Fusion Method Based on Low Temporal Sensitivity. *IEEE Transactions on Geoscience and Remote Sensing*, *1*, 1–16. Advance online publication. DOI: 10.1109/TGRS.2024.3416152

Li, B., Peng, X., Wang, Z., Xu, J., & Feng, D. (2017). AOD-Net: All-in-One Dehazing Network. *2017 IEEE International Conference on Computer Vision (ICCV)*, (pp. 4780-4788). DOI: 10.1109/ICCV.2017.511

Li, J., Shen, Y., & Yang, C. (2020). An adversarial generative network for crop classification from remote sensing timeseries images. *Remote Sensing (Basel)*, *13*(1), 65. DOI: 10.3390/rs13010065

Lim, W., Kelvin, S. C. Y., Lau, B. T., & Colin, C. L. T. (2024), Future of generative adversarial networks (GAN) for anomaly detection in network security: A review, Published by Elsevier Ltd, ISSN 0167-4048.

Linaza, M. T., Posada, J., Bund, J., Eisert, P., Quartulli, M., Döllner, J., Pagani, A. G., Olaizola, I., Barriguinha, A., Moysiadis, T., & Lucat, L. (2021). Data-Driven Artificial Intelligence Applications for Sustainable Precision Agriculture. *Agronomy (Basel)*, *11*, 1227. DOI: 10.3390/agronomy11061227

Lin, D., Xu, G., Wang, X., Wang, Y., Sun, X., & Fu, K. (2019). A Remote Sensing Image Dataset for Cloud Removal. CoRR, abs/1901.00600. Retrieved from http://arxiv.org/abs/1901.00600

Lin, Q., Fang, Z., Chen, Y., Tan, K. C., & Li, Y. (2022). Evolutionary architectural search for generative adversarial networks. *IEEE Transactions on Emerging Topics in Computational Intelligence*, *6*(4), 783–794. DOI: 10.1109/TETCI.2021.3137377

Lipton, Z. C. (2016). The mythos of model interpretability. *Communications of the ACM*, *59*(10), 36–43. DOI: 10.1145/3233231

Li, S., Dragicevic, S., Castro, F. A., Sester, M., Winter, S., Coltekin, A., Pettit, C., Jiang, B., Haworth, J., Stein, A., & Cheng, T. (2016). Geospatial big data handling theory and methods: A review and research challenges. *ISPRS Journal of Photogrammetry and Remote Sensing*, *115*, 119–133. DOI: 10.1016/j.isprsjprs.2015.10.012

Little, C., Elliot, M., Allmendinger, R., & Samani, S. S. (2021). Generative adversarial networks for synthetic data generation: a comparative study. *arXiv preprint arXiv:2112.01925*.

Liu, H., Gu, X., & Samaras, D. (2019). Wasserstein gan with quadratic transport cost. In *Proceedings of the IEEE/CVF international conference on computer vision* (pp. 4832-4841).

Liu, J. J., Yao, J. P., Liu, J. H., Wang, Z. Y., & Huang, L. (2024). Missing data imputation and classification of small sample missing time series data based on gradient penalized adversarial multi-task learning. *Applied Intelligence*, *54*(3), 2528–2550. DOI: 10.1007/s10489-024-05314-3

Liu, J., Hou, W., Luo, X., Su, J., Hou, Y., & Wang, Z. (2022). SI-SA GAN: A Generative Adversarial Network Combined With Spatial Information and Self-Attention for Removing Thin Cloud in Optical Remote Sensing Images. *IEEE Access : Practical Innovations, Open Solutions*, *10*, 114318–114330. DOI: 10.1109/ACCESS.2022.3213354

Liu, L., Lei, S., Shi, Z., Zhang, N., & Zhu, X. (2021). Hyperspectral Remote Sensing Imagery Generation From RGB Images Based on Joint Discrimination. *IEEE Journal of Selected Topics in Applied Earth Observations and Remote Sensing*, *14*, 7624–7636. DOI: 10.1109/JSTARS.2021.3099242

Liu, P. (2023). A review on remote sensing data fusion with generative adversarial networks (gan). *Authorea Preprints*.

Liu, P. (2023). A review on remote sensing data fusion with generative adversarial networks (GAN). *Authorea Preprints*.

Liu, P., Li, J., Wang, L., & He, G. (2022). Remote sensing data fusion with generative adversarial networks: State-of-the-art methods and future research directions. *IEEE Geoscience and Remote Sensing Magazine*, *10*(2), 295–328. DOI: 10.1109/MGRS.2022.3165967

Liu, T., Liu, H., Li, Y.-F., Chen, Z., Zhang, Z., & Liu, S. (2019). Flexible ftir spectral imaging enhancement for industrial robot infrared vision sensing. *IEEE Transactions on Industrial Informatics*, *16*(1), 544–554. DOI: 10.1109/TII.2019.2934728

Liu, Y., Liu, Y., & Wang, J. (2023). RFM-GAN: Robust feature matching with GAN-based neighborhood representation for agricultural remote sensing image registration. *IEEE Geoscience and Remote Sensing Letters*, *20*, 1–5.

Liu, Y., Wang, W., Fang, F., Zhou, L., Sun, C., Zheng, Y., & Chen, Z. (2021). Cscgan: Conditional scale-consistent generation network for multi-level remote sensing image to map translation. *Remote Sensing (Basel)*, *13*(10), 1936. DOI: 10.3390/rs13101936

Li, X., Yuan, Y., Liu, G., Han, Z., & Stouffs, R. (2024). A predictive model for daylight performance based on multimodal generative adversarial networks at the early design stage. *Energy and Building*, *305*, 113876. DOI: 10.1016/j.enbuild.2023.113876

Li, X., Zhang, H., Yang, H., & Li, T.-Q. (2023). Cs-mri reconstruction using an improved gan with dilated residual networks and channel attention mechanism. *Sensors (Basel)*, *23*(18), 7685. DOI: 10.3390/s23187685 PMID: 37765747

Li, Y., Ma, J., & Zhang, Y. (2021). Image retrieval from remote sensing big data: A survey. *Information Fusion*, *67*, 94–115. DOI: 10.1016/j.inffus.2020.10.008

Li, Y., Zhang, Y., & Wang, Y. (2020). A GAN-based approach for cloud removal in remote sensing images. *Remote Sensing*, *12*(3), 456.

Li, Z., Shi, S., Wang, L., Xu, M., & Li, L. (2022). Unsupervised Generative Adversarial Network with Background Enhancement and Irredundant Pooling for Hyperspectral Anomaly Detection. *Remote Sensing (Basel)*, *14*(5), 1265. Advance online publication. DOI: 10.3390/rs14051265

Lu, B., Dao, P. D., Liu, J., He, Y., & Shang, J. (2020). Recent advances of hyperspectral imaging technology and applications in agriculture. *Remote Sensing (Basel)*, *12*(16), 2659. DOI: 10.3390/rs12162659

Luleci, F., Catbas, F. N., & Avci, O. (2022). Generative adversarial networks for data generation in structural health monitoring. *Frontiers in Built Environment, 8*, 816644. DOI: 10.3389/fbuil.2022.816644

Luo, J., Zhu, L., Li, Q., Liu, D., & Chen, M. (2021). Imbalanced fault diagnosis of rotating machinery based on deep generative adversarial networks with gradient penalty. *Processes (Basel, Switzerland), 9*(10), 1751. DOI: 10.3390/pr9101751

Luo, Y., Guo, X., Wang, L., Zheng, J., Liu, J., & Liao, F. (2023). Unsupervised structural damage detection based on an improved generative adversarial network and cloud model. *Journal of Low Frequency Noise, Vibration and Active Control, 42*(3), 1501–1518. DOI: 10.1177/14613484221150804

Lu, Y., Chen, D., Olaniyi, E., & Huang, Y. (2022). Generative adversarial networks (gans) for image augmentation in agriculture: A systematic review. *Computers and Electronics in Agriculture, 200*, 107208. DOI: 10.1016/j.compag.2022.107208

Lyu, H., Qian, J., & Wang, Y. (2024, July). Study on the Unified Theory of Thin Cloud Detection and Removal based on Physical Model. In IGARSS 2024-2024 IEEE International Geoscience and Remote Sensing Symposium (pp. 5549-5552). IEEE.

Mahanta, D. K., Bhoi, T. K., Komal, J., Samal, I., & Mastinu, A. (2024). Spatial, spectral and temporal insights: Harnessing high-resolution satellite remote sensing and artificial intelligence for early monitoring of wood boring pests in forests. *Plant Stress, 11*, 100381. DOI: 10.1016/j.stress.2024.100381

Maldonado-Romo, J., Aldape-Perez, M., & Rodriguez-Molina, A. (2021). Analysis of depth and semantic mask for perceiving a physical environment using virtual samples generated by a gan. *IEEE Access : Practical Innovations, Open Solutions, 10*, 5595–5607. DOI: 10.1109/ACCESS.2021.3137797

Manikandababu, C. S., Preethi, V., Kanna, M. Y., Vedhathiri, K., & Kumar, S. S. (2024, May). Enhancing Crop Yield Prediction with IoT and Machine Learning in Precision Agriculture. In *2024 International Conference on Advances in Computing, Communication and Applied Informatics (ACCAI)* (pp. 1-6). IEEE.

Mansourifar, H., Moskovitz, A., Klingensmith, B., Mintas, D., & Simske, S. J. (2022). Gan-based satellite imaging: A survey on techniques and applications. *IEEE Access : Practical Innovations, Open Solutions, 10*, 118123–118140. DOI: 10.1109/ACCESS.2022.3221123

Mao, X., Li, Q., Xie, H., Lau, R. Y., Wang, Z., & Paul Smolley, S. (2017). Least squares generative adversarial networks. In *Proceedings of the IEEE international conference on computer vision* (pp. 2794-2802).

Maqsood, M. H., Mumtaz, R., Haq, I. U., Shafi, U., Zaidi, S. M. H., & Hafeez, M. (2021). Super resolution generative adversarial network (Srgans) for wheat stripe rust classification. *Sensors (Basel), 21*(23), 7903. DOI: 10.3390/s21237903 PMID: 34883905

Mash, H. E., & Weavers, L. K. (2011). Water resources. *Water Research,* 45(1), 1-10. Shorten, C., & Khoshgoftaar, T. M. (2019). A survey on image data augmentation for deep learning. *Journal of Big Data, 6*(1), 1–48.

Ma, T., Ma, J., Yu, K., Zhang, J., & Fu, W. (2021). Multispectral Remote Sensing Image Matching via Image Transfer by Regularized Conditional Generative Adversarial Networks and Local Feature. *IEEE Geoscience and Remote Sensing Letters, 18*(2), 351–355. DOI: 10.1109/LGRS.2020.2972361

Ma, X. (2024, February). A comparison of art style transfer in Cycle-GAN based on different generators. [). IOP Publishing.]. *Journal of Physics: Conference Series, 2711*(1), 012006. DOI: 10.1088/1742-6596/2711/1/012006

Ma, Y., Zhang, Y., & Wang, Y. (2020). Generative adversarial networks for remote sensing applications: A review. *Remote Sensing, 12*(1), 1–20.

McAmis, R., Sim, M., Bennett, M., & Kohno, T. (2024). Over fences and into yards: Privacy threats and concerns of commercial satellites. Proceedings on Privacy Enhancing Technologies.

Meadows, M., & Wilson, M. (2021). A comparison of machine learning approaches to improve free topography data for flood modelling. *Remote Sensing (Basel), 13*(2), 275. DOI: 10.3390/rs13020275

Mgendi, G. (2024). Unlocking the potential of precision agriculture for sustainable farming. *Discov Agric., 2*(1), 87. DOI: 10.1007/s44279-024-00078-3

Miftahushudur, T., Grieve, B., & Yin, H. (2024). Permuted KPCA and SMOTE to Guide GAN-Based Oversampling for Imbalanced HSI Classification. *IEEE Journal of Selected Topics in Applied Earth Observations and Remote Sensing, 17*, 489–505. DOI: 10.1109/JSTARS.2023.3326963

Miletic, M., & Sariyar, M. (2024). Challenges of Using Synthetic Data Generation Methods for Tabular Microdata. *Applied Sciences (Basel, Switzerland), 14*(14), 5975. DOI: 10.3390/app14145975

Millière, R. (2022). Deep learning and synthetic media. *Synthese, 200*(3), 231. DOI: 10.1007/s11229-022-03739-2

Mirza, M. (2014). Conditional generative adversarial nets. arXiv preprint arXiv:1411.1784.

Mirza, M., & Osindero, S. (2014). Conditional generative adversarial nets. *arXiv preprint arXiv:1411.1784.*

Mirza, M., & Osindero, S. (2014). Conditional Generative Adversarial Nets. Conditional Generative Adversarial Nets. Retrieved from https://arxiv.org/abs/1411.1784

Misra, V., & Mall, A. K. (2024). Harnessing image processing for precision disease diagnosis in sugar beet agriculture. *Crop Design.*, *3*(4), 100075. DOI: 10.1016/j.cropd.2024.100075

Modanwal, G., Vellal, A., & Mazurowski, M. A. (2021). Normalization of breast MRIs using cycle-consistent generative adversarial networks. *Computer Methods and Programs in Biomedicine*, *208*, 106225. DOI: 10.1016/j.cmpb.2021.106225 PMID: 34198016

Mohan, S., Manke, A., Verma, S., & Baskar, K. (2024). Machine Learning at the Edge: GANs for Anomaly Detection in Wireless Sensor Networks. In *Enhancing Security in Public Spaces Through Generative Adversarial Networks (GANs)* (pp. 305-317). IGI Global. https://doi.org/DOI: 10.4018/979-8-3693-3597-0.ch021

Monteiro, A., Santos, S., & Gonçalves, P. (2021). Precision agriculture for crop and livestock farming—Brief review. *Animals (Basel)*, *11*(8), 2345. DOI: 10.3390/ani11082345 PMID: 34438802

Morales-García, J., Bueno-Crespo, A., Terroso-Sáenz, F., Arcas-Túnez, F., Martínez-España, R., & Cecilia, J. M. (2023). Evaluation of synthetic data generation for intelligent climate control in greenhouses. *Applied Intelligence*, *53*(21), 24765–24781. DOI: 10.1007/s10489-023-04783-2

Moreno-Martínez, Á., Izquierdo-Verdiguier, E., Maneta, M. P., Camps-Valls, G., Robinson, N., Muñoz-Marí, J., & Running, S. W. (2020). Multispectral high resolution sensor fusion for smoothing and gap-filling in the cloud. *Remote Sensing of Environment*, *247*, 111901.

Mouniane, Y., Chriqui, A., El-Khadir, I., Hbyaj, K., El-Moustaqim, K., Mezouara, M., Zirari, M., Mabrouki, J., & Hmouni, D. (2024). Impact of Remote Sensing on Irrigation Management in Precision Agriculture. In Mabrouki, J., & Mourade, A. (Eds.), *Technical and Technological Solutions Towards a Sustainable Society and Circular Economy. World Sustainability Series*. Springer., DOI: 10.1007/978-3-031-56292-1_19

Nair, R. R., & Singh, T. (2019). Multi-sensor medical image fusion using pyramid-based DWT: A multi-resolution approach. *IET Image Processing*, *13*(9), 1447–1459. DOI: 10.1049/iet-ipr.2018.6556

Nair, R. R., Singh, T., Sankar, R., & Gunndu, K. (2021). Multi-modal medical image fusion using lmf-gan-a maximum parameter infusion technique. *Journal of Intelligent & Fuzzy Systems*, *41*(5), 5375–5386. DOI: 10.3233/JIFS-189860

Namboodiri, A. S., Kumar Sanodiya, R., & Arun, P. V. (2023). Remote Sensing Cloud Removal using a Combination of Spatial Attention and Edge Detection. 2023 11th International Symposium on Electronic Systems Devices and Computing (ESDC), 1, pp. 1-6. DOI: 10.1109/ESDC56251.2023.10149875

Naqvi, S. M. A., Shabaz, M., Khan, M. A., & Hassan, S. I. (2023). Adversarial attacks on visual objects using the fast gradient sign method. *Journal of Grid Computing*, *21*(4), 52. DOI: 10.1007/s10723-023-09684-9

Neethirajan, S. (2024). Net Zero Dairy Farming—Advancing Climate Goals with Big Data and Artificial Intelligence. *Climate (Basel)*, *12*(2), 15. DOI: 10.3390/cli12020015

Neff, T., Payer, C., Stern, D., & Urschler, M. (2017, May). Generative adversarial network based synthesis for supervised medical image segmentation. In *Proc. OAGM and ARW joint Workshop* (Vol. 3, p. 4).

Nguyen, A., ... (2015), Deep Neural Networks Are Easily Fooled: High Confidence Predictions for Unrecognizable Images, Computer Vision and Pattern Recognition, http://bit.ly/2ZKclwW

Noguchi, A., & Harada, T. (2019). Image generation from small datasets via batch statistics adaptation. In *Proceedings of the IEEE/CVF International Conference on Computer Vision* (pp. 2750-2758). DOI: 10.1109/ICCV.2019.00284

Nunn, E. J., Khadivi, P., & Samavi, S. (2021). Compound frechet inception distance for quality assessment of gan created images. *arXiv preprint arXiv:2106.08575.*

Nyberg, D. (2021). *Exploring the Capabilities of Generative Adversarial Networks in Remote Sensing Applications.* Linkoping University Electronic Press.

Öcal, A., & Özbakır, L. (2021). Supervised deep convolutional generative adversarial networks. *Neurocomputing*, *449*, 389–398. DOI: 10.1016/j.neucom.2021.03.125

Oh, J., Hong, J., & Baek, J. (2019). Oversampling method using outlier detectable generative adversarial network. *Expert Systems with Applications*, *133*, 1–8. DOI: 10.1016/j.eswa.2019.05.006

Onakpojeruo, E. P., Mustapha, M. T., Ozsahin, D. U., & Ozsahin, I. (2024). A comparative analysis of the novel conditional deep convolutional neural network model, using conditional deep convolutional generative adversarial network-generated synthetic and augmented brain tumor datasets for image classification. *Brain Sciences*, *14*(6), 559. DOI: 10.3390/brainsci14060559 PMID: 38928561

Oubara, A., Wu, F., Amamra, A., & Yang, G. (2022, May). Survey on remote sensing data augmentation: Advances, challenges, and future perspectives. In *International Conference on Computing Systems and Applications* (pp. 95-104). Cham: Springer International Publishing. DOI: 10.1007/978-3-031-12097-8_9

Pan, H. (2020). Cloud Removal for Remote Sensing Imagery via Spatial Attention Generative Adversarial Network. Cloud Removal for Remote Sensing Imagery via Spatial Attention Generative Adversarial Network.

Pan, S. J., & Yang, Q. (2009). A survey on transfer learning. *IEEE Transactions on Knowledge and Data Engineering*, *22*(10), 1345–1359. DOI: 10.1109/TKDE.2009.191

Pantelidakis, M., Panagopoulos, A. A., Mykoniatis, K., Ashkan, S., Eravi, R. C., Pamula, V., Verduzco, E. C., Babich, O., Panagopoulos, O. P., & Chalkiadakis, G. (2022). Identifying sunlit leaves using Convolutional Neural Networks: An expert system for measuring the crop water stress index of pistachio trees. *Expert Systems with Applications*, *209*, 118326. DOI: 10.1016/j.eswa.2022.118326

Pan, X., Xie, F., Jiang, Z., & Yin, J. (2015). Haze Removal for a Single Remote Sensing Image Based on Deformed Haze Imaging Model. *IEEE Signal Processing Letters*, *22*(10), 1806–1810. DOI: 10.1109/LSP.2015.2432466

Pan, X., Yang, F., Gao, L., Chen, Z., Zhang, B., Fan, H., & Ren, J. (2019). Building extraction from high-resolution aerial imagery using a generative adversarial network with spatial and channel attention mechanisms. *Remote Sensing (Basel)*, *11*(8), 917. DOI: 10.3390/rs11080917

Paolanti, M., Tiribelli, S., Giovanola, B., Mancini, A., Frontoni, E., & Pierdicca, R. (2024). Ethical framework to assess and quantify the trustworthiness of artificial intelligence techniques: Application case in remote sensing. *Remote Sensing (Basel)*, *16*(23), 4529. DOI: 10.3390/rs16234529

Paola, Z. L., Jesús, L. S., Christian, A. H., & Sonia, R. U. (2023). Correction of Banding Errors in Satellite Images With Generative Adversarial Networks (GAN). *IEEE Access : Practical Innovations, Open Solutions*, *11*, 51960–51970. DOI: 10.1109/ACCESS.2023.3279265

Park, T., Liu, M. Y., Wang, T. C., & Zhu, J. Y. (2019). Semantic image synthesis with spatially-adaptive normalization. In *Proceedings of the IEEE/CVF Conference on Computer Vision and Pattern Recognition* (pp. 2337-2346).

Patel, D., Raut, G., Cheetirala, S. N., Nadkarni, G. N., Freeman, R., Glicksberg, B. S., Klang, E., & Timsina, P. (2024). Cloud platforms for developing generative ai solutions: A scoping review of tools and services. arXiv preprint arXiv:2412.06044.

Patibandla, R., Rao, B. T., & Malla, R. (2024). *Precision Agriculture and AI*. Informa., DOI: 10.1201/9781003451648-7

Patil, A. (2021, December). DCGAN: Deep convolutional GAN with attention module for remote view classification. In *2021 International Conference on Forensics, Analytics, Big Data, Security (FABS)* (Vol. 1, pp. 1-10). IEEE. DOI: 10.1109/FABS52071.2021.9702655

Pavan Kumar, M. R., & Jayagopal, P. (2021). Generative adversarial networks: A survey on applications and challenges. *International Journal of Multimedia Information Retrieval, 10*(1), 1–24. DOI: 10.1007/s13735-020-00196-w

Pedro, J. S. V. (2021). Gilson Alexandre Ostwald Pedro da, Raul Queiroz Feitosa, Mabel Ximena, Claudio Aparecido de Almeida, Christian Heipke, Franz Rottensteiner, (2021), An unsupervised domain adaptation approach for change detection and its application to deforestation mapping in tropical biomes. *ISPRS Journal of Photogrammetry and Remote Sensing*, *181*(November), 113–128.

Pengqiang, Z. H. A. N. G., Bing, L. I. U., Xuchu, Y. U., Xiong, T. A. N., Fan, Y. A. N. G., & Zenghua, Z. H. O. U. (2020). Generative adversarial networks for hyperspectral image classification. *Bulletin of Surveying and Mapping*, (3), 29. http://tb.chinasmp.com/EN/10.13474/j.cnki.11-2246.2020.0073

Perarnau, G., Van De Weijer, J., Raducanu, B., & Álvarez, J. M. (2016). Invertible conditional gans for image editing. *arXiv preprint arXiv:1611.06355.*

Peterson, K. T., Sagan, V., Sidike, P., Cox, A., & Martinez, M. (2018). Suspended sediment concentration estimation from land sat imagery along the lower missouri and middle mississippi rivers using an extreme learning machine. *Remote Sensing (Basel), 10*(10), 1503. DOI: 10.3390/rs10101503

Peterson, K. T., Sagan, V., & Sloan, J. J. (2020). Deep learning-based water quality estimation and anomaly detection using Landsat-8/Sentinel-2 virtual constellation and cloud computing. *GIScience & Remote Sensing*, *57*(4), 510–525. DOI: 10.1080/15481603.2020.1738061

Petrović, B., Bumbálek, R., Zoubek, T., Kuneš, R., Smutný, L., & Bartoš, P. (2024). Application of precision agriculture technologies in Central Europe-review. *Journal of Agriculture and Food Research, 15*, 101048. DOI: 10.1016/j.jafr.2024.101048

Piao, S., Liu, Q., Chen, A., Janssens, I. A., Fu, Y., Dai, J., Liu, L., Lian, X., Shen, M., & Zhu, X. (2019). Plant phenology and global climate change: Current progresses and challenges. *Global Change Biology, 25*(6), 1922–1940. DOI: 10.1111/gcb.14619 PMID: 30884039

Pingi, S. T., Zhang, D., Bashar, M. A., & Nayak, R. (2024). Joint representation learning with generative adversarial imputation network for improved classification of longitudinal data. *Data Science and Engineering, 9*(1), 5–25. DOI: 10.1007/s41019-023-00232-9

Pomar, C., & Remus, A. (2023). Review: Fundamentals, limitations and pitfalls on the development and application of precision nutrition techniques for precision livestock farming. *Animal, 100763*, 100763. Advance online publication. DOI: 10.1016/j.animal.2023.100763 PMID: 36966025

Potts, J., Jangra, S., Michael, V. N., & Wu, X. (2023). Speed breeding for crop improvement and food security. *Crops (Basel), 3*(4), 276–291. DOI: 10.3390/crops3040025

Prabha, D., Subramanian, R. S., Dinesh, M. G., & Girija, P. (2024). Sustainable Farming Through AI-Enabled Precision Agriculture. In *Artificial Intelligence for Precision Agriculture* (pp. 159–182). Auerbach Publications. DOI: 10.1201/9781003504900-8

Prinsa, K., & Saritha, E. (2018). Detection and Removal of Clouds on Remote Sensing Images. 2018 4th International Conference for Convergence in Technology (I2CT), (pp. 1-5). DOI: 10.1109/I2CT42659.2018.9057879

Pulakurthi, P. R., Mozaffari, M., Dianat, S. A., Rabbani, M., Heard, J., & Rao, R. (2024). Enhancing GAN performance through neural architecture search and tensor decomposition. *ICASSP 2022 - 2022 IEEE International Conference on Acoustics, Speech and Signal Processing (ICASSP)*, 7280–7284. DOI: 10.1109/ICASSP48485.2024.10446488

Purwono, P., Wulandari, A. N. E., Ma'arif, A., & Salah, W. A. (2025). Understanding Generative Adversarial Networks (GANs): A Review. *Control Systems and Optimization Letters, 3*(1), 36–45.

Pyroistech (n.d.). Hyperspectral and multispectral imaging. Accessed: 2024-12 22.

Qiao, L., Wang, X., Smith, P., Fan, J., Lu, Y., Emmett, B., Li, R., Dorling, S., Chen, H., Liu, S., Benton, T. G., Wang, Y., Ma, Y., Jiang, R., Zhang, F., Piao, S., Müller, C., Yang, H., Hao, Y., & Fan, M. (2022). Soil quality both increases crop production and improves resilience to climate change. *Nature Climate Change*, *12*(6), 574–580. DOI: 10.1038/s41558-022-01376-8

Qi, C., Gao, J., Chen, K., Shu, L., & Pearson, S. (2022). Tea chrysanthemum detection by leveraging generative adversarial networks and edge computing. *Frontiers in Plant Science*, *13*, 850606. DOI: 10.3389/fpls.2022.850606 PMID: 35463441

Qin, Z., Liu, Z., Zhu, P., & Xue, Y. (2020). A GAN-based image synthesis method for skin lesion classification. *Computer Methods and Programs in Biomedicine*, *195*, 105568. DOI: 10.1016/j.cmpb.2020.105568 PMID: 32526536

Radford, A. (2015). Unsupervised representation learning with deep convolutional generative adversarial networks. *arXiv preprint arXiv:1511.06434*.

Radford, A., Metz, L., & Chintala, S. (2015). Unsupervised representation learning with deep convolutional generative adversarial networks. *arXiv preprint arXiv:1511.06434*.

Radford, A., Metz, L., & Chintala, S. (2016). Unsupervised representation learning with deep convolutional generative adversarial networks. In Y. Bengio & Y. LeCun (Eds.), 4th International Conference on Learning Representations, ICLR 2016, San Juan, Puerto Rico, May 2-4, 2016, Conference Track Proceedings.

Radford, A., Metz, L., & Chintala, S. (2016). Unsupervised Representation Learning with Deep Convolutional Generative Adversarial Networks. Unsupervised Representation Learning with Deep Convolutional Generative Adversarial Networks. Retrieved from https://arxiv.org/abs/1511.06434

Rani, V., Nabi, S. T., Kumar, M., Mittal, A., & Kumar, K. (2023). Self-supervised Learning: A Succinct Review. *Archives of Computational Methods in Engineering*, *2023*(30), 2761–2775. DOI: 10.1007/s11831-023-09884-2 PMID: 36713767

Rastogi, R., Rawat, V., & Kaushal, S. (2025). Advancements in Image Restoration Techniques: A Comprehensive Review and Analysis through GAN. *Generative Artificial Intelligence and Ethics: Standards, Guidelines, and Best Practices*, 53-90.

Ratnaparkhi, N. S., Nagne, A. D., & Gawali, B. (2014). A land use land cover classification system using remote sensing data. changes, 7(8).

Ray, A., Kolekar, M. H., Balasubramanian, R., & Hafiane, A. (2023). Trans fer learning enhanced vision-based human activity recognition: A decade-long analysis. *International Journal of Information Management Data Insights*, *3*(1), 100142. DOI: 10.1016/j.jjimei.2022.100142

Reed, S., Akata, Z., Yan, X., Logeswaran, L., Schiele, B., & Lee, H. (2016). Generative adversarial text to image synthesis. In *International conference on machine learning* (pp. 1060-1069). PMLR.

Reed, S., Akata, Z., Yan, X., Logeswaran, L., Schiele, B., & Lee, H. (2016). Generative adversarial text to image synthesis. In *International Conference on Machine Learning* (pp. 1060-1069). PMLR.

Rewhel, E. M., Li, J., Hamed, A. A., Keshk, H. M., Mahmoud, A. S., Sayed, S. A., Samir, E., Zeyada, H. H., Mohamed, S. A., Moustafa, M. S., Nasr, A. H., & Helmy, A. K. (2023). Deep Learning Methods Used in Remote Sensing Images: A Review. *Journal of Environment and Earth Science*, *5*(1), 33–64. DOI: 10.30564/jees.v5i1.5232

Riaz, K., McAfee, M., & Gharbia, S. S. (2023). Management of climate resilience: Exploring the potential of digital twin technology, 3D city modelling, and early warning systems. *Sensors (Basel)*, *23*(5), 2659. DOI: 10.3390/s23052659 PMID: 36904867

Rocchini, D., Luque, S., Pettorelli, N., Bastin, L., Doktor, D., Faedi, N., Feilhauer, H., Féret, J.-B., Foody, G. M., Gavish, Y., Godinho, S., Kunin, W. E., Lausch, A., Leitão, P. J., Marcantonio, M., Neteler, M., Ricotta, C., Schmidtlein, S., Vihervaara, P., & Nagendra, H. (2018). Measuring β-diversity by remote sensing: A challenge for biodiversity monitoring. *Methods in Ecology and Evolution*, *9*(8), 1787–1798. DOI: 10.1111/2041-210X.12941

Rodríguez-Suárez, B., Quesada-Barriuso, P., & Argüello, F. (2022). Design of CGAN models for multispectral reconstruction in remote sensing. *Remote Sensing (Basel)*, *14*(4), 816. DOI: 10.3390/rs14040816

Rodriguez-Vazquez, J., Fernandez-Cortizas, M., Perez-Saura, D., Molina, M., & Campoy, P. (2023). Overcoming domain shift in neural networks for accurate plant counting in aerial images. *Remote Sensing (Basel)*, *15*(6), 1700. DOI: 10.3390/rs15061700

Rožanec, J. M., Zajec, P., Theodoropoulos, S., Koehorst, E., Fortuna, B., & Mladenić, D. (2023). Synthetic data augmentation using GAN for improved automated visual inspection. *IFAC-PapersOnLine*, *56*(2), 11094–11099. DOI: 10.1016/j.ifacol.2023.10.817

Rui, X., Cao, Y., Yuan, X., Kang, Y., & Song, W. (2021). Disastergan: Generative adversarial networks for remote sensing disaster image generation. *Remote Sensing (Basel), 13*(21), 4284. DOI: 10.3390/rs13214284

Saad, M. M., Rehmani, M. H., & O'Reilly, R. (2022a, December). A self-attention guided multi-scale gradient GAN for diversified x-ray image synthesis. In *Irish Conference on Artificial Intelligence and Cognitive Science* (pp. 18-31). Cham: Springer Nature Switzerland.

Saad, M. M., Rehmani, M. H., & O'Reilly, R. (2022b, July). Addressing the intra-class mode collapse problem using adaptive input image normalization in gan-based x-ray images. In *2022 44th annual international conference of the IEEE Engineering in Medicine & Biology Society (EMBC)* (pp. 2049-2052). IEEE. DOI: 10.1109/EMBC48229.2022.9871260

Saad, M. M., O'Reilly, R., & Rehmani, M. H. (2024). A survey on training challenges in generative adversarial networks for biomedical image analysis. *Artificial Intelligence Review, 57*(2), 19. DOI: 10.1007/s10462-023-10624-y

Sabins, F. F.Jr, & Ellis, J. M. (2020). *Remote sensing: Principles, interpretation, and applications*. Waveland Press.

Sabnam, S., & Rajagopal, S. (2024). Application of generative adversarial networks in image, face reconstruction and medical imaging: Challenges and the current progress. *Computer Methods in Biomechanics and Biomedical Engineering. Imaging & Visualization, 12*(1), 2330524.

Sachithra, V., & Subhashini, L. D. C. S. (2023). How artificial intelligence uses to achieve the agriculture sustainability: Systematic review. *Artificial Intelligence in Agriculture., 8*, 46–59. DOI: 10.1016/j.aiia.2023.04.002

Sagan, V., Peterson, K. T., Maimaitijiang, M., Sidike, P., Sloan, J., Greeling, B. A., & Adams, C. (2020). Monitoring inland water quality using remote sensing: Potential and limitations of spectral indices, bio-optical simulations, machine learning, and cloud computing. *Earth-Science Reviews, 205*, 103187. DOI: 10.1016/j.earscirev.2020.103187

Sahoo, G. S., Dass, S., & Prakash, F. (2023, November). The Synergy of Advanced Language Models, Sequential Skeleton Features, and GAN-Enhanced Data Augmentation for Unprecedented Precision. In *2023 International Conference on System, Computation, Automation and Networking (ICSCAN)* (pp. 1-6). IEEE.

Saini, A. K., Yadav, A. K., & Dhiraj, . (2025). A Comprehensive review on technological breakthroughs in precision agriculture: IoT and emerging data analytics. *European Journal of Agronomy*, *163*, 127440. DOI: 10.1016/j.eja.2024.127440

Saleem, S. R., Levison, J., & Haroon, Z. (2023). Environment: Role of precision agriculture technologies. *Precision Agriculture*, •••, 211–229. DOI: 10.1016/B978-0-443-18953-1.00012-X

Salehi, P., Chalechale, A., & Taghizadeh, M. (2020). Generative adversarial networks (GANs): An overview of theoretical model, evaluation metrics, and recent developments. *arXiv preprint arXiv:2005.13178.*

Salimans, T., Goodfellow, I., Zaremba, W., Cheung, V., Radford, A., & Chen, X. (2016). Improved techniques for training gans. *Advances in Neural Information Processing Systems*, •••, 29.

Salimans, T., Goodfellow, I., Zaremba, W., Cheung, V., Radford, A., & Chen, X. (2016). Improved techniques for training GANs. *Advances in Neural Information Processing Systems*, •••, 29.

Sanders, C. E., Mayfield-Smith, K. A., & Lamm, A. J. (2021). Exploring Twitter discourse around the use of artificial intelligence to advance agricultural sustainability. *Sustainability (Basel)*, *13*(21), 12033. DOI: 10.3390/su132112033

Santoro, M. A., & Chakrabarti, A. (2002). The role of research in the development of new technologies: A case study of the telecommunications industry. *Technological Forecasting and Social Change*, *69*(5), 487–507.

Sara, D., Mandava, A. K., Kumar, A., Duela, S., & Jude, A. (2021). Hyperspectral and multispectral image fusion techniques for high resolution applications: A review. *Earth Science Informatics*, *14*(4), 1685–1705. DOI: 10.1007/s12145-021-00621-6

Saranathan, A. M., Smith, B., & Pahlevan, N. (2023). Per-pixel uncertainty quantification and reporting for satellite-derived chlorophyll-a estimates via mixture density networks. *IEEE Transactions on Geoscience and Remote Sensing*, *61*, 1–18. DOI: 10.1109/TGRS.2023.3234465

Sarker, I. H. (2021). Deep Learning: A Comprehensive Overview on Techniques, Taxonomy, Applications and Research Directions. *SN Computer Science*, *2*(6), 420. Advance online publication. DOI: 10.1007/s42979-021-00815-1 PMID: 34426802

Sarmah, B., Baruah, V. J., & Sarmah, T. (2024). Remote sensing–aided disaster risk mitigation in agriculture. In *Remote Sensing in Precision Agriculture* (pp. 487–499). Academic Press.

Sattar, Y., & Oymak, S. (2020). Quickly finding the best linear model in high dimensions via projected gradient descent. *IEEE Transactions on Signal Processing*, *68*, 818–829. DOI: 10.1109/TSP.2020.2964216

Saxena, D., & Cao, J. (2021). Generative adversarial networks (GANs) challenges, solutions, and future directions. *ACM Computing Surveys*, *54*(3), 1–42. DOI: 10.1145/3446374

Schnell, T., Bott, K., Puck, L., Buettner, T., Roennau, A., & Dillmann, R. (2022, October). RoBiGAN: A bidirectional Wasserstein GAN approach for online robot fault diagnosis via internal anomaly detection. In *2022 IEEE/RSJ International Conference on Intelligent Robots and Systems (IROS)* (pp. 4332-4337). IEEE. DOI: 10.1109/IROS47612.2022.9982240

Segarra, J. (2024). Satellite Imagery in Precision Agriculture. In Priyadarshan, P. M., Jain, S. M., Penna, S., & Al-Khayri, J. M. (Eds.), *Digital Agriculture*. Springer., DOI: 10.1007/978-3-031-43548-5_10

Segato, A., Corbetta, V., Di Marzo, M., Pozzi, L., & De Momi, E. (2020). Data augmentation of 3D brain environment using deep convolutional refined auto-encoding alpha GAN. *IEEE Transactions on Medical Robotics and Bionics*, *3*(1), 269–272. DOI: 10.1109/TMRB.2020.3045230

Shafik, W. (2025). Quantum Computing and Generative Adversarial Networks (GANs): Ethics, Privacy, and Security. In *Quantum AI and its Applications in Blockchain Technology* (pp. 111-156). IGI Global Scientific Publishing. https://doi.org/DOI: 10.4018/979-8-3373-1657-4.ch007

Shafik, W. (2024a). Deep Learning Impacts in the Field of Artificial Intelligence. In *Deep Learning Concepts in Operations Research* (pp. 9–26). Auerbach Publications., DOI: 10.1201/9781003433309-2

Shafik, W. (2024b). Introduction to ChatGPT. In *Advanced applications of generative AI and natural language processing models* (pp. 1–25). IGI Global., DOI: 10.4018/979-8-3693-0502-7.ch001

Shafique, A., Cao, G., Khan, Z., Asad, M., & Aslam, M. (2022). Deep learning-based change detection in remote sensing images: A review. *Remote Sensing (Basel)*, *14*(4), 871. DOI: 10.3390/rs14040871

Shamsolmoali, P., Zareapoor, M., Shen, L., Sadka, A., & Yang, J. (2020). Imbalanced Data Learning by Minority Class Augmentation using Capsule Adversarial Networks. *ArXiv*, abs/2004.02182. https://doi.org/.DOI: 10.1016/j.neucom.2020.01.119

Sharma, S., & Namboodiri, V. (2018). No Modes left behind: Capturing the data distribution effectively using GANs. *ArXiv*, abs/1802.00771. https://doi.org/.DOI: 10.1609/aaai.v32i1.11790

Sharma, A., Singh, P., & Chandra, R. (2021). SMOTified-GAN for Class Imbalanced Pattern Classification Problems. *IEEE Access : Practical Innovations, Open Solutions*, *10*, 30655–30665. DOI: 10.1109/ACCESS.2022.3158977

Sharma, K., & Shivandu, S. K. (2024). Integrating artificial intelligence and Internet of Things (IoT) for enhanced crop monitoring and management in precision agriculture. *Sensors International*, *5*, 100292. DOI: 10.1016/j.sintl.2024.100292

Sharma, P., Kumar, M., Sharma, H. K., & Biju, S. M. (2024). Generative adversarial networks (gans): Introduction, taxonomy, variants, limitations, and applications. *Multimedia Tools and Applications*, *83*(41), 1–48. DOI: 10.1007/s11042-024-18767-y

Sharmilan, S., & Chaminda, H. T. (2017). Generate bioinformatics data using Generative Adversarial Network: a review. In 2nd international conference on information technology research. Colombo.

Shekhar, S., Durgam, M., Khose, S. B., Pohshna, C., & Bhalekar, D. G. (2024). Advancement and Challenges of Implementing Artificial Intelligence of Things in Precision Agriculture. In Chouhan, S. S., Saxena, A., Singh, U. P., & Jain, S. (Eds.), *Artificial Intelligence Techniques in Smart Agriculture*. Springer., DOI: 10.1007/978-981-97-5878-4_13

Shendryk, Y., Davy, R., & Thorburn, P. (2021). Integrating Satellite Imagery and Environmental Data to Predict Field-Level Cane and Sugar Yields in Australia Using Machine Learning. *Field Crops Research*, *260*, 107984. DOI: 10.1016/j.fcr.2020.107984

Shen, Y., Gu, J., Tang, X., & Zhou, B. (2020). Interpreting the latent space of GANs for semantic face editing. In *Proceedings of the IEEE/CVF Conference on Computer Vision and Pattern Recognition* (pp. 9243-9252). DOI: 10.1109/CVPR42600.2020.00926

Shi, C., Dang, Y., Fang, L., Zhao, M., Lv, Z., Miao, Q., & Pun, C. M. (2022). Multifeature collaborative adversarial attack in multimodal remote sensing image classification. *IEEE Transactions on Geoscience and Remote Sensing*, *60*, 1–15. DOI: 10.1109/TGRS.2022.3208337

Shorten, C., & Khoshgoftaar, T. M. (2019). A survey on image data augmentation for deep learning. *Journal of Big Data*, *6*(1), 1–48. DOI: 10.1186/s40537-019-0197-0

Shukla, J. S., & Pandya, R. J. (2023). Deep learning-oriented c-GAN models for vegetative drought prediction on peninsular india. *IEEE Journal of Selected Topics in Applied Earth Observations and Remote Sensing*.

Sidheekh, S., Aimen, A., & Krishnan, N. C. (2021, July). On characterizing GAN convergence through proximal duality gap. In *International Conference on Machine Learning* (pp. 9660-9670). PMLR.

Si, H. (2023, November). On the convergence acceleration of vanilla Generative Adversarial Network. [). IOP Publishing.]. *Journal of Physics: Conference Series, 2634*(1), 012047. DOI: 10.1088/1742-6596/2634/1/012047

Sihan, R. (2024). *Applications and challenges of GAN in AI-powered artistry*. Applied and Computational Engineering., DOI: 10.54254/2755-2721/41/20230713

Singh, P., & Komodakis, N. (2018). Cloud-Gan: Cloud Removal for Sentinel-2 Imagery Using a Cyclic Consistent Generative Adversarial Networks. IGARSS 2018 - 2018 IEEE International Geoscience and Remote Sensing Symposium, (pp. 1772-1775). DOI: 10.1109/IGARSS.2018.8519033

Singh, P., & Komodakis, N. (2018, July). Cloud-gan: Cloud removal for sentinel-2 imagery using a cyclic consistent generative adversarial networks. In *IGARSS 2018-2018 IEEE International Geoscience and Remote Sensing Symposium* (pp. 1772-1775). IEEE.

Singla, K., Pandey, R., & Ghanekar, U. (2022). A review on single image super resolution techniques using generative adversarial network. *Optik (Stuttgart), 266*, 169607. DOI: 10.1016/j.ijleo.2022.169607

Sishodia, R. P., Ray, R. L., & Singh, S. K. (2020). Applications of remote sensing in precision agriculture: A review. *Remote Sensing (Basel), 12*(19), 3136. DOI: 10.3390/rs12193136

Song, C. H., Han, H. J., & Avrithis, Y. (2021). All the attention you need: Global-local, spatial-channel attention for image retrieval. CoRR, abs/2107.08000. Retrieved from https://arxiv.org/abs/2107.08000

Sood, A. (2023). An overview of precision farming. *Int. J. Environ. Clim. Change, 13*(12), 441–456.

Soon, Ae. (2000). *Chun., Vijayalakshmi, Atluri*. Protecting Privacy from Continuous High-resolution Satellite Surveillance., DOI: 10.1007/0-306-47008-X_21

Soori, M., Arezoo, B., & Dastres, R. (2023). Artificial intelligence, machine learning and deep learning in advanced robotics, a review. *Cognitive Robotics*, *3*, 54–70. DOI: 10.1016/j.cogr.2023.04.001

Sortino, R., Magro, D., Fiameni, G., Sciacca, E., Riggi, S., DeMarco, A., Spampinato, C., Hopkins, A. M., Bufano, F., Schillirò, F., Bordiu, C., & Pino, C. (2023). Radio astronomical images object detection and segmentation: A benchmark on deep learning methods. *Experimental Astronomy*, *56*(1), 293–331. DOI: 10.1007/s10686-023-09893-w

Spyrou, E. D., Tsoulos, I., & Stylios, C. (2022). Applying and comparing LSTM and ARIMA to predict CO levels for a time-series measurements in a port area. *Signals*, *3*(2), 235–248. DOI: 10.3390/signals3020015

Stengel, K., Glaws, A., Hettinger, D., & King, R. N. (2020). Adversarial super-resolution of climatological wind and solar data. *Proceedings of the National Academy of Sciences of the United States of America*, *476*(2240), 20190493. DOI: 10.1073/pnas.1918964117 PMID: 32631993

Strelet, E., Peng, Y., Castillo, I., Rendall, R., Wang, Z., Joswiak, M., Braun, B., Chiang, L., & Reis, M. S. (2023). Multi-source and multimodal data fusion for improved management of a wastewater treatment plant. *Journal of Environmental Chemical Engineering*, *11*(6), 111530. DOI: 10.1016/j.jece.2023.111530

Suh, S., Lee, H., Lukowicz, P., & Lee, Y. (2020). CEGAN: Classification Enhancement Generative Adversarial Networks for unraveling data imbalance problems. *Neural networks: the official journal of the International Neural Network Society*, 133, 69-86 . https://doi.org/.DOI: 10.1016/j.neunet.2020.10.004

Su, L., Sui, Y., & Yuan, Y. (2023). An Unmixing-Based Multi-Attention GAN for Unsupervised Hyperspectral and Multispectral Image Fusion. *Remote Sensing (Basel)*, *15*(4), 936. DOI: 10.3390/rs15040936

Sumaiya, T. (2024, August 30). Kazi, Aminul, Islam. (2024). Improve Adversarial Robustness of AI Models in Remote Sensing via Data-Augmentation and Explainable-AI Methods. *Remote Sensing (Basel)*, *16*(17), 3210. Advance online publication. DOI: 10.3390/rs16173210

Suresh Babu, C. V., Suruthi, G., Indhumathi, C., & Sushruth, S. (2024b). Integrated approach for network intrusion detection: Combining KNN algorithm, Karhunen-Loève transform, and genetic algorithm optimization. In Metaheuristic and machine learning optimization strategies for complex systems (p. 21). DOI: 10.4018/979-8-3693-7842-7.ch002

Suresh Babu, C. V. (2003). *Artificial Intelligence and Expert Systems.* Anniyappa Publicaitons.

Suresh Babu, C. V., Abhinaba Pal, A., Vinith, A., Muralirajan, V., & Gunasekaran, S. (2024). Enhancing Cloud and IoT Security: Leveraging IoT Technology for Multi-Factor User Authentication. In Ahmed Nacer, A., & Abdmeziem, M. (Eds.), *Emerging Technologies for Securing the Cloud and IoT* (pp. 258–282). IGI Global., DOI: 10.4018/979-8-3693-0766-3.ch011

Suresh Babu, C. V., & Adhithya, S. (2023). ChatGPT and Other Generative AI Tools in Education: Transformative Potential and Ethical Considerations. In Keengwe, J. (Ed.), *Creative AI Tools and Ethical Implications in Teaching and Learning* (pp. 135–152). IGI Global., DOI: 10.4018/979-8-3693-0205-7.ch007

Suresh Babu, C. V., Sowmi Saltonya, M., Ganapathi, S., & Gunasekar, A. (2024). AIoT Revolution: Transforming Networking Productivity for the Digital Age. In Rezaei, S., & Ansary, A. (Eds.), *Artificial Intelligence of Things (AIoT) for Productivity and Organizational Transition* (pp. 108–143). IGI Global., DOI: 10.4018/979-8-3693-0993-3.ch005

Suresh Babu, C. V., & Yadavamuthiah, K. (2023). Precision Agriculture and Farming Using Cyber-Physical Systems: A Systematic Study. In Karthick, G. (Ed.), *Contemporary Developments in Agricultural Cyber-Physical Systems* (pp. 184–203). IGI Global., DOI: 10.4018/978-1-6684-7879-0.ch010

Suryawati, E., Pardede, H. F., Zilvan, V., Ramdan, A., Krisnandi, D., Heryana, A., Yuwana, R. S., Kusumo, R. B. S., Arisal, A., & Supianto, A. A. (2021, December). R. Sandra Yuwana, R. Budiarianto Suryo Kusumo, Andria Arisal & Ahmad Aff Supianto, Unsupervised feature learning-based encoder and adversarial networks. *Journal of Big Data*, *8*(1), 118. Advance online publication. DOI: 10.1186/s40537-021-00508-9

Swadi, L. A., & Haider, A. M. (2025). A Review of Generative Adversarial Networks for Addressing Data Imbalance and Enhancing Model Performance. [IJICI]. *Iraqi Journal of Intelligent Computing and Informatics*, *4*(1), 18–32.

Tai, C. Y., Wang, W. J., & Huang, Y. M. (2023). Using time-series generative adversarial networks to synthesize sensing data for pest incidence forecasting on sustainable agriculture. *Sustainability (Basel)*, *15*(10), 7834. DOI: 10.3390/su15107834

Tang, L. (2024). Motoaki, Asai., Yoichiro, Kato., Yuya, Fukano., Wei, Guo. (2024). Channel Attention GAN-based Synthetic Weed Generation for Precise Weed Identification. *Plant Phenomics (Washington, D.C.)*, *6*, 0122. Advance online publication. DOI: 10.34133/plantphenomics.0122

Tang, Y., Dananjayan, S., Hou, C., Guo, Q., Luo, S., & He, Y. (2021). A survey on the 5G network and its impact on agriculture: Challenges and opportunities. *Computers and Electronics in Agriculture, 180*, 105895. DOI: 10.1016/j.compag.2020.105895

Taseer, A., & Han, X. (2024). Advancements in variable rate spraying for precise spray requirements in precision agriculture using Unmanned aerial spraying Systems: A review. *Computers and Electronics in Agriculture, 219*, 108841. DOI: 10.1016/j.compag.2024.108841

Thirumagal, E., & Saruladha, K. (2021). GAN models in natural language processing and image translation. In *Generative Adversarial Networks for Image-to-Image Translation* (pp. 17–57). Academic Press. DOI: 10.1016/B978-0-12-823519-5.00001-4

Tiwari, T., Tiwari, T., & Tiwari, S. (2018), How Artificial Intelligence, Machine Learning and Deep Learning are Radically Different? International Journals of Advanced Research in Computer Science and Software Engineering ISSN: 2277-128X (Volume-8, Issue-2).

Toizumi, T., Zini, S., Sagi, K., Kaneko, E., Tsukada, M., & Schettini, R. (2019). Artifact-Free Thin Cloud Removal Using Gans. *2019 IEEE International Conference on Image Processing (ICIP)*, (pp. 3596-3600). DOI: 10.1109/ICIP.2019.8803652

Tran, N. T., Tran, V. H., Nguyen, N. B., Nguyen, T. K., & Cheung, N. M. (2021). On data augmentation for GAN training. *IEEE Transactions on Image Processing, 30*, 1882–1897. DOI: 10.1109/TIP.2021.3049346 PMID: 33428571

Tran, N.-T., Tran, V.-H., Nguyen, N.-B., Yang, L., & Cheung, N.-M. (2019), Self-supervised GAN: Analysis and Improvement with Multi-class Minimax Game,*33rd Conference on Neural Information Processing Systems (NeurIPS 2019)*, Vancouver, Canada.

Trinh, L. T., & Hamagami, T. (2024). Latent Denoising Diffusion GAN: Faster sampling, Higher image quality. *IEEE Access : Practical Innovations, Open Solutions, 12*, 78161–78172. DOI: 10.1109/ACCESS.2024.3406535

Tsatsaris, A., Kalogeropoulos, K., Stathopoulos, N., Louka, P., Tsanakas, K., Tsesmelis, D. E., Krassanakis, V., Petropoulos, G. P., Pappas, V., & Chalkias, C. (2021). Geoinformation technologies in support of environmental hazards monitoring under climate change: An extensive review. *ISPRS International Journal of Geo-Information, 10*(2), 94. DOI: 10.3390/ijgi10020094

Tzeng, E., Hoffman, J., Saenko, K., & Darrell, T. (2017). Adversarial discriminative domain adaptation. In *Proceedings of the IEEE Conference on Computer Vision and Pattern Recognition* (pp. 7167-7176).

Ullo, S. L., & Sinha, G. R. (2020). Advances in smart environment monitoring systems using IoT and sensors. *Sensors (Basel), 20*(11), 3113. DOI: 10.3390/s20113113 PMID: 32486411

Ullo, S. L., & Sinha, G. R. (2021). Advances in IoT and Smart Sensors for Remote Sensing and Agriculture Applications. *Remote Sensing (Basel), 13*(13), 2585. DOI: 10.3390/rs13132585

Vellingiri, A., Kokila, R., Nisha, P., Kumar, M., Chinnusamy, S., & Boopathi, S. (2025). Harnessing GPS, Sensors, and Drones to Minimize Environmental Impact: Precision Agriculture. In Designing Sustainable Internet of Things Solutions for Smart Industries (pp. 77-108). IGI Global.

Vibhute, A. D., Kale, K. V., & Gaikwad, S. V. (2024). Machine learning-enabled soil classification for precision agriculture: A study on spectral analysis and soil property determination. *Appl Geomat, 16*(1), 181–190. DOI: 10.1007/s12518-023-00546-3

Vinod Chandra, S. S., Hareendran, A., & Albaaji, G. F. (2024). Precision farming for sustainability: An agricultural intelligence model. *Computers and Electronics in Agriculture, 226*, 109386. DOI: 10.1016/j.compag.2024.109386

Vuletić, M., Prenzel, F., & Cucuringu, M. (2024). Fin-gan: Forecasting and classifying financial time series via generative adversarial networks. *Quantitative Finance, 24*(2), 175–199. DOI: 10.1080/14697688.2023.2299466

Vyawahare, H. R., Rathod, S. B., Dhande, S., Bajaj, K. A., & Khandelwal, S. (2024). Exploring Generative Adversarial Networks (GANs) in the Context of Public Space Protection. In *Enhancing Security in Public Spaces Through Generative Adversarial Networks (GANs)* (pp. 231-245). IGI Global.

Wang, J., Yi, G., Zhang, S., & Wang, Y. (2020), An Unsupervised Generative Adversarial Network-Based Method for Defect Inspection of Texture Surfaces. Appl. Sci. 2021, 11, 283. https://doi.org/doi:10.3390/app11010283

Wang, K., Gou, C., Duan, Y., Lin, Y., Zheng, X., & Wang, F.-Y. (2017), Generative Adversarial Networks: Introduction and Outlook, IEEE/CAA JOURNAL OF AUTOMATICA SINICA, VOL. 4, NO. 4, OCTOBER 2017.

Wang, C., Chen, B., Zou, Z., & Shi, Z. (2023). Remote sensing image synthesis via semantic embedding generative adversarial networks. *IEEE Transactions on Geoscience and Remote Sensing, 61*, 1–11. DOI: 10.1109/TGRS.2023.3335484

Wang, C., & Zhu, H. (2024). Addressing imbalance data for online fraud detection. In *Universal Behavior Computing for Security and Safety* (pp. 61–84). Springer. DOI: 10.1007/978-981-97-9014-2_3

Wang, D., Zhao, M., Li, Z., Xu, S., Wu, X., Ma, X., & Liu, X. (2025). A survey of unmanned aerial vehicles and deep learning in precision agriculture. *European Journal of Agronomy*, *164*, 127477. DOI: 10.1016/j.eja.2024.127477

Wang, J., Shi, Y., Yu, H., Zhang, K., Wang, X., Yan, Z., & Li, H. (2023). Temporal density-aware sequential recommendation networks with contrastive learning. *Expert Systems with Applications*, *211*, 118563. DOI: 10.1016/j.eswa.2022.118563

Wang, J., Wang, S., Lv, M., & Jiang, H. (2024). Forecasting VaR and ES by using deep quantile regression, GANs-based scenario generation, and heterogeneous market hypothesis. *Financial Innovation*, *10*(1), 36. DOI: 10.1186/s40854-023-00564-5 PMID: 35251896

Wang, J., Wang, Y., Li, G., & Qi, Z. (2024). Integration of Remote Sensing and Machine Learning for Precision Agriculture: A Comprehensive Perspective on Applications. *Agronomy (Basel)*, *14*(9), 1975. DOI: 10.3390/agronomy14091975

Wang, Q., Li, Q., & Li, X. (2020a). A fast neighborhood grouping method for hyperspectral band selection. *IEEE Transactions on Geoscience and Remote Sensing*, *59*(6), 5028–5039. DOI: 10.1109/TGRS.2020.3011002

Wang, S., Li, W., Hou, S., Guan, J., & Yao, J. (2022). STA-GAN: A spatio-temporal attention generative adversarial network for missing value imputation in satellite data. *Remote Sensing (Basel)*, *15*(1), 88. DOI: 10.3390/rs15010088

Wang, S., Wang, Y., Liu, Y., & Li, L. (2021a). Sar image change detection based on sparse representation and a capsule network. *Remote Sensing Letters*, *12*(9), 890–899. DOI: 10.1080/2150704X.2021.1946199

Wang, W. (2022). Multi-scale attention generative adversarial network for single image rain removal. *Pattern Recognition and Image Analysis*, *32*(2), 436–447. DOI: 10.1134/S1054661822020201

Wang, W., Lai, Q., Fu, H., Shen, J., Ling, H., & Yang, R. (2021b). Salient object detection in the deep learning era: An in-depth survey. *IEEE Transactions on Pattern Analysis and Machine Intelligence*, *44*(6), 3239–3259. DOI: 10.1109/TPAMI.2021.3051099 PMID: 33434124

Wang, X., Sun, L., Chehri, A., & Song, Y. (2023). A review of GAN-based super-resolution reconstruction for optical remote sensing images. *Remote Sensing (Basel)*, *15*(20), 5062. DOI: 10.3390/rs15205062

Wang, X., Wang, K., & Lian, S. (2020b). A survey on face data augmentation for the training of deep neural networks. *Neural Computing & Applications*, *32*(19), 15503–15531. DOI: 10.1007/s00521-020-04748-3

Wang, Y., Wu, C., Herranz, L., van de Weijer, J., Gonzalez-Garcia, A., & Raducanu, B. (2018). Transferring GANs: Generating images from limited data. In *Proceedings of the European Conference on Computer Vision (ECCV)* (pp. 218-234). Springer, Cham. DOI: 10.1007/978-3-030-01231-1_14

Wang, Z., She, Q., & Ward, T. E. (2021). Generative adversarial networks in computer vision: A survey and taxonomy. *ACM Computing Surveys*, *54*(2), 1–38.

War, K. (2019), Strengthing Deep Neural Netwoks: Making AI Less Susceptible to Adversarial Trickery, O'Reilly publications, August 2019, ISBN:978-93-5213-873-9.

Weidman, S. (2019), Deep Learning from Scratch, O'Reilly publications, September 2019, ISBN:978-93-5213-902-6.

Wi, H., Shin, Y., & Park, N. (2024, March). Continuous-time Autoencoders for Regular and Irregular Time Series Imputation. In *Proceedings of the 17th ACM International Conference on Web Search and Data Mining* (pp. 826-835). https://doi.org/DOI: 10.1145/3616855.3635831

Woo, S., Park, J., Lee, J.-Y., & Kweon, I. S. (2018). CBAM: Convolutional Block Attention Module. CoRR, abs/1807.06521. Retrieved from http://arxiv.org/abs/1807.06521

Wu, Y., Yue, Y., Tan, X., Wang, W., & Lu, T. (2018, October). End-to-end chromosome Karyotyping with data augmentation using GAN. In *2018 25th IEEE international conference on image processing (ICIP)* (pp. 2456-2460). IEEE. DOI: 10.1109/ICIP.2018.8451041

Wu, A. N., Stouffs, R., & Biljecki, F. (2022). Generative Adversarial Networks in the built environment: A comprehensive review of the application of GANs across data types and scales. *Building and Environment*, *223*, 109477. DOI: 10.1016/j.buildenv.2022.109477

Wu, C., Zhang, H., Chen, J., Gao, Z., Zhang, P., Muhammad, K., & Del Ser, J. (2022). Vessel-gan: Angiographic reconstructions from myocardial ct per fusion with explainable generative adversarial networks. *Future Generation Computer Systems*, *130*, 128–139. DOI: 10.1016/j.future.2021.12.007

Wu, Z., Li, J., Wang, Y., Hu, Z., & Molinier, M. (2020). Self-Attentive Generative Adversarial Network for Cloud Detection in High Resolution Remote Sensing Images. *IEEE Geoscience and Remote Sensing Letters*, *17*(10), 1792–1796. DOI: 10.1109/LGRS.2019.2955071

Wu, Z., Lischinski, D., & Shechtman, E. (2020). StyleSpace analysis: Disentangled controls for StyleGAN image generation. In *Proceedings of the IEEE/CVF Conference on Computer Vision and Pattern Recognition* (pp. 12863-12872).

Xiao, H., Wang, X., Wang, J., Cai, J. Y., Deng, J. H., Yan, J. K., & Tang, Y. D. (2024). Single image super-resolution with denoising diffusion GANs. *Scientific Reports, 14*(1), 4272. DOI: 10.1038/s41598-024-52370-3 PMID: 38383573

Xiao, J., Li, J., Yuan, Q., Jiang, M., & Zhang, L. (2021). Physics-Based GAN With Iterative Refinement Unit for Hyperspectral and Multispectral Image Fusion. *IEEE Journal of Selected Topics in Applied Earth Observations and Remote Sensing, 14*, 6827–6841. DOI: 10.1109/JSTARS.2021.3075727

Xia, X., Pan, X., Li, N., He, X., Ma, L., Zhang, X., & Ding, N. (2022). Gan based anomaly detection: A review. *Neurocomputing, 493*, 497–535. DOI: 10.1016/j.neucom.2021.12.093

Xu, H. (2024). Rotated single-stage detector with transformer in remote sensing rotating object detection. In 2024 5th International Conference on Computer Engineering and Application (ICCEA), pages 1298–1303. IEEE. DOI: 10.1109/ICCEA62105.2024.10603764

Xue, W., Chen, Z., Tian, W., Wu, Y., & Hua, B. (2022). A cascade defense method for multidomain adversarial attacks under remote sensing detection. *Remote Sensing (Basel), 14*(15), 3559. DOI: 10.3390/rs14153559

Xue, Y., Zhou, Q., Ye, J., Long, L. R., Antani, S., Cornwell, C., Xue, Z., & Huang, X. (2019). Synthetic augmentation and feature-based filtering for improved cervical histopathology image classification. In *Medical Image Computing and Computer Assisted Intervention–MICCAI 2019: 22nd International Conference, Shenzhen, China, October 13–17, 2019* [Springer International Publishing.]. *Proceedings, 22*(Part I), 387–396.

Xu, L., Zeng, X., Huang, Z., Li, W., & Zhang, H. (2020). Low-dose chest X-ray image super-resolution using generative adversarial nets with spectral normalization. *Biomedical Signal Processing and Control, 55*, 101600. DOI: 10.1016/j.bspc.2019.101600

Xu, M., Wu, M., Chen, K., Zhang, C., & Guo, J. (2022). The eyes of the gods: A survey of unsupervised domain adaptation methods based on remote sensing data. *Remote Sensing (Basel), 14*(17), 4380. DOI: 10.3390/rs14174380

Xu, T., Zhao, W., Meng, X., Ma, Y., & Zhou, J. (2023). Adversarial learning based method for recognition of bionic and highly contextual underwater tar gets. *Journal of Electronic Imaging*, *32*(2), 023027–023027. DOI: 10.1117/1.JEI.32.2.023027

Xu, Y., Du, B., & Zhang, L. (2020). Assessing the threat of adversarial examples on deep neural networks for remote sensing scene classification: Attacks and defenses. *IEEE Transactions on Geoscience and Remote Sensing*, *59*(2), 1604–1617. DOI: 10.1109/TGRS.2020.2999962

Xu, Y., & Ghamisi, P. (2022). Universal adversarial examples in remote sensing: Methodology and benchmark. *IEEE Transactions on Geoscience and Remote Sensing*, *60*, 1–15. DOI: 10.1109/TGRS.2022.3156392

Yang, G., Lu, H., Li, C., Zhou, G., Wu, H., & Lu, Z. (2024, April). Progressive Image Synthesis from Semantics to Details with Denoising Diffusion GAN. In *ICASSP 2024-2024 IEEE International Conference on Acoustics, Speech and Signal Processing (ICASSP)* (pp. 7495-7499). IEEE. https://doi.org/DOI: 10.1109/ICASSP48485.2024.10445885

Yang, Q., Wang, G., Zhao, Y., Zhang, X., Dong, G., & Ren, P. (2020). Multi-Scale Deep Residual Learning for Cloud Removal. IGARSS 2020 - 2020 IEEE International Geoscience and Remote Sensing Symposium, (pp. 4967-4970). DOI: 10.1109/IGARSS39084.2020.9323261

Yang, H., Huang, X., & Liu, Y. (2023). *InfoGAN-Enhanced Underwater Acoustic Target Recognition Method Based on Deep Learning*. ISBN. DOI: 10.1007/978-981-99-0479-2_249

Yang, J., Li, C., Zhang, P., Dai, X., Xiao, B., Yuan, L., & Gao, J. (2021). Focal attention for long-range interactions in vision transformers. *Advances in Neural Information Processing Systems*, *34*, 30008–30022.

Yang, S., Zhou, Y., Zheng, Z., Wang, Y., Zhu, L., & Wu, Y. (2023). Towards unified text-based person retrieval: A large-scale multi-attribute and language search benchmark. In *Proceedings of the 31st ACM International Conference on Multimedia*, pages 4492–4501. DOI: 10.1145/3581783.3611709

Yican, H. (2023, December). Zhengguang, Chen., Jinming, Liu. (2023). Limited agricultural spectral dataset expansion based on generative adversarial networks. *Computers and Electronics in Agriculture*, *215*, 108385. Advance online publication. DOI: 10.1016/j.compag.2023.108385

Yin, Y., Yuan, Z., Tanvir, I. M., & Bao, X. (2024). Electronic medical records imputation by temporal Generative Adversarial Network. *BioData Mining*, *17*(1), 19. DOI: 10.1186/s13040-024-00372-2 PMID: 38926718

Yu, W., Zhang, X., Pun, M.-O., & Liu, M. (2021). A Hybrid Model-Based and Data-Driven Approach for Cloud Removal in Satellite Imagery Using Multi-Scale Distortion-Aware Networks. 2021 IEEE International Geoscience and Remote Sensing Symposium IGARSS, (pp. 7160-7163). DOI: 10.1109/IGARSS47720.2021.9554963

Yu, M., Zhang, W., Chen, X., Liu, Y., & Niu, J. (2022). An End-to-End Atrous Spatial Pyramid Pooling and Skip-Connections Generative Adversarial Segmentation Network for Building Extraction from High-Resolution Aerial Images. *Applied Sciences (Basel, Switzerland)*, *12*(10), 5151. DOI: 10.3390/app12105151

Yun, J., Kim, D., Kim, D. M., Song, T., & Woo, J. (2023). GAN-based sensor data augmentation: Application for counting moving people and detecting directions using PIR sensors. *Engineering Applications of Artificial Intelligence*, *117*, 105508. DOI: 10.1016/j.engappai.2022.105508

Yu, W., Xu, Y., & Ghamisi, P. (2024). Universal adversarial defense in remote sensing based on pre-trained denoising diffusion models. *International Journal of Applied Earth Observation and Geoinformation*, *133*, 104131. DOI: 10.1016/j.jag.2024.104131

Yu, Y., Chen, W., Chen, F., Jia, W., & Lu, Q. (2024). Night-time vehicle model recognition based on domain adaptation. *Multimedia Tools and Applications*, *83*(4), 9577–9596. DOI: 10.1007/s11042-023-15447-1

Zakari, R. Y., Lawal, Z. K., Kalinaki, K., & Maiwada, U. D. (2024). Time Series Forecasting Techniques for Climate Trend Prediction. In *Reshaping Environmental Science Through Machine Learning and IoT* (pp. 282–310). IGI Global., DOI: 10.4018/979-8-3693-2351-9.ch014

Zakari, R. Y., Shafik, W., Kalinaki, K., & Iheaturu, C. J. (2024). Internet of forestry things (IoFT) technologies and applications in Forest management. In *Advanced IoT technologies and applications in the industry 4.0 digital economy* (pp. 275–295). CRC Press. DOI: 10.1201/9781003434269-15

Zhan, C. (2024, May). Tan., Wang, Man., Gui, Song, Wang., Qin, Nie. (2024). Unsupervised remote sensing image thin cloud removal method based on contrastive learning. *IET Image Processing*, *18*(7), 1844–1861. Advance online publication. DOI: 10.1049/ipr2.13067

Zhang, C., Zhang, X., Yu, Q., & Ma, C. (2022). An Improved Method for Removal of Thin Clouds in Remote Sensing Images by Generative Adversarial Network. IGARSS 2022 - 2022 IEEE International Geoscience and Remote Sensing Symposium, (pp. 6706-6709). DOI: 10.1109/IGARSS46834.2022.9884704

Zhang, B., Wu, Y., Zhao, B., Chanussot, J., Hong, D., Yao, J., & Gao, L. (2022). Progress and challenges in intelligent remote sensing satellite systems. *IEEE Journal of Selected Topics in Applied Earth Observations and Remote Sensing*, *15*, 1814–1822. DOI: 10.1109/JSTARS.2022.3148139

Zhang, B., Zhang, Y., Li, Y., Wan, Y., Guo, H., Zheng, Z., & Yang, K. (2022a). Semi-supervised deep learning via transformation consistency regularization for remote sensing image semantic segmentation. *IEEE Journal of Selected Topics in Applied Earth Observations and Remote Sensing*, *16*, 5782–5796. DOI: 10.1109/JSTARS.2022.3203750

Zhang, D., & Khoreva, A. (2019). Progressive augmentation of gans. *Advances in Neural Information Processing Systems*, •••, 32.

Zhang, D., Ren, L., Shafiq, M., & Gu, Z. (2022). A lightweight privacy-preserving system for the security of remote sensing images on iot. *Remote Sensing (Basel)*, *14*(24), 6371. DOI: 10.3390/rs14246371

Zhang, D., Shafiq, M., Wang, L., Srivastava, G., & Yin, S. (2023). Privacy-preserving remote sensing images recognition based on limited visual cryptography. *CAAI Transactions on Intelligence Technology*, *8*(4), 1166–1177. DOI: 10.1049/cit2.12164

Zhang, H., Goodfellow, I., Metaxas, D., & Odena, A. (2019). Self-attention generative adversarial networks. In *International conference on machine learning* (pp. 7354-7363). PMLR.

Zhang, H., Goodfellow, I., Metaxas, D., & Odena, A. (2019). Self-attention generative adversarial networks. In *International Conference on Machine Learning* (pp. 7354-7363). PMLR.

Zhang, H., Song, Y., Han, C., & Zhang, L. (2020). Remote sensing image spatiotemporal fusion using a generative adversarial network. *IEEE Transactions on Geoscience and Remote Sensing*, *59*(5), 4273–4286. DOI: 10.1109/TGRS.2020.3010530

Zhang, L., Huang, G., Li, Y., Yang, S., Lu, L., & Huo, W. (2023). A robust insar phase unwrapping method via improving the pix2pix network. *Remote Sensing (Basel)*, *15*(19), 4885. DOI: 10.3390/rs15194885

Zhang, L., Long, C., Zhang, X., & Xiao, C. (2020). Ris-gan: Explore residual and illumination with generative adversarial networks for shadow removal. *Proceedings of the AAAI Conference on Artificial Intelligence*, *34*(7), 12829–12836. DOI: 10.1609/aaai.v34i07.6979

Zhang, L., & Zhang, L. (2022). Artificial intelligence for remote sensing data analysis: A review of challenges and opportunities. *IEEE Geoscience and Remote Sensing Magazine*, *10*(2), 270–294. DOI: 10.1109/MGRS.2022.3145854

Zhang, L., Zhang, L., & Du, B. (2016). Deep Learning for Remote Sensing Data: A Technical Tutorial on the State of the Art. *IEEE Geoscience and Remote Sensing Magazine*, *2016*(4), 22–40. DOI: 10.1109/MGRS.2016.2540798

Zhang, M., Zhou, Y., Liu, B., Zhao, J., Yao, R., Shao, Z., & Zhu, H. (2022b). Weakly supervised few-shot semantic segmentation via pseudo mask enhancement and meta learning. *IEEE Transactions on Multimedia*, *25*, 7980–7991. DOI: 10.1109/TMM.2022.3232037

Zhang, N., Wang, Y., Zhang, X., Xu, D., & Wang, X. (2020). An unsupervised remote sensing single-image super-resolution method based on generative adversarial network. *IEEE Access : Practical Innovations, Open Solutions*, *8*, 29027–29039. DOI: 10.1109/ACCESS.2020.2972300

Zhang, Q., & Wang, T. (2024). Deep learning for exploring landslides with remote sensing and geo-environmental data: Frameworks, progress, challenges, and opportunities. *Remote Sensing (Basel)*, *16*(8), 1344. DOI: 10.3390/rs16081344

Zhang, X., Feng, G., & Sun, X. (2024). Advanced technologies of soil moisture monitoring in precision agriculture: A Review. *Journal of Agriculture and Food Research*, *18*, 101473. DOI: 10.1016/j.jafr.2024.101473

Zhang, Y., Peng, C., Wang, Q., Song, D., Li, K., & Zhou, S. K. (2024). Unified multi-modal image synthesis for missing modality imputation. *IEEE Transactions on Medical Imaging*. Advance online publication. DOI: 10.1109/TMI.2024.3424785 PMID: 38976465

Zhang, Y., Wang, Y., & Li, X. (2018). A review of deep learning in remote sensing: Current applications and future directions. *ISPRS Journal of Photogrammetry and Remote Sensing*, *145*, 1–16.

Zhao, P., Liao, W., Huang, Y., & Lu, X. (2023). Intelligent design of shear wall layout based on attention-enhanced generative adversarial network. *Engineering Structures*, *274*, 115170. DOI: 10.1016/j.engstruct.2022.115170

Zheng, B., Zhang, R., Diao, S., Zhu, J., Yuan, Y., Cai, J., Shao, L., Li, S., & Qin, W. (2024). Dual domain distribution disruption with semantics preservation: Unsupervised domain adaptation for medical image segmentation. *Medical Image Analysis, 103275*, 103275. Advance online publication. DOI: 10.1016/j.media.2024.103275 PMID: 39032395

Zheng, C., Jiang, X., & Liu, X. (2019). Semi-supervised sar atr via multi-discriminator generative adversarial network. *IEEE Sensors Journal*, *19*(17), 7525–7533. DOI: 10.1109/JSEN.2019.2915379

Zheng, Y., Su, J., Zhang, S., Tao, M., & Wang, L. (2022). Dehaze-AGGAN: Unpaired remote sensing image dehazing using enhanced attention-guide generative adversarial networks. *IEEE Transactions on Geoscience and Remote Sensing*, *60*, 1–13. DOI: 10.1109/TGRS.2022.3225843

Zhou, J., Xu, T., Guo, W., Zhao, W., & Cai, L. (2022). Underwater occlusion object recognition with fusion of significant environmental features. *Journal of Electronic Imaging*, *31*(2), 023016–023016. DOI: 10.1117/1.JEI.31.2.023016

Zhou, M., Liu, X., Yi, T., Bai, Z., & Zhang, P. (2023). A superior image in painting scheme using transformer-based self-supervised attention gan model. *Expert Systems with Applications*, *233*, 120906. DOI: 10.1016/j.eswa.2023.120906

Zhu, C., Deng, S., Li, J., Zhang, Y., Gong, L., Gao, L., Ta, N., Chen, S., & Wu, Q. (2023). Hyperspectral and multispectral remote sensing image fusion using SwinGAN with joint adaptive spatial-spectral gradient loss function. *International Journal of Digital Earth*, *16*(1), 3580–3600. DOI: 10.1080/17538947.2023.2253206

Zhu, C., Deng, S., Zhou, Y., Deng, L., & Wu, Q. (2023). QIS-GAN: A Lightweight Adversarial Network With Quadtree Implicit Sampling for Multispectral and Hyperspectral Image Fusion. *IEEE Transactions on Geoscience and Remote Sensing*, *61*, 1–15. DOI: 10.1109/TGRS.2023.3332176

Zhu, J. Y., Park, T., Isola, P., & Efros, A. A. (2017). Unpaired image-to-image translation using cycle-consistent adversarial networks. In *Proceedings of the IEEE international conference on computer vision* (pp. 2223-2232). DOI: 10.1109/ICCV.2017.244

Zhu, L., Chen, Y., Ghamisi, P., & Benediktsson, J. (2018). Generative Adversarial Networks for Hyperspectral Image Classification. *IEEE Transactions on Geoscience and Remote Sensing*, *56*(9), 5046–5063. DOI: 10.1109/TGRS.2018.2805286

Zi, Y., Xie, F., Zhang, N., Jiang, Z., Zhu, W., & Zhang, H. (2021). Thin Cloud Removal for Multispectral Remote Sensing Images Using Convolutional Neural Networks Combined With an Imaging Model. *IEEE Journal of Selected Topics in Applied Earth Observations and Remote Sensing*, *14*, 3811–3823. DOI: 10.1109/JSTARS.2021.3068166

Zualkernan, I., Abuhani, D. A., Hussain, M. H., Khan, J., & ElMohandes, M. (2023). Machine Learning for Precision Agriculture Using Imagery from Unmanned Aerial Vehicles (UAVs): A Survey. *Drones (Basel)*, *2023*(7), 382.

About the Contributors

Amol D. Vibhute received his PhD in Computer Science under the domain of Geospatial Technology, M. Phil and M. Sc. in Computer Science from the Department of Computer Science and Information Technology, Dr. Babasaheb Ambedkar Marathwada University, Chhatrapati Sambhaji Nagar (formarly known as Aurangabad), MH-India. He is an Assistant Professor (Senior Scale) and Program Head of M. Sc. (CA), MBA (IT), and MBA (DT) courses at the Symbiosis Institute of Computer Studies and Research (SICSR), Symbiosis International (Deemed University), Pune, Maharashtra, India. He has more than 10+ years of academic experience in research, innovation, and UG/PG-level teaching. His six (06) Indian patents are under the granting stage, two (02) International and one (01) national patent have been granted, and he authored/co-authored over 85+ referred journals, book chapters, and conference papers in reputed international journals and conferences indexed by Scopus/SCIE/UGC. The current publication citations of his research are more than 1250+, 700+, and 350+, with an h-index of 17, 12, and 7 as per Google Scholar, Scopus, and Web of Science, respectively. His research interests include Geospatial Technology, Digital Image Processing, Pattern Recognition, Big Data Analysis, the Internet of Things (IoT), Machine Learning, and Artificial Intelligence. He is an ISPRS, IAEng, CSTA, IACSIT, and ISCA member. He is an editorial board member for Scientific Reports (Sci Rep) published by "Nature Portfolio" ISSN: 2045-2322 (online). Also, he is an active reviewer of the Taylor and Francis, Elsevier, Wiley Online, IEEE Access and Springer journals indexed by SCI, SCIE, and Scopus.

Rajesh Kumar Dhanaraj is a distinguished Professor at Symbiosis International (Deemed University) in Pune, India. Before joining Symbiosis International University, he served as a Professor at the School of Computing Science & Engineering at Galgotias University in Greater Noida, India. His academic and research achievements have earned him a place among the top 2% of scientists globally, a recognition bestowed upon him by Elsevier and Stanford University. He earned his B.E. degree in Computer Science and Engineering from Anna University Chennai, India, in 2007. Subsequently, he obtained his M.Tech degree from Anna University Coimbatore, India, in 2010. His relentless pursuit of knowledge culminated in a Ph.D. in Computer Science from Anna University, Chennai, India, in 2017. He has authored and edited over 100 books on various cutting-edge technologies and holds 21 patents. Furthermore, he has contributed over 200 articles and papers to esteemed refereed journals and international conferences, in addition to providing chapters for several influential books. Dr. Dhanaraj has shared his insights with the academic community by delivering numerous techs talks on disruptive technologies. He has forged meaningful partnerships with esteemed professors from top QS-ranked universities around the world, fostering a global network of academic excellence. His research interests encompass Machine Learning, Cyber-Physical Systems, and Wireless Sensor Networks. Dr. Dhanaraj's expertise in these areas has led to numerous research talks on Applied AI and Cyber Physical Systems at various esteemed institutions. Dr. Dhanaraj has earned the distinction of being a Senior Member of the Institute of Electrical and Electronics Engineers (IEEE). He is also a member of the Computer Science Teacher Association (CSTA) and the International Association of Engineers (IAENG). Dr. Dhanaraj's commitment to academic excellence extends to his role as an Associate Editor and Guest Editor for renowned journals, including Elsevier Computers and Electrical Engineering, Human-centric Computing and Information Sciences, Emerald - International Journal of Pervasive Computing and Communications, and Hindawi - Mobile Information Systems. His expertise has earned him a position as an Expert Advisory Panel Member of Texas Instruments Inc., USA.

K. V. Kale is currently working as a Vice-Chancellor of Dr Babasaheb Ambedkar Technological University, Lonere, Raigad, India. He worked as a Professor, Program Coordinator of UGC SAP and DST FIST, in the Department of Computer Science and IT as well as Coordinator of University RUSA project & School of Computational and Physical Sciences. He worked as a Director, Board of College and University Development (BCUD) of Dr. Babasaheb Ambedkar Marathwada University, Aurangabad, Maharashtra. He was Head of the Department of Computer Science and IT at University more than 9 years. He also served as Director of University Network & Information Center (UNIC) of Dr. BAMU. He has acted as President of ICT Section the Indian Science Congress Association, Kolkata in the year 2015-2016 under the Ministry of Science & Technology, Government of India. He is a pioneer of Multimodal Biometrics and Remote Sensing Data Analysis in the region and established & developed state-of-art Multimodal Biometrics Research Laboratory (MBRL) as a unique national Laboratory, RS and

GIS, Pattern Recognition and as a center of excellence in teaching and research in biometrics and advance Image processing technology in university. He has more than 30 years of experience in research, innovation and teaching at UG, PG and PhD level. He has published more than 350+National and International research articles, 06 books and 23 book chapters. The current total impact factor of his research is more than 24, with h-index of 25 and i10-index of 57 and 8 Indian Patent 2- Australian patents. He has given a number of plenary lectures, invited talks and Chaired Sessions at Symposia, workshop, national &international conferences. He has successfully supervised 36 Ph.D., 08 M. Phil and 22M.Tech students and currently 07 PhD students, 06 M.Phil. Students and 14M. Tech (CSE) scholars are working with him. His area of research includes Remote Sensing and GIS, Image Processing, Pattern Recognition, Computer Vision, Software Engineering, Artificial Intelligence, Neural Network, Big Data Analytic, Internet of Things (IoT) etc. He is Editor in Chief, Editors, Reviewers of various International, National Journals. Professor Kale has several academic honours and professional distinctions to his credits. He has served on various bodies and authorities of the university like Senate Member, Management Council Member, Academic Council Member, BOS & RRC's chairman, BUTER Member, BCUD member, etc. time to time. He has also been working on various statutory authorities of other universities, in different capacities such as; Chairman and Member.

***

Jayakrishnan Anandakrishnan received bachelor's degree in computer science and engineering from Kerala University, Kerala, India, in 2011, and master's degree in computer science with specialization in image processing from the Cochin University of Science and Technology, Kerala, in 2013. He is a Researcher from the National Institute of Technology Puducherry, Karaikal, India. His research interests include machine learning, remote sensing, geospatial analytics, IoT.

Bikie Gerald Anicet was born on January 5th, 1991 in Esse, Cameroon. He is from Cameroon. Born of Ndoumou Senguena Pierre Lotie and Atebete Mengue Catherine. EDUCATION: BIKIE attended the Central Government English Nursery and Primary School, GBHS Essos, and the Faculty of Sciences of the University of Yaounde I in Yaoundé-Cameroon, and then in 2018 obtained a scholarship from UNOOSA(United Nation's Office of Outer space affaires) to study Space Science and Technology at the African Regional Centre For Space Science and Technology education in the French Language Affiliated with the united Nation's Organisation at the Mohammedia school of engineers in Rabat Morocco he equally obtained a second Masters in GIS and Remote Sensing at the Faculty of Sciences of Rabat/Department of Geology in Mohammed V University a then began his journey as a PhD Researcher in the domain of Geospatial AI and Oceanography at the Geobiodiversity Laboratory of the Scientific Institute of Rabat/Department of Geology and Remote Sensing . WORK: BIKIE is best known for his research publications in space science and technologies, some of which include: Mbou et al. (2024): The Contribution of Remote Sensing to Groundwater Exploration: A Case Study in the Far North Region of Cameroon. DOI: 10.38124/ijisrt/IJISRT24AUG127/ResearchGate Bikie et al. (2024): The Contribution of Remote Sensing to Groundwater Exploration: A Case Study in the Far North Region of Cameroon. DOI: https://doi.org/10 .38124/ijisrt Bikie et al. (2024): Mapping areas suitable for agriculture and their accessibility: The case of the centre region of Cameroon. publication/DOI: 10.54254/2977-3903/6/2024048 /ResearchGate Bikie et al. (2024): Green Water and Hyacinth Invasion Monitoring in Lake Lagdo North Region of Cameroon a Space Observation Based Approach. https://ssrn.com/author=6786603/ELSVIER Bikie et al. (2024): Implementation of a Real Estate Management System using Photogrammetry Method Allowing 3d Navigation of Buildings. DOI:10.38124/ijisrt/ijisrt24mar2159/SEMANTIC SCHOLAR Bikie et al. (2024): multi-criteria analysis and identification of soils suitable for wheat cultivation in the Adamaoua region of Cameroon. ResearchGate Bikie et al. (2024): Elaboration of the Land use Land Cover Plan of the Subdivision of Monatélé in Cameroon. DOI: https://doi.org/10 .38124/ijisrt Bikie et al. (2024): Optimisation de la Gestion des Incendies à Yaoundé 4 grâce à l'Implémentation Novatrice des Systèmes d'Information Géographique (SIG). ResearchGate Bikie et al.(2023): Implementation of a GIS for management effectiveness of data relating to land property in the department of the ocean Kribi location. publication/DOI: 10.54254/2977-3903/4/2023049 /ResearchGate Bikie Gerald Anicet. (2022): Status Trends and Gaps of Oceanography in the Gulf of Guinea " A case study on Cameroon's Coastal region". California: Public Lab. He also made important contributions to international conferences such as: TOPIC 1: Modeling And Spatiotemporal Monitoring of Marine and Coastal Ecosystems of The Moroccan Margin of the Gulf of Cadiz and The Gulf of Guinea A Hybride Approach For Spaceborn Oceanography And Artificial Intelligence. -The 67th International Conference on Great Lakes Research from May 20 to May 24, 2024. Caesars Convention Center, Windsor, Ontario, Canada:(Presented already) -24th Annual Meeting May 29-31, 2024 Virginia Tech - Blacksburg, Virginia, Hosted by Virginia Tech Department of Biological Systems Engineering: -Mediterranean Geological Union's 4th Annual Meeting November 25th -28th 2024 Barcelona, Hosted by the University of Barcelona. -British Society for Geomorphology Annual Conference 2024. 2-5 September, Hosted by the Loughborough University of the United Kingdom. TOPIC 2: Contribution Of Remote Sensing to The Structural Study Of The Ngovayang Region In South-Central Cameroon: Metallogenic

Implications. -MANAGEM DOCTORAL DAYS 2022: 7th,8th And 9th December, MANAGEM ACADEMY, Marrakech, MOROCCO. IMPACT: BIKIE 's work had a major impact on the development of space science and technology as well as environmental preservation in coastal areas. As he contributed to successful research projects such as: -The Status Trends and Gaps of Oceanography in the Gulf of Guinea " A case study on Cameroon's Coastal region (2021). - -The creation of Geospatial infrastructures such as: • IDoGek (Infrastructure de Données Geospatiale de Kribi) for coastal monitoring: https://idogek-14be.onrender.com/ • Trace cocoa by Mane a traceability system for sustainable cocoa:https://tracecocoa.onrender.com PROFESSIONAL LIFE: BIKIE is a geospatial expert and teacher of GIS, Remote Sensing and Photogrammetry at the Department of Land Surveying of National Advanced School of Public Works in Cameroon. He served as examiner and reporter in several research juries and as a Geospatiale consultant on several projects.

Susanta Das is a seasoned educator with a proven record of teaching and mentoring diverse students, working at multiple institutions, and providing service to the various initiatives of organizations. He received his Ph.D. and M.A. degrees from Western Michigan University (WMU), USA, and M.Sc. degree from Banaras Hindu University (BHU), India all in Physics. He continued his research as a Marie-Curie post-doctoral fellow at Stockholm University, Stockholm, Sweden on beam diagnostics for the DESIREE (Double ElectroStatic Ion Ring ExpEriment) in the project DITANET (Diagnostic Techniques for particle Accelerators – a Marie-Curie initial training NETwork), at the Indian Institute of Science Education and Research-Kolkata, India on high-pressure physics as a Project Scientist-B, and at the University of Electro-Communications, Tokyo, Japan on ion-surface interactions as a Post-doctoral fellow. Throughout his career, Dr. Das worked and collaborated with many researchers, Ph.D., master, and visiting students from different countries. He further visited many countries to discuss research and an international conference participant (UK, Italy, Germany, Greece, Belgium, Bulgaria, Romania, Brazil etc.). He co-authored several research articles in WoS/Scopus indexed international journals, conference proceedings, and scholarly book chapters published by renowned international publishing houses. Beyond his research endeavors, Dr. Das has a storied history of service to academic and administrative committees, exhibiting his commitment to the growth and development of educational institutions. His experience includes tenure at Central University South Bihar, Sri Sri University, and P.K. University before assuming his current role at Ajeenkya DY Patil University. He received the Marie-Curie post-doctoral fellowship in Sweden, Science Academies' Summer Research Fellowship in India, Gwen Frostic Doctoral Fellowship, Department Graduate Research and Creative Scholar Award by WMU, and Leo R. Parpart Doctoral Fellowship by Dept. of Physics, WMU, among many others, throughout his academic journey. At present, Dr. Das continues to delve into cutting-edge fields, with a keen interest in nanotechnology, quantum computing, and data science. His multifaceted contributions, spanning teaching, research, and administrative leadership, showcase a dedicated professional who is instrumental in advancing the vision and mission of the institutions he serves.

Pooja Dehankar is working as an Assistant Professor at Ajeenkya DY Patil School of Engineering, Savitribai Phule Pune University, Pune. She has completed BE in Computer Engineering from Priyadarshini College of Engineering, Nagpur, and ME in Information Technology from Sinhgad College of Engineering, Pune. She has done Post Graduate Diploma in Advance Computing. She has 15 years of experience in academics. Her areas of interest are Artificial Intelligence, Cyber Security, and Data Mining. She has published papers in Scopus Journals. She actively participates in workshops, faculty development programs, short-term training programs, conferences, and webinars. She is pursuing Ph.D. from Symbiosis Center for Research and Innovation, Symbiosis International (Deemed University).

Udayakumar K is a senior assistant professor of computer science and engineering at Vel Tech Rangarajan Dr.Sagunthala R&D Institute of Science and Engineering. With a Ph.D. in Computer Science from SRM Institute of Science and Technology, his research focuses on deep reinforcement learning, multi-agent systems, and applications in industry, autonomous systems, and robotics. He has published 20+ articles extensively in reputed journals and conferences. He also serves as a reviewer for prestigious journals and conferences. Dedicated to teaching, he has accumulated extensive teaching experience over 15 years in the field of computer science and engineering. During this time, he has developed and taught a wide range of undergraduate and postgraduate courses, including machine learning, theory of computation, and functional languages.

Kavita Khobragade is currently working an Associate Professor in Computer Science at Fergusson College, Pune, Maharashtra, India with more than 28 years of teaching experience. She is a well-known teacher and she has written 26 textbooks for undergraduate students and a reference book published by Lambert Academic Publishing, Germany. Her research interests include Image Processing, Machine Learning, Deep Learning, Pattern Recognition and Computer Vision.

Yukesh Kumar is a Student at Hindustan institute of technology and science.

Venkatesan M Sundaram received the bachelor's degree in computer science and engineering from the University of Madras, Chennai, India, in 1999, the master's degree in information technology, and the Ph.D. degree in computer science and engineering from VIT University, Vellore, India, in 2006 and 2014, respectively., He is currently an Assistant Professor and the Head of the Department of Computer Science and Engineering, National Institute of Technology Puducherry, Karaikal, India. He has 23 years of teaching experience in various prestigious institutions, such as Adhiparasakthi Engg. College, Vellore Institute of Technology, and now VIT University and NIT Surathkal. In connection with his research work, he has authored or coauthored 50 articles in various reputed journals and international conference proceedings book chapters, including SCI, ESCI and Scopus. He has successfully run and executed several government-funded projects of 4 300 000 INR from ISRO, SAC, Ahmedabad, India, and DST-ICPS. His research areas include AI, machine learning, data science, data analytics, geo-spatial analysis, computational intelligence, blockchain, information retrieval, and social network analysis.

Alkha Mohan is an Assistant Professor in the Department of Computer Science and Engineering at IIIT Tiruchirappalli. She earned her PhD from NIT Surathkal in 2022, and her M.Tech and B.Tech from CUSAT, Kerala. Her research focuses on machine learning, deep learning, data science, remote sensing, and precision agriculture. With extensive teaching experience at institutions like VIT Vellore, GITAM Bangalore, and CE Karunagappally, Dr. Mohan has also completed certifications in deep learning and cloud computing, enhancing her expertise in these cutting-edge fields.

Prabhavathy Paneer received the M.Tech. and the Ph.D. degrees in computer science from VIT University, Vellore, India, in 2008 and 2016, respectively. She is currently a Professor with the School of Information Technology and Engineering, VIT University, Vellore, India. She is also a part of various school activity committees. She had completed funded project from ISRO, SAC, Ahmedabad, India. She has authored or coauthored around 20 journal papers in her research field, and number of papers in international conferences. Her research interests include computational intelligence, data mining, data science, machine learning, and deep learning.

Poonam Ponde is currently working an Associate Professor in Computer Science at Nowrosjee Wadia College, Pune, Maharashtra, India with more than 26 years of academic and corporate training experience in leading IT companies. She holds an M.Phil. in Information Technology and a Doctorate in Computer Science. She has published more than 20 research papers in National and International journals. She has also authored several text books and two reference books in various subjects in Computer Science. Her research interests include Computer Networks, Algorithms and Machine Learning.

Auwal Sagir is a PhD scholar at the Fujian Key Laboratory of Sensing and Computing for Smart Cities, Xiamen University, Xiamen 361005, China. My research interest focuses on spatial computing, ubiquitous computing and intelligent transport systems.

Wasswa Shafik (Member, IEEE) received a Bachelor of Science degree in information technology from Ndejje University, Uganda, a Master of Engineering degree in information technology engineering (Computer and communication networks options) from Yazd University, Iran, and a PhD in Computer Science with the School of Digital Science, Universiti Brunei Darussalam, Brunei Darussalam. He is also the Founder and Lead Investigator of the Dig Connectivity Research Laboratory (DCRLab) after serving as a Research Associate at the Network Interconnectivity Research Laboratory at Yazd University. Prior to this, he worked as a Community Data Analyst at Population Services International (PSI-Uganda), Community Data Officer at Programme for Accessible Health Communication (PACE-Uganda), Research Assistant at the Socio-Economic Data Centre (SEDC-Uganda), Uganda, an Assistant Data Officer at TechnoServe, Kampala, IT Support at Thurayya Islam Media, Uganda, and Asmaah Charity Organization. He has over hundreds of publications in renowned publishers. His research interests include Computer Vision, Explainable AI, Smart Agriculture, Health and ecological informatics, Security and Privacy.

C.V. Suresh Babu is a pioneer in content development. A true entrepreneur, he founded Anniyappa Publications, a company that is highly active in publishing books related to Computer Science and Management. Dr. C.V. Suresh Babu has also ventured into SB Institute, a center for knowledge transfer. He holds a Ph.D. in Engineering Education from the National Institute of Technical Teachers Training & Research in Chennai, along with seven master's degrees in various disciplines such as Engineering, Computer Applications, Management, Commerce, Economics, Psychology, Law, and Education. Additionally, he has UGC-NET/SET qualifications in the fields of Computer Science, Management, Commerce, and Education. Currently, Dr. C.V. Suresh Babu is a Professor in the Department of Information Technology at the School of Computing Science, Hindustan Institute of Technology and Science (Hindustan University) in Padur, Chennai, Tamil Nadu, India. For more information, you can visit his personal blog at https://sites.google .com/view/cvsureshbabu/.

Rohini Tarade is working as an Assistant Professor in the School of Engineering, Ajinkya D Y Patil University, Pune, Maharashtra, India. She qualified NET and SET and is pursuing a Ph.D. in Computer Science. She has 18 years of experience in academics. Her areas of interest are machine learning and data mining. She has published research papers. She actively participates in outreach activities, workshops, conferences, short-term training courses, and faculty development programs. She is deeply committed to education and has taught a variety of courses at both the undergraduate and graduate levels. She is known for her engaging teaching style, which combines rigorous theoretical foundations with practical, hands-on experience. Mrs. Rohini actively mentors students, guiding them in their academic and research pursuits. She has also worked on various Examination committees such as Paper-Setter, Moderator, and Senior Supervisor. Outside of academia, she enjoys singing and exploring new technologies.

Rufai Yusuf Zakari obtained his bachelor's and master's degrees from Bayero University in Kano, Nigeria, and Jodhpur National University in India, respectively. Currently, he is pursuing a Ph.D. in data science at the School of Digital Science, Universiti Brunei Darussalam. His present research focuses on Natural Language Processing (NLP), Computer Vision, Deep Learning, and Graph Neural Networks.

Index

E

F

G

H

I

L

M

N

O

P

R

S

T

U

W

Z

www.ingramcontent.com/pod-product-compliance
Ingram Content Group UK Ltd.
Pitfield, Milton Keynes, MK11 3LW, UK
UKHW050948290926
15470UKWH00005B/1717

9 798369 369005